Harmonization of Seismic Hazard in Vrancea Zone

NATO Science for Peace and Security Series

This Series presents the results of scientific meetings supported under the NATO Programme: Science for Peace and Security (SPS).

The NATO SPS Programme supports meetings in the following Key Priority areas: (1) Defence Against Terrorism; (2) Countering other Threats to Security and (3) NATO, Partner and Mediterranean Dialogue Country Priorities. The types of meeting supported are generally "Advanced Study Institutes" and "Advanced Research Workshops". The NATO SPS Series collects together the results of these meetings. The meetings are co-organized by scientists from NATO countries and scientists from NATO's "Partner" or "Mediterranean Dialogue" countries. The observations and recommendations made at the meetings, as well as the contents of the volumes in the Series, reflect those of participants and contributors only; they should not necessarily be regarded as reflecting NATO views or policy.

Advanced Study Institutes (ASI) are high-level tutorial courses intended to convey the latest developments in a subject to an advanced-level audience

Advanced Research Workshops (ARW) are expert meetings where an intense but informal exchange of views at the frontiers of a subject aims at identifying directions for future action

Following a transformation of the programme in 2006 the Series has been re-named and re-organised. Recent volumes on topics not related to security, which result from meetings supported under the programme earlier, may be found in the NATO Science Series.

The Series is published by IOS Press, Amsterdam, and Springer, Dordrecht, in conjunction with the NATO Public Diplomacy Division.

Sub-Series

A.	Chemistry and Biology	Springer
B.	Physics and Biophysics	Springer
C.	Environmental Security	Springer
D.	Information and Communication Security	IOS Press
E.	Human and Societal Dynamics	IOS Press

http://www.nato.int/science
http://www.springer.com
http://www.iospress.nl

Series C: Environmental Security

Harmonization of Seismic Hazard in Vrancea Zone

with Special Emphasis on Seismic Risk Reduction

Edited by

Anton Zaicenco

Institute of Geology and Seismology
Academy of Sciences
Chisinau, Moldova

Iolanda Craifaleanu

National Institute for Building Research
Bucharest, Romania

and

Ivanka Paskaleva

Bulgarian Academy of Sciences
Sofia, Bulgaria

Springer

Published in cooperation with NATO Public Diplomacy Division

Proceedings of the NATO Science for Peace Project on
Harmonization of Seismic Hazard and Risk Reduction in Countries
Influenced by Vrancea Earthquakes
Chisinau, Moldova
20 May 2008

Library of Congress Control Number: 2008937478

ISBN 978-1-4020-9241-1 (PB)
ISBN 978-1-4020-9240-4 (HB)
ISBN 978-1-4020-9242-8 (e-book)

Published by Springer,
P.O. Box 17, 3300 AA Dordrecht, The Netherlands.

www.springer.com

Printed on acid-free paper

PREFACE

The NATO Science for Peace Project SfP-980468 *Harmonization of Seismic Hazard and Risk Reduction in Countries Influenced by Vrancea Earthquakes* was an ambitious attempt to harmonize the seismic-hazard assessment in Bulgaria, Moldova and Romania, and provide the guidelines for seismic risk reduction in the target countries. Related to the study of intermediate-depth Vrancea earthquakes, it became operational in 2005.

The project co-coordinators were as follows:

- Prof. Güney Özcebe, Ankara, Turkey;

- Dr. Anton Zaicenco, Chisinau, Moldova;

- Dr. Iolanda Craifaleanu, Bucharest, Romania;

- Prof. Ivanka Paskaleva, Sofia, Bulgaria.

The project has brought together leading research personalities in the area of earthquake engineering, seismology and earth physics from several countries for brainstorming sessions, informal discussions, and exchanges of ideas. One of its key components was an upgrade of the strong-motion seismic networks of the countries-participants, which created a foundation for a long-term collaboration. A number of papers have been published as a result of the work conducted under this project.

The present book contains the Proceedings of the Closing Workshop for Project SfP-980468, which was organized in Chisinau, Moldova on May 20, 2008.

From hazard analyses to protection of the historical buildings, from study of the dynamic properties of the soft soils to paleoseismology, there are few areas of interest that remain untouched. Research from the NATO members and partner countries in South-Eastern Europe that forms the components of NATO Project SfP-980468 has made solid contributions to the Workshop theme.

Lungu and Craifaleanu provide an extensive study involving the mapping of key ground motion characteristics and of linear/nonlinear spectral ordinates for all strong Vrancea earthquakes that occurred in the last three decades. The second paper by these authors presents an evaluation of the damage potential of Vrancea earthquakes based on the use of damage spectra.

Bonjer, Ionescu, Sokolov, Radulian, Grecu, Popa, and Popescu present the ground motion patterns of the October 27, 2004 Vrancea earthquake. Ground motion maps for the main shock are presented using the recordings of the accelerometer networks installed in Romania and the macroseismic intensities as observed by the Internet community provided by the United States Geological Survey.

In her paper Craifaleanu gives comprehensive material on the required overstrength, expressed by the ratio between the demands imposed by Vrancea earthquakes and those imposed by the Romanian seismic design codes.

Borcia interprets instrumental data obtained in the Republic of Moldova (Chisinau and Cahul) and Romania (Vaslui and Birlad) during the 1986 and 1990 Vrancea earthquakes. The response spectra of strong-motion records, the global parameters that characterize an individual component of a record and instrumental intensity are the main numerical results obtained for the seismic records.

Alkaz, Zaicenco, and Isiciko investigate the dynamic properties of soft soils in Chişinău City. Authors propose a methodology for seismic microzonation based on a complex study of soil dynamic properties, structural damage observations, and numerical methods in seismic wave propagation.

A combination of different methods in the assessments of recurrence and probability of Vrancea intermediate-depth earthquakes is offered by Ginsari. She performs an analysis of recurrence relationships for two types of magnitudes, for different time intervals and by assigning alternative values of maximum possible earthquake magnitude.

Grecu, Radulian, Mandrescu, and Panza investigate the validity and relevance of the H/V spectral ratio technique in the particular case of Bucharest City, which was strongly affected by the 4 March 1977 (M_w=7.4) damaging earthquake.

Dineva, Paskaleva, La Mura, and Panza present and solve the 2-D elastodynamic model for seismic in-plane wave propagation in laterally inhomogeneous geological profiles embedded in a vertically inhomogeneous half-space in which an earthquake source is buried. The model and hybrid computational tool developed are applied in order to contribute to the seismic risk analysis of the Bulgarian capital Sofia.

An upper limit for horizontal peak ground acceleration generated by Paleolithic earthquakes is investigated by Gribovszki, Paskaleva, Kostov, Varga, and Nikolov. To pursue their goals authors examine broken and slim intact speleothems in Snezanka and Eminova caves in southwestern Bulgaria in the Rhodope Massif.

An acceleration data logger incorporating MEMS sensor for seismic arrays and structural health monitoring is presented by Mohniuc, Zaicenco, Kuendig, and Kurmann. This low-power class C device resolves accelerations at sub milli-g level on a bandwidth of 100 Hz, incorporates 16-bit ADC and has a GPS sensor.

Analysis of focal mechanisms of Vrancea earthquakes is provided by Sandu and Zaicenco. Seismic events with magnitude $M_w>3.0$ were studied for the time span 1967-2006. Ten independent catalogues were examined by the authors. They recomputed about 250 focal mechanism solutions from ISC database containing information on P-wave polarities, and made thorough comparative analyses.

In their paper, Zaicenco and Alkaz discuss numerical solutions of an elastic wave equation using spectral method based on Chebyshev grids.

Koleva, Ilie, and Akkar investigate maximum interstory drift ratio for shear-wall systems with fundamental periods ranging from 0.5 seconds to 1.25 seconds. Predictive equations, derived from a recently compiled ground-motion dataset, include focal mechanism and site class as explanatory variables.

A snapshot of the development of a strong motion registration system in Russe, Bulgaria is offered by Kouteva and Paskaleva, and the benefits of seismic monitoring in the region are discussed.

A parametric model for the August 30, 1986 Vrancea earthquake is offered by Zaicenco, Gavin, and Dickinson. They use Gabor wavelet to capture the coherent component of ground motion, and describe the high-frequency part by a stochastic method.

Paskaleva, Simeonov, Koleva, Kouteva, and Hadjiiski present results of applying a seismic microzonation procedure for designing and constructing a 34-storey administrative building in Sofia.

Paskaleva, Koleva, Vaccari, Zuccolo, and Panza describe an advanced modeling technique that allows computation of realistic synthetic seismograms in a set of selected sites in the Sofia urban area. Afterwards, they use a 3-D finite element modeling to assess a building structural performance.

A comprehensive study that includes catalogues of Vrancea earthquakes, earthquake records in Romania, the evolution of seismic zonation and seismic design codes, a seismic hazard map of Romania and design spectra in the new seismic design code P100-1/2006 following the Eurocode 8 format is proposed by Lungu, Aldea, and Arion. Another paper authored by this team tackles current challenges involved in conserving historical buildings in the center of Bucharest. The study includes historical earthquake damage and lessons learned, strengthening procedures for tall reinforced concrete fragile buildings in central Bucharest, and international projects for seismic risk reduction in Romania.

The authors of the papers comprised in the present volume express deep gratitude to John Ebel, John Evans, Klaus Hinzen, Luis Esteva Maraboto, Dieter Mayer-Rosa, Imtiyaz Parvez, Peter Rangelow, Yevgeny Rogozhin, Joachim Ritter, and Wolfgang Wirth for reviewing the articles.

The assistance of Judith Goldman in editing and proofreading the papers for this volume is highly appreciated.

The success of the project would not be possible without continuous support of the NATO Science Committee, and particularly Dr. Chris De Wispelaere and Dr. Susanne Michaelis.

August 2008

Guney Ozcebe
Anton Zaicenco

LIST OF CONTRIBUTORS WITH COMPLETE ADDRESSES

Akkar Sinan
Middle East Technical University
Department of Civil Engineering
06531 Ankara
Turkey

Aldea Alexandru
Technical University of Civil Engineering
Bucharest, 124 Lacul Tei
RO-020396
Romania

Alkaz Vasile
Institute of Geology and Seismology
Academy of Sciences of Moldova
str. Academiei 3, MD-2028 Chisinau
Moldova

Arion Cristian
Technical University of Civil Engineering
Bucharest, 124 Lacul Tei
RO-020396
Romania

Bonjer Klaus-Peter
Geophysical Institute University of Karlsruhe
Hertzstr. 16
D 76187 Karlsruhe
Germany

Borcia Ioan Sorin
National Institute for Building Research, INCERC
Earthquake Engineering Division
266 Pantelimon St., Sector 2, RO-021652, Bucharest, Romania

Craifaleanu Iolanda-Gabriela
Technical University of Civil Engineering Bucharest
Faculty of Civil, Industrial and Agricultural Buildings
Department of Reinforced Concrete Structures
124 Lacul Tei Blvd., Sector 2, RO-020396 Bucharest, Romania
National Institute for Building Research, INCERC
Earthquake Engineering Division
266 Pantelimon St., Sector 2, RO-021652, Bucharest, Romania

Dickinson Bruce
Department of Civil and Environmental Engineering
Edmund T. Pratt School of Engineering
Duke University
Durham, NC
USA

Dineva Petia
Institute of Mechanics
Bulgarian Academy of Sciences (BAS)
Acad. G. Bonchev block 4, 1113 Sofia
Bulgaria

Gavin Henri
Department of Civil and Environmental Engineering
Edmund T. Pratt School of Engineering
Duke University
Durham, NC
USA

Ginsari Victoria
Institute of Geology and Seismology
Academy of Sciences of Moldova
str. Academiei 3, MD-2028 Chisinau
Moldova

Grecu Bogdan
National Institute of Research and Development
for Earth Physics (NIEP)
Bucharest-Magurele (PO Box: MG-2)
12 Calugareni st., Ilfov
Romania

Gribovszki Katalin
Hungarian Academy of Sciences
Geodetic and Geophysical Research Institute
MTA-ELTE Geological, Geophysical and Space Science Research Group
Hungary

Hadjiiski Kiril
Central Laboratory Seismic Mechanics and Earthquake Engineering
Bulgarian Academy of Science (BAS)
Acad. G. Bonchev block 3, 1113 Sofia
Bulgaria

Ionescu Constantin
National Institute of Research and Development for Earth Physics (NIEP)
Bucharest-Magurele (PO Box: MG-2)
12 Calugareni st., Ilfov
Romania

Isiciko Evgheniy
Institute of Geology and Seismology
Academy of Sciences of Moldova
str. Academiei 3, MD-2028 Chisinau
Moldova

Koleva Gergana
Central Laboratory Seismic Mechanics and Earthquake Engineering
Bulgarian Academy of Science (BAS)
Acad. G. Bonchev block 3, 1113 Sofia
Bulgaria

Koleva Nikolina
Central Laboratory Seismic Mechanics and Earthquake Engineering
Bulgarian Academy of Science (BAS)
Acad. G. Bonchev block 3, 1113 Sofia
Bulgaria

Kostov Marin
Central Laboratory Seismic Mechanics and Earthquake Engineering
Bulgarian Academy of Science (BAS)
Acad. G. Bonchev block 3, 1113 Sofia
Bulgaria

Kouteva – Guentcheva Mihaela
Central Laboratory Seismic Mechanics and Earthquake Engineering
Bulgarian Academy of Science (BAS)
Acad. G. Bonchev block 3, 1113 Sofia
Bulgaria

Kuendig Christoph
GeoSIG Ltd.
Europastrasse 11
8152 Glattbrugg
Switzerland

Kurmann Thomas
GeoSIG Ltd.
Europastrasse 11
8152 Glattbrugg
Switzerland

La Mura Cristina
DST– Department of Earth Sciences
University of Trieste
Via E. Weiss 4, I-34142 Trieste
Italy

Lungu Dan
Technical University of Civil Engineering
Bucharest, 124 Lacul Tei
RO-020396
Romania

Mandrescu Neculai
National Institute of Research and Development
for Earth Physics (NIEP)
Magurele-Bucharest (PO Box: MG-2)
12 Calugareni st., Ilfov
Romania

Mohniuc Ruslan
Institute of Geology and Seismology
Academy of Sciences of Moldova
Str. Academiei 3, MD-2028 Chisinau
Moldova

Nikolov Gabriel
Geological Institute
Sofia 1113
Acad. G. Bonchev, str. block 4
Bulgaria

Panza Giuliano
DST – Department of Earth Sciences
University of Trieste
Via E. Weiss 4, I-34142 Trieste
Italy
and
The ICTP – Abdus Salam International Centre for Theoretical Physics
Strada Costiera 11, I-34100 Trieste
Italy

Paskaleva Ivanka
Central Laboratory Seismic Mechanics and Earthquake Engineering
Bulgarian Academy of Science (BAS)
Acad. G. Bonchev block 3, 1113 Sofia
Bulgaria

Popa Mihaela
National Institute of Research and Development for Earth Physics (NIEP)
Bucharest-Magurele (PO Box: MG-2)
12 Calugareni st., Ilfov
Romania

Popescu Emilia
National Institute of Research and Development for Earth Physics (NIEP)
Bucharest-Magurele (PO Box: MG-2)
12 Calugareni st., Ilfov
Romania

Radulian Mircea
National Institute of Research and Development for Earth Physics (NIEP)
Bucharest-Magurele (PO Box: MG-2)
12 Calugareni st., Ilfov
Romania

Sandu Ilie
Institute of Geology and Seismology
Academy of Sciences of Moldova
Str. Academiei 3, MD-2028 Chisinau
Moldova

Simeonov Svetoslav
Central Laboratory Seismic Mechanics and Earthquake Engineering
Bulgarian Academy of Science (BAS)
Acad. G. Bonchev block 3, 1113 Sofia
Bulgaria

Sokolov Vladimir
Geophysical Institute University of Karlsruhe
Hertzstr. 16
D 76187 Karlsruhe
Germany

Vaccari Franco
DST – Department of Earth Sciences
University of Trieste
Via E. Weiss 4, I-34142 Trieste
Italy

Varga Peter
Geodetic and Geophysical Research
Institute of Hungarian Academy of Sciences (HAS)
6-8 Csatkai 6-8, 9400 Sopron
Hungary

Zaicenco Anton
Institute of Geology and Seismology
Academy of Sciences of Moldova
str. Academiei 3, MD-2028 Chisinau
Moldova

Zuccolo Elisa
DST – Department of Earth Sciences
University of Trieste
Via E. Weiss 4, I-34142 Trieste
Italy

CONTENTS

ROMANIA'S SEISMICITY AND SEISMIC HAZARD: FROM HISTORICAL RECORDS TO DESIGN CODES

D. LUNGU[*], A. ALDEA, C. ARION
Technical University of Civil Engineering, Bucharest

Abstract. The paper presents (i) catalogues of Vrancea earthquakes (ii) past strong earthquakes, (iii) earthquake records in Romania, (iv) the evolution of seismic zonation and seismic design codes, (v) a seismic hazard map of Romania and design spectra in the new seismic design code P100-1/2006 following the Eurocode 8 format.

Keywords: Earthquakes, Vrancea, seismic hazard, seismic design codes

1. Introduction

The Vrancea region is located where the Carpathian Mountain arch bends at about 135 kilometers (km) epicentral distance from Bucharest. It is a source of subcrustal seismic activity that affects more than two thirds of the territory of Romania and significant parts of the Republic of Moldova, Bulgaria and Ukraine. The Vrancea source induces a high seismic risk in the densely built zones of southeastern Romania.

2. Vrancea Earthquake Catalogues

Catalogues of earthquakes that occurred in Romania have been compiled by Radu (1979 and 1994 manuscripts published Lungu et al., 1997) and Constantinescu and Marza (1980 and 1995), Figure 1. In 1997, *Romplus*, a version of the Constantinescu and Marza catalogue, was presented at the First International Seminar on Vrancea earthquakes and has been updated daily on the web site www.infp.ro by the National Institute of Earth Physics (INFP) in Magurele. The

---

[*]Technical University of Civil Engineering, Bucharest, 124 Lacul Tei, RO-020396, Romania

A. Zaicenco et al. (eds.), *Harmonization of Seismic Hazard in Vrancea Zone,*
© Springer Science + Business Media B.V. 2008

most complete Vrancea historical catalogue is, however, the Radu Catalogue, even though significant events are also included in the Constantinescu and Marza Catalogue. The magnitude in the Radu catalogue is the Gutenberg-Richter magnitude M.

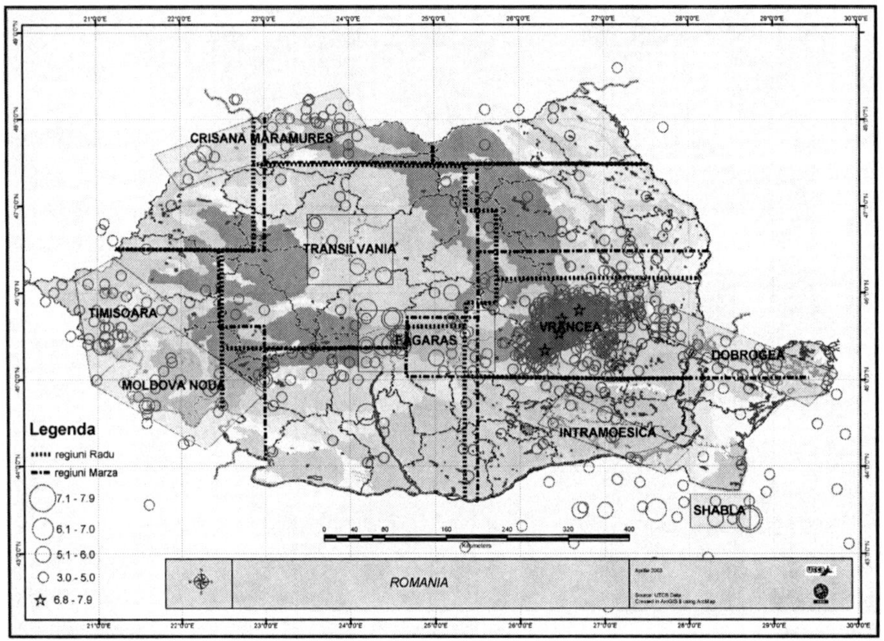

Figure 1: Romania: Location of the epicenters from 984 to 2003 and seismic regions according to Radu and Marza

From existing catalogues, one can note one event/century with epicentral intensity $I_0 \geq 9.0$ from 984 to 1900 and two events per century with intensity $I_0 \geq 9.0$ (or magnitude $M \geq 7.2$) from 1901 to 2000.

The most powerful Vrancea earthquake is generally accepted to be the 26 October 1802 event ($M_{G-R} \geq 7.5$) (Table 1). It was felt on a surface area of over 2,000,000 km^2 (Popescu, 1938). Strong earthquakes followed on 26 November 1829 and on 23 January 1838 (maximum seismic intensity $8 \div 9$) (Table 1). The Vrancea earthquake with the highest magnitude in the 20th century was the 10 November 1940 event ($M_{G-R} = 7.4$), and the most destructive one that century was the one on 4 March 1977 ($M_{G-R} = 7.2$, moment magnitude $M_w = 7.5$) (Table 2). On this occasion, the first strong ground motion in Romania was obtained on a SMAC-B Japanese instrument at the National Institute for Building Research (INCERC) seismic station, Eastern Bucharest.

TABLE 1: Radu's historical catalogue of major Vrancea subcrustal earthquakes, $I_0 \geq 9$ (Radu manuscripts, 1994 published Lungu et al., 1997[a,b,c])

Nr.	Date		Time (GMT) h:m:s	Max MSK intensity, I_o		Magnitude			Radu's source
						M_{G-R}		M_s	
				Radu	Others	Radu	Others		
1	1196	Feb. 13	07:	(8) 8–9	9/CM	(6.7) 7.2	7.0/KS	7.3/CM	RT, R
2	1230	May 10	07:	(8–9) 9[+]	8.5/CM	(6.9) 7.4	7.1/KS	7.1/CM	RT, R, N
3	1446	Oct. 10	04:	8	8.5/CM, 8-9/RT	6.7	7.3/KS	7.3/CM	RT, R
4	1471	Aug. 29	10–11:	(8) 9	9/CM, 8-9 KS	(6.9) 7.4	7.1/KS	7.3/CM	RT, R
5	1516	Nov. 8	12:	9	9/CM, 8/KS	7.2		6.8/KS	RT, R
6	1620	Nov. 8	13–14:	(8–9) 9	9/CM, 8/KS	(6.9) 7.2	6.5/KS	7.3/CM	RT
7	1679	Jan. 29/Aug. 9		(8) 6	9/CM	(6.7) 5.5	6.8/KS	7.3/CM	RT
8	1681	Aug. 18	(00)01:	9	8/CM	(6.7) 7.4	6.8/KS	6.8/CM	RT
9	1738	May 31/ June	10–11:	(8–9) 9	9.5/CM	(6.9) 7.4	7.0/KS	7.5/CM	RT, R
10	1802	Oct. 26	10:55	9	10/CM	7.5	7.4/KS	7.7/CM	R
11	1838	Jan. 23	18:45	8	9/CM	6.7	6.9/KS	7.3/CM	R

[a]Source abbreviations: R – Radu (1971, 1974) catalogues; RT – Radu and Torro (1986) catalogue; N – Nikonov catalogues; CM – Constantinescu and Marza (1980) catalogue; KS – Kondorskaya and Shebalin (1977) catalogue; SKH – Shebalin et al. (1974) catalogue.

[b]Focus depth h is considered as intermediate depth: $60 \leq h \leq 170$ km.

[c]The intensity in parenthesis indicates the previous estimations made by Radu (1980).

TABLE 2: Twentieth century strong Vrancea earthquakes, M_{G-R} or $M_s \geq 6.0$ (Lungu et al., 2000)

No.	Date	Lat. N°	Long. E°	Focus depth h km			Magnitude[a]			
							M_S	M_w	M_{G-R}	M_w
				Radu	Marza	www.infp.ro	Marza	www.infp.ro	Radu	[b]
1	1903 Sep. 13	45.7	26.6	≥60	70	70	5.7	6.3	6.3	6.6
2	1904 Feb. 6	45.7	26.6	≥60	75	75	6.3	6.6	5.7	–
3	1908 Oct. 6	45.7	26.5	150	125	125	6.8	7.1	6.8	7.1
4	1912 May 25	45.7	27.2	80	90	90	6.4	6.7	6.0	6.3
5	1934 Mar. 29	45.8	26.5	90	90	90	6.3	6.6	6.3	6.6
6	1939 Sep. 5	45.9	26.7	115	120	120	6.1	6.2	5.3	–
7	1940 Oct. 22	45.8	26.4	122	125	125	6.2	6.5	6.5	6.8
8	1940 Nov. 10	45.8	26.7	133[c]	135[c]	150	7.4	7.7	7.4	7.7
9	1945 Sep. 7	45.9	26.5	75	80	80	6.5	6.8	6.5	6.8
10	1945 Dec. 9	45.7	26.8	80	80	80	6.2	6.5	6.0	6.3
11	1948 May 29	45.8	26.5	140	130	130	6.0	6.3	5.8	–
12	1977 Mar. 4	45.34	26.30	109	–	94	7.2	7.4	7.2	7.5
13	1986 Aug.	45.53	26.47	133	–	131	–	7.1	7.0	7.3
14	1990 May 30	45.82	26.90	91	–	91	–	6.9	6.7	7.0
15	1990 May 31	45.83	26.89	79	–	87	–	6.4	6.1	6.4

Notes:

[a]Magnitudes: M_{G-R} – Gutenberg-Richter magnitude, M_S – surface waves magnitude, M_W – moment magnitude.

[b]$M_w \cong M_{G-R} + 0.3$ for $6.0 \leq M_{G-R} \leq 7.4$ (Lungu, 1999).

[c]In present the value accepted for focus depth for Nov. 10, 1940 event is $h = 140$ km.

2.1. PAST STRONG EARTHQUAKES

2.1.1. *1802 and 1838*

The 26 October 1802 (M_S = 7.4 – 7.7) earthquake is considered to be the strongest Vrancea subcrustal event; there is no precise information on causalities but there is some on damages. The earthquake was felt in Ukraine, Russia, Poland, Bulgaria and Turkey. As mentioned previously, the event was felt on a total estimated area of more than 2,000,000 km^2. The 23 January 1838 (M_S = 6.9 – 7.3) earthquake was also felt over an extended area in Europe including Ukraine, Poland, Bulgaria and Turkey up to (then) Constantinople and in the northeast of Italy.

The following description of the effects of 1838 earthquake in Bucharest is on page 144 in *Voyage dans la Russie Méridionale et la Crimée par la Hongrie, la Valachie et la Moldavie* by M. A. de Démidoff, illustrated by Raffet and edited by E. Bourdin in Paris in 1841 and 1854 (Figure 2):

Figure 2: M. Anatole de Démidoff, "Voyage dans la Russie Méridionale et la Crimée par la Hongrie, la Valachie et la Moldavie". Illustré par Raffet. Ernest Bourdin éditeur, Paris, 1841

Chaque année, le sol de la Valachie est ébranlé par deux ou trois secousses de tremblement de terre plus ou moins sensibles; mais, malheureusement, on a à noter, tous les huit ou dix ans, quelque atteinte réellement désastreuse de ce fléau. On conserve encore le souvenir du tremblement de terre de 1802, qui renversa la tour du monastère de Koltza; de celui de 1829, qui ébranla fortement la plupart des édifices de Bukharest. Depuis que ces lignes sont écrites, une secousse plus violente que toutes celles dont le souvenir attriste encore le pays, a pensé engloutir Bukharest. Tout à coup, le 11-23 janvier 1838, c'était le soir, la ville s'ébranle; les plus solides monuments chancellent; plusieurs maisons s'écroulent; toutes son endommagées, et, dans tout ces ravages, plusieurs hommes perdent la vie.

2.1.2. *1940*

According to *Comptes Rendus des Séances de l'Académie des Sciences de Roumanie,* "The November 10, 1940 earthquake caused damage all around Romania and threw the people into mourning," (Figure 3) (*Numéro consacré aux recherches sur le tremblement de terre du 10 Novembre 1940 en Roumanie,* Tome V, numéro 3, mai-juin, Cartea Romaneasca, Bucuresti, pp. 177–288).

Figure 3: Comptes Rendus des Séances de l'Academie des Sciences de Roumanie, 1941

Figure 4: The Carlton building in Bucharest collapsed during the 1940 earthquake

In Bucharest, the most significant loss from that earthquake (M_{G-R} = 7.4) was the complete collapse of the Carlton Building (Figure 4), the highest reinforced concrete building (47 m, 12 stories) in Romania at the time. The earthquake was also felt in an area of about 2,000,000 km^2. In Odessa, Cracovia and Moscow the estimated intensity was V–VI. It was also felt north up to (then) Leningrad, to the west up to the Tissa River and to the south in (then) Yugoslavia and all over Bulgaria to Istanbul.

Two zones of maximum intensity were identified In Romania: one was in the area of Panciu and Focsani and the second one was from Campina to Bucharest. In those areas, the seismic intensity was over VIII on the Mercalli-Sieberg scale, i.e., close to IX in Campina, Focsani, Tecuci and Beresti and about X in Panciu.

No realistic estimation of the number of deaths was reported. Some authors indicated the following numbers: >1,000 (Beles, Ifrim 1962; Coburn and

Spence 1992), 980 (The Emergency Disasters Data Base (EM-DAT): The Office of Foreign Disaster Assistance/Centre for Research on the Epidemiology of Disasters International Disaster Database; Univ.Cath.Louvain, Brussels) and 1,000 (Gutenberg and Richter, 1956; United States Geological Survey web site).

2.1.3. *1977*

The event on 4 March 1977 (M_{G-R} = 7.2) was the most destructive earthquake in the history of Romania (Figure 5). International experts dispatched to Romania in the aftermath of the earthquake reported as follows.

"The unusual nature of the ground motion and the extent and distribution of the structural damage have important bearing on earthquake engineering efforts in the United States." Jennings and Blume, National Research Council and the Eearthquake Engineering Research Institute, Washington.

"It was felt on an area of 1.3 x 106 km^2 and caused damage over an area of about 80,000 km^2 within which the most frequently occurring intensity did not exceed VII (MM). It mostly affected the densely populated and rapidly developing centres of Craiova, Pitesti, Zimnicea, Bucharest, Ploiesti, Focsani, Barlad." Ambraseys, N.N., 1977. "Long-period effects in the Romanian earthquake of March 1977" *Nature*, Vol. 268, July, pp. 324–325.

"About 108 people died and three multistorey reinforced concrete apartment buildings (having natural period of vibration of the order 0.8-0.9 s) collapsed completely in the town of Svistov, northern Bulgaria, approximately 270 km to the south of the epicenter." Tezcan, S.S., Yerlici V., Durgunoglu, H.T., 1978. "A reconnaissance report for the Romanian earthquake of 4 March 1977". *Earthquake Engineering and Structural Dynamics*, Vol. 6, pp. 397–421, Wiley.

Figure 5: Bucharest 4 March 1977. Photo © UTCB D. Lungu (Computer Center of Ministry of Transportation – *left*; Dunarea building – *right*)

The 4 March 1977 earthquake killed 1,578 people, including 1,424 in Bucharest, and injured 11,221 people including 7,598 in Bucharest. According to the World Bank (Report 16.P-2240-RO, 1978) the total damage in Romania was worth 2.05 billion 1977 United States dollars. After that earthquake, the Munich Re World Map of Natural Hazards in 1998 indicated that Bucharest was a "large city with a Mexico City effect."

3. Seismic Records

As stated previously, the first strong ground motion recorded in Romania was the 1977 record in eastern Bucharest at the seismic station of INCERC on a Japanese SMAC-B instrument. The ground motion was digitized and analyzed by the Building Research Institute of the Ministry of Construction in Japan in 1978 in "Digitized data of strong-motion earthquake accelerograms in Romania (March 4, 1977)" by the Observational Committee of Strong Motion Earthquake, *Kenchiku Kenkyu Shiro* No. 20, January.1.4. The unusual 1977 record which was characterized by a long predominant period of ground vibration, $T_p \cong 1.6$ s, was used for calibrating design response spectra in the Romanian seismic code from 1977 to 1992 when almost 40% of the buildings in Bucharest were built.

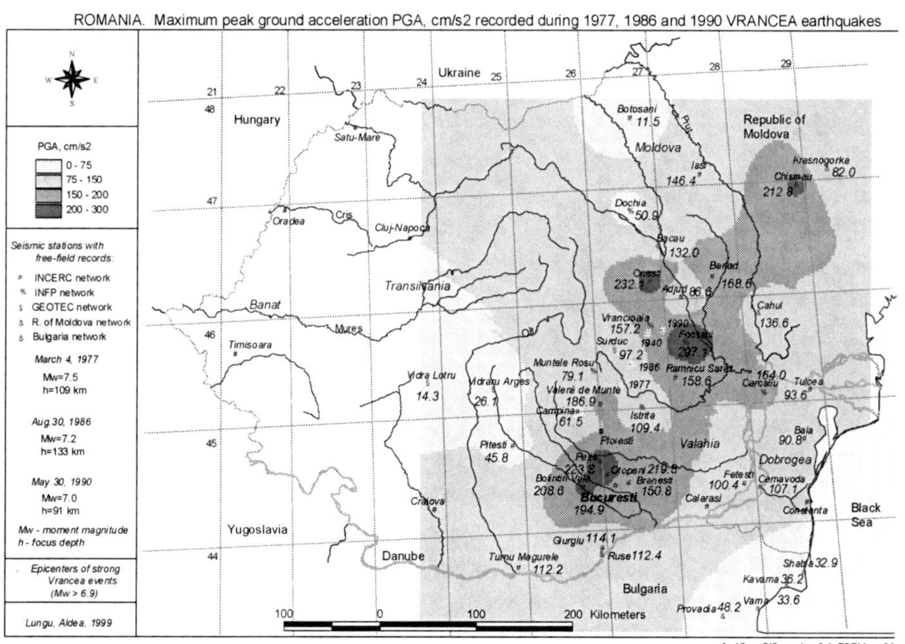

Figure 6: Maximum recorded peak ground acceleration during the last Vrancea strong events

After 1977, a significant ground-motion database of more than 40 records was collected in Bucharest during the 1986 and 1990 earthquakes.

The envelope of peak ground acceleration (PGA) recorded during the strongest three recent Vrancea events in Romania are in Figure 6. Note the NE-SW directivity of the subcrustal Vrancea source as was first observed by Montessus de Ballore in 1906.

Seismic microzonation maps for the city of Bucharest during the 1986 earthquake ($M_w = 7.0$) in terms of PGA and control period of response spectra $T_C = 2\pi\ EPV/EPA$ are shown in Figures 7 and 8.

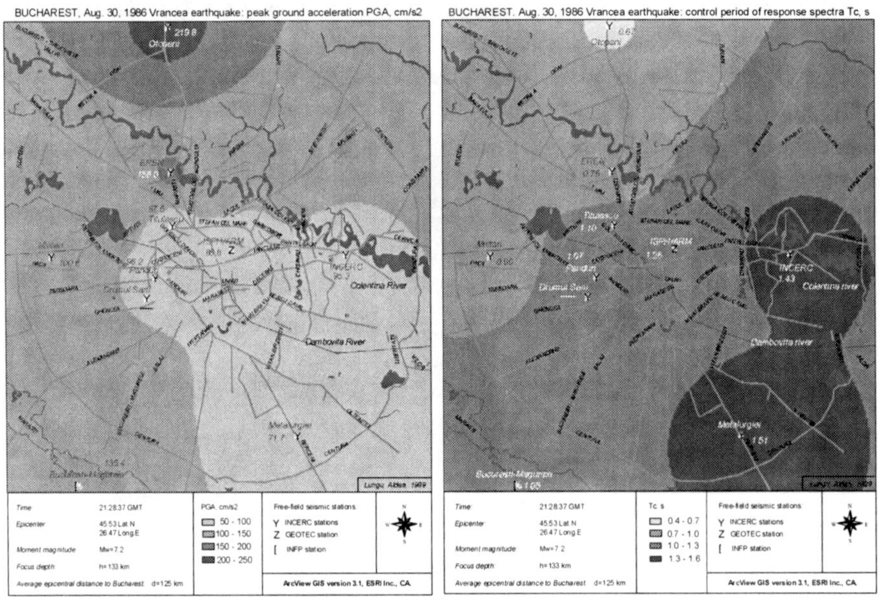

Figure 7: Bucharest Aug. 30, 1986 event: microzonation of PGA

Figure 8: Bucharest Aug. 30, 1986 event: microzonation T_C

Effective peak acceleration (EPA) and effective peak velocity (EPV) are defined as follows (Lungu ct al., 1997): $EPA = \max \overline{SA}_{0.4s}/2.5$ where $\overline{SA}_{0.4s}$ means the maximum value of peak acceleration ordinates averaged on a 0.4 s mobile window on a period axis and $EPV = \max \overline{SV}_{0.4s}/2.5$ where $\overline{SV}_{0.4s}$ means the maximum value of peak velocity ordinates averaged on a 0.4 s mobile window on a period axis.

Based on the very important conclusions from the 1977, 1986 and 1990 earthquakes, the seismic instrumentation of Bucharest has been recently extended and improved by various national and international efforts. Presently Romania has more than 100 digital K2 and ETNA, Kinemetrics instruments, about half of them in Bucharest, distributed in three seismic networks: INCERC, INFP and the National Center for Seismic Risk Reduction (NCSRR).

NCSRR received seismic equipment (Kinemetrics) from the Japan International Cooperation Agency (JICA) (Figure 9), and OYO Seismic Instrumentation Corp. and NCSRR installed the equipment in 2003. In 2005 and 2006, the NCSRR network was enlarged with Romanian funding from the MTCT budget, and additional sites received Geosig equipment and technical support. The NCSRR network contains three types of instrumentation: seven free-field stations outside Bucharest, five buildings with instruments in Bucharest and eight stations with ground surface and borehole sensors in Bucharest with the deepest borehole at 153 m.

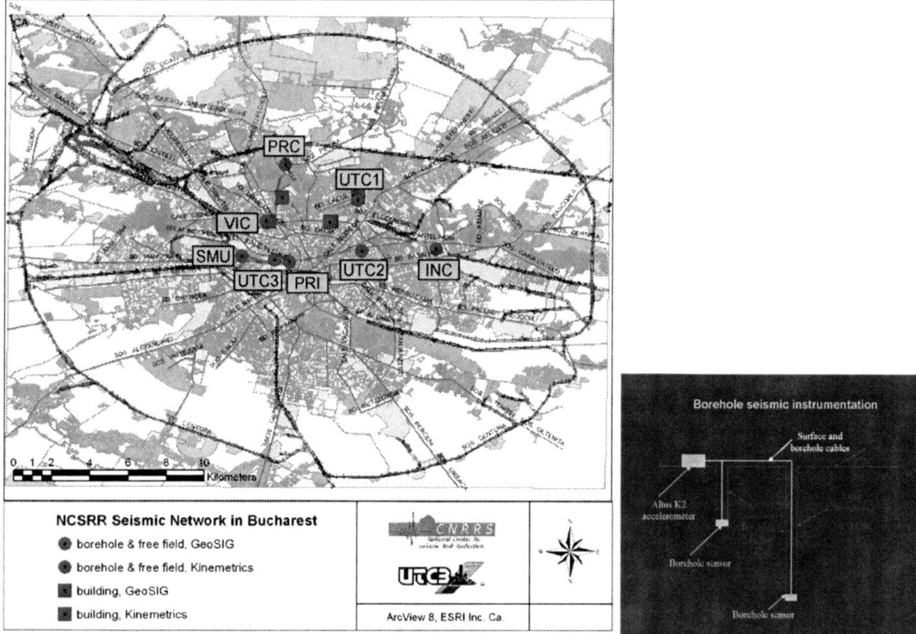

Figure 9: NCSRR seismic instrumentation in Bucharest within JICA Project (2002–2007) on Seismic Risk reduction in Romania

The database of Vrancea strong ground motion contains records from 47 free-field stations in Romania distributed by networks and events. The "free field" accelerograms were obtained either at ground level or in the basement of one to two story buildings. The distribution of the recorded accelerograms is given in Table 3.

The maximum of the two horizontal components of the ground motion was considered for *PGA* attenuation. The following attenuation model was selected for the analysis:

TABLE 3: Distribution of the free field accelerograms used in the attenuation analysis

Seismic Network	Romania			Rep of Moldova	Bulgaria	Total
Earthquake	*INCERC*[a]	*INFP*[b]	*GEOTEC*[c]	*IGG*[d]		
Mar. 4, 1977	1	–	–	–	–	1
Aug. 30, 1986	24	8	3	2	–	37
May 30, 1990	23	10	2	2	5	42
Total	48	18	5	4	5	80

[a]INCERC, National Building Research Institute, Bucharest.
[b]INFP, National Institute for Earth Physics, Bucharest-Magurele.
[c]GEOTEC, Institute for Geotechnical and Geophysical Studies, Bucharest.
[d]IGG, Institute of Geophysics and Geology, Moldovan Academy of Science, Chisinau.

$$\ln PGA = c_0 + c_1 M_w + c_2 \ln R + c_3 R + c_4 h + \varepsilon \tag{1}$$

where: *PGA* is peak ground acceleration at the site, M_w is moment magnitude, R is hypocentral distance to the site $R = \sqrt{h^2 + \Delta^2}$, h is focal depth, c_0, c_1, c_2, c_3, c_4 are data dependent coefficients and ε is a random variable with zero mean and standard deviation $\sigma_\varepsilon = \sigma_{\ln PGA}$. The coefficients obtained from the regression are as follows (Lungu et al., 1999): $c_0 = 3.098$, $c_1 = 1.053$, $c_2 = -1.000$, $c_3 = -0.0005$, $c_4 = -0.006$, $\sigma_{\ln PGA} = 0.502$.

The seismic hazard assessment for a Vrancea subcrustal source took magnitude recurrence into consideration. The classical recurrence relationship was modified into a truncated Gutenberg-Richter relationship that for Vrancea sources gives (Elnashai and Lungu, 1995):

$$n\left(\geq M_w\right) = e^{8.654-1.687M_w} \frac{1-e^{-1.687(8.1-M_w)}}{1-e^{-1.687(8.1-6.3)}} \tag{2}$$

In Equation (2), the threshold lower magnitude is $M_w = 6.3$, the maximum credible magnitude is $M_{w,max} = 8.1$, and $\alpha = 3.76 \ln 10 = 8.654$, $\beta = 0.73 \ln 10 = 1.687$.

The relationship between the magnitude of a potentially destructive Vrancea earthquake ($M_w \geq 6.3$) and the corresponding focal depth shows that the higher the magnitude, the deeper the focus (Lungu et al., 1997):

$$\ln h = -0.866 + 2.846 \ln M_w - 0.18 P \tag{3}$$

where P is a binary variable: $P = 0$ for the mean relationship and $P = 1.0$ for mean minus one standard deviation.

4. Zonation Standards and Codes for Earthquake Resistance of Structures in Romania

The first seismic zonation of Romania was done after the 1940 Vrancea earthquake. The 1941 zonation map contains two areas: one seismic area (Moldova, Valachia and Brasov) and a second one considered not to be seismic (the remaining parts of Romania). A. Cismigiu and E. Titaru wrote the first modern Romanian code for the seismic design of buildings in 1954, but the first official regulation appeared only in 1963. Table 4 presents the evolution of the seismic zonation and seismic design codes for buildings in Romania during the last 50 years. The P100-1/2006 code for the earthquake-resistant design of buildings Part 1 (244 p.), was approved by Ministry of Transportation and Construction order No. 1711/19.09.2006, and it entered into force on 1 January 2007. After the 1977 event, ductility rules for reinforced concrete structures were imported from the American Concrete Institute (ACI) code of practice. Those ductility rules were improved and enlarged in 1990 according to Eurocode eight requirements.

TABLE 4: Classification of codes for design of earthquake resistant buildings and of standards for seismic zonation of Romania (1940–2008)

Period		Code for earthquake resistant buildings	Seismic zonation standard
Pre-code, before 1963	Prior to the 1940 earthquake and prior to the 1963 code	*P.I. - 1941* *I - 1945*	*P.I. – 1941* *I – 1945* *STAS 2923 - 52*
Low-code, 1963–1977	Inspired by the Russian seismic practice	*P 13 - 63* *P 13 - 70*	*STAS 2923 - 63*
Moderate-code, 1977–1990	After the great 1977 earthquake	*P 100 - 78* *P 100 - 81*	*STAS 11100/1 - 77*
Moderate-code to High-code, after 1990	After the 1986 and the 1990 earthquakes	*P 100 - 90* *P 100 - 92*	*STAS 11100/1 - 91* *SR 11100/1 - 93*
High code, after 2006	Inspired by Eurocode 8	*P100 –1/2006*	*In P100 –1/2006 code*

5. New Romanian Seismic Design Code

Based on the results of probabilistic seismic hazard assessment from Vrancea source and taking into account the macroseismic fields from the crustal seismic sources in Romania, Figure 10 presents the seismic hazard map of the P100-1/2006 earthquake resistant design code in force since 1 January 2007. The map presents the design ground acceleration (PGA) for an event with a mean recurrence interval MRI = 100 years, which represents a transitional MRI towards the MRI = 475 years recommended by Eurocode 8 and American Society of Civil Engineers 7. Certainly the next version of the P100 code after 2010 will definitely introduce the 475 year PGA in Romania. For Bucharest, the PGA with a 475 year recurrence interval is about 0.36 g.

In the P100-1/2006 code, ground conditions are characterized by the control period of response spectra T_C. The T_C macrozonation in the P100-1/2006 code is presented in Figure 11 based on ground motion processed from 1977, 1986 and 1990 event records. The studies made on the complete data base of strong motion accelerograms in Romania showed that the control period of response spectra T_C is the most reliable and stable indicator for characterizing the frequency content of ground motions and indirectly of ground conditions.

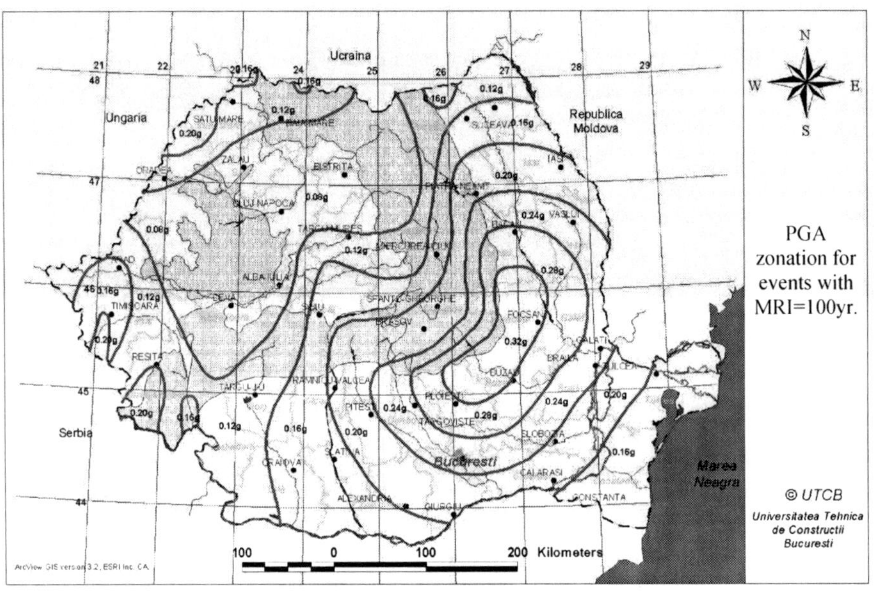

Figure 10: P100–1/2006 design ground acceleration a_g, for an event with *MRI* = 100 years

A direct characterization/classification of ground conditions based on soil data (soil profile and shear velocity profile) is not yet possible due to the lack of significant soil data correlated with strong recorded ground motions. The few cases where such information exists (i.e., soil data and earthquake records) showed that imported criteria and corresponding design spectra from other countries are not valid for areas under the influence of Vrancea source where deep sediment deposits exist (Lungu et al., 2002, Aldea, 2002).

Using recent PS logging results (from NCSRR), the average shear wave velocities in the top 30 m of soil in Bucharest are presented in Figure 12.

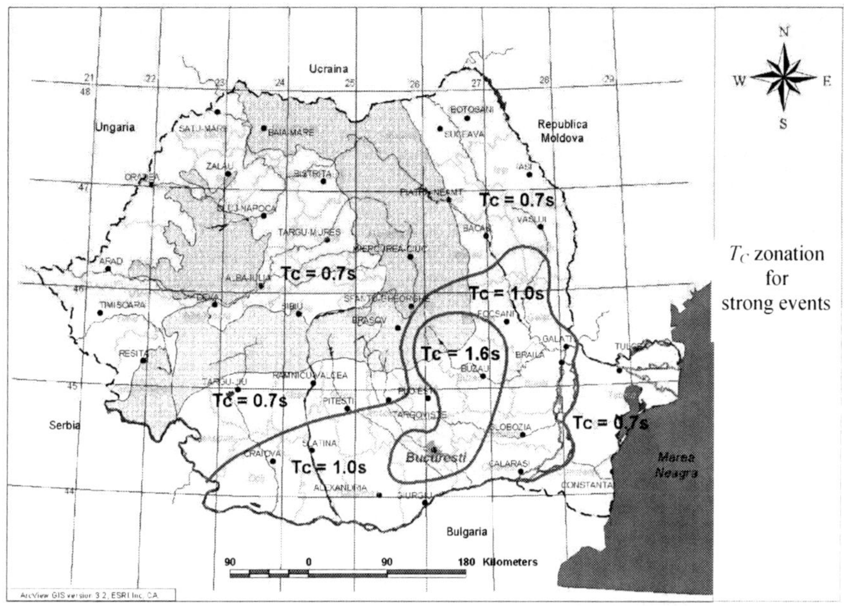

Figure 11: P100-1/2006 zonation of control period of response spectra, T_C

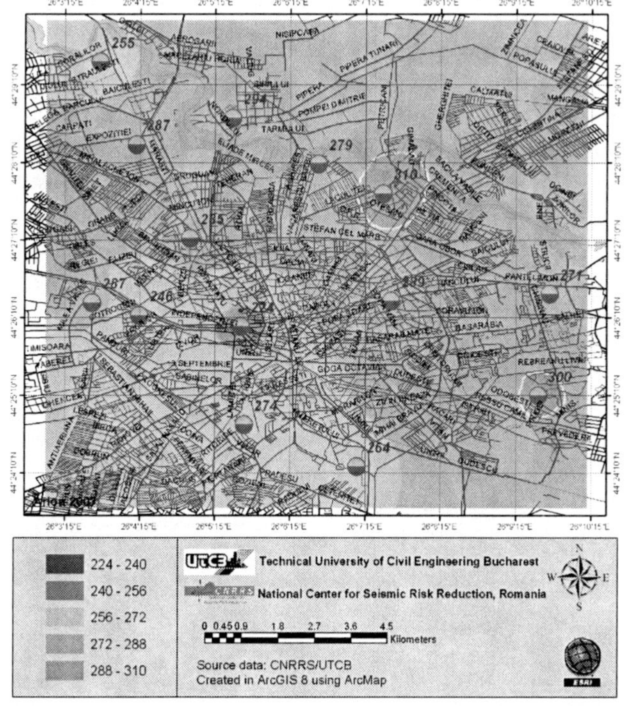

Figure 12: Bucharest. Microzonation map for shear wave velocity (m/s) averaged on 30 m

5.1. DESIGN SPECTRA

For each of the three zones in Figure 11, a normalized acceleration elastic response spectrum is recommended according to the Eurocode 8 format (Figure 13). Based on records obtained during the 1991 events (MRI $\approx$ 50 years) in the Banat region (western Romania with local shallow earthquakes), the design spectra in Figure 14 is recommended.

The seismic action in the P100-1/2006 design spectra is to be used for ultimate limit state design in connection with three other new Romanian codes: basis of design, wind action and snow action.

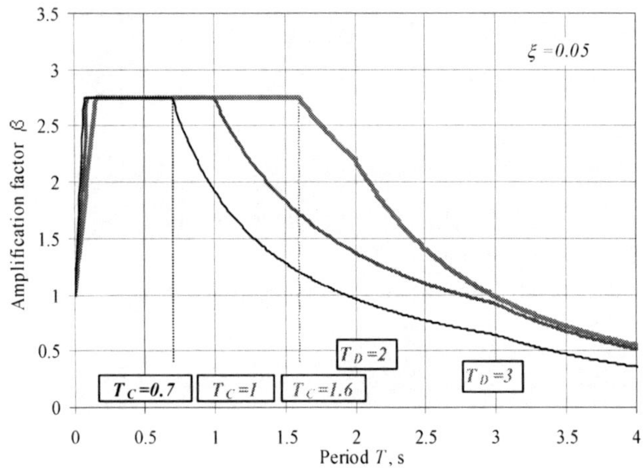

Figure 13: P100-1/2006 Normalized acceleration response spectra for horizontal ground motion

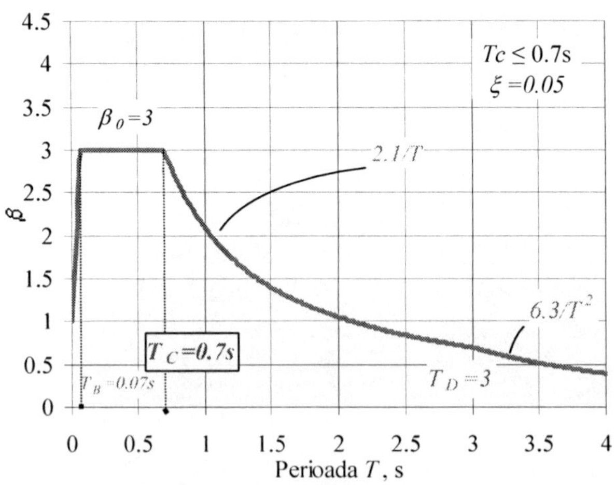

Figure 14: P100-1/2006 Normalized acceleration response spectrum for crustal sources

The new set of codes was prepared in 2005 at the Technical University of Civil Engineering in Bucharest (UTCB) following Eurocode 0 and Eurocode 1 format and contents. They were approved by the Ministry of Development, Public Works and Housing in 2007.

6. Conclusions

The P100-1/2006 code was written at UTCB in 2005 as part of a contract with the Ministry of Transport, Construction and Tourism and was published in the *Official Gazette*. The code uses a transitional definition of seismic action with MRI = 100 years instead of MRI = 475 years (as recommended by Eurocode 8), and it includes design spectra for various zones characterized by a control (corner) period of response spectra T_C = 0.7, 1.0 and 1.6 s. However, the code comes after decades of using a seismic design load having MRI = 50 years and T_C equal to either 0.7 s (pre 1977) or 1.5 s (between 1977 and 2007). In coming years, 10% in the 50-year exceedance hazard map will be introduced in the seismic code of Romania.

Acknowledgements

The authors acknowledge the JICA project accomplishments, and they are deeply indebted to their Japanese colleagues over the last 6 years for the $7.5 million Japan International Cooperation Project on Seismic Risk Reduction, implemented by NCSRR with support from UTCB and the National Institute for Building Research.

References

Aldea, A., 2002. Vrancea source seismic hazard assessment and site effects, Ph.D. thesis Technical University of Civil Engineering Bucharest, Bucharest, 256 p.

Ambraseys, N.N., 1977. Long-period effects in the Romanian earthquake of March 1977, *Nature*, Vol. 268, July, pp. 324–325.

Beles, A., Ifrim, M., 1962. *Elemente de seismologie inginerească*, Bucureşti, Editura Tehnică, 247 p.

Coburn, A., Spence, R., 1992. Earthquake protection. Wiley, 355 pp.

Constantinescu, L., Marza, V., 1980 (updated and revised 1995): A computer-compiled and computer-oriented catalogue of Romania's earthquakes duing a millennium (AD 984-1979), *Rev. Roum. GeoL, Geophys., Geogr., Ser Geophys.*, 24, 171–191, Bucharest.

Elnashai, A., Lungu, D., 1995. Zonation as a tool for retrofit and design of new facilities. Report of the Session A.1.2. 5th International Conference on Seismic Zonation. Proceedings Vol. 3, Quest Editions, Preses Academiques.

Gutenberg, B., Richter, C.F., 1956. Earthquake magnitude, intensity, energy and acceleration. *Bull. Seism. Soc. Am.*, 46, 105–145.

Lungu, D., Cornea, T., Aldea, A., Zaicenco, A., 1997. Basic representation of seismic action. In: *Design of structures in seismic zones: Eurocode 8 – Worked examples. TEMPUS PHARE* CM Project 01198: Implementing of structural Eurocodes in Romanian civil engineering standards. Editors D. Lungu, F. Mazzolani and S. Savidis. Bridgeman Ltd., Timisoara, pp. 9–60.

Lungu, D., Arion, C., Aldea, A., Demetriu, S., 1999. Assessment of seismic hazard in Romania based on 25 years of strong ground motion instrumentation, *NATO ARW Conference on strong motion instrumentation for civil engineering structures*, Istanbul, Turkey, 1999, 12 p.

Lungu, D., Aldea, A., Arion, C., Demetriu, S., Cornea, T., 2000. Microzonage Sismique de la ville de Bucarest – Roumanie, Cahier Technique de l'Association Française du Génie Parasismique, No. 20, pp. 31–63.

Lungu, D., Vacareanu, R., Aldea, A., Arion, C., 2002. RISK-U.E., Work Package 1: European Distinctive Features, Inventory Data base and Typology. International Conference Earthquake Loss Estimation and Risk Reduction, Bucharest, October 24–26, 2002.

Lungu, D., Demetriu, S., Aldea, A., Arion, C., 2006. Probabilistic Seismic Hazard Assessment for Vrancea Earthquakes and Seismic Action in the New Seismic Code of Romania, Proceedings of the *First European Conference on Earthquake Engineering and Seismology*, Geneva, September 3–8, Paper No. 1003.

Popescu, I., 1938. Cutremurele de pamint din Dobrogea. Analele Dobregei. Cernauti, pp. 22–24.

Radu, C., 1971. Catalogue of earthquakes in Romania: $1801–1900/I_0 \geq VII$, prior to $1800/I_0 \gtrsim VIII$. Manuscript (RCR), UNESCO, REM 70/172, Bucharest/Skopje.

Radu, C., 1979. *Catalogue of strong earthquakes originated on the Romanian territory. Part I-before 1901, Part II-1901-1979* (in Romanian) in "Seismological research on the earthquake of March 4, 1977 (Coordinators Cornea I., Radu C.,)", 723–752, ICEFIZ, Bucharest, 1979.

Radu, C., 1994. The revised and completed catalogue of historical earthquakes occurred in Romania before 1801. *European Seismological Commission*, XXIV General Assembly, Sept. 19–24, Athens, Greece, Proceedings (and enlarged manuscript).

Radu, C., Toro E., 1996. Two strong historical earthquakes in Transylvania (Romania): November 19, 1523 and October 3, 1880. Annali di Geofisica, 39, 5, pp. 1069–1078.

Shebalin, N.V. (ed.), 1977. Atlas of isoseismal maps for the "New Catalogue of Strong Earthquakes on the USSR territory from Ancient Times through 1975". Unpublished, Archive of the Institute of Physics of the Earth, Moscow.

Новый каталог сильных землетрясений на территории СССР (ред. Н.В.Кондорская, Н.В.Шебалин). М.: Наука. 1977. 536 с

Шебалин Н.В. Очаги сильных землетрясений на территории СССР. М.: Наука. 1974.

FOCAL MECHANISM SOLUTIONS FOR VRANCEA SEISMIC AREA

I. SANDU[*], A. ZAICENCO

Institute of Geology and Seismology, Academy of Sciences of Moldova

Abstract. The current study is focused on the problems of the analysis of focal mechanisms for the region of South-Eastern Carpathians, particularly of the Vrancea zone. The data from the ISC Bulletin for the period 1967–2006 were used in the study, to which the series of supplementary catalogues were added: HVD, M&P, ONC, ANSS, MED, ROM+, SBL, FSU, USGS in order to carry out a thorough complete analysis. As a result the focal mechanisms for about 250 catalogued events were obtained, which were projected on a uniform magnitude scale (M_w) for the geographic area in the limits: Latit. 44–50° and Longit. 24–30° for the continuum reference period of 40 years. Uncertainties in the data influencing focal mechanism solutions are discussed at the end of the paper.

Keywords: Focal mechanism, Vrancea earthquakes

1. Introduction

Reports of seismic events based on data in historical catalogues could be considered the first investigations of seismic source parameters (Ambraseys et al., 2003). Cataloguing source parameters on a systematic basis revealed the dynamic character of the physical process of mechanical energy release by the source and initiated the study of recurrence intervals of strong events (Kramer, 1996; Purcaru, 1979; Popescu et al., 2003). As a result, it has been established that in an average century, Vrancea seismic sources have generated at least five strong earthquakes with magnitudes $M_w > 6.0$ (Wenzel et al., 1999). The same rate of earthquake occurrence was observed in the 20th century with strong events on November 10, 1940, September 7, 1945, March 4, 1977, August 31, 1986 and May 30, 1990 (Engdahl and Villasenor, 2003). Statistical information aout seismic activity in the Vrancea region dates back to AD 1000 (Radu, 2003) and shows the unique peculiarities of this seismic source which limits its epicenters

---

[*] Institute of Geology and Seismology, Academy of Sciences of Moldova, e-mail: ilie_sandu@ yahoo.com

A. Zaicenco et al. (eds.), *Harmonization of Seismic Hazard in Vrancea Zone,* to an area of about 60 × 30 km and to a maximum hypocentral depth of about 180 km as is show in Figure 1. The most severe subcrustal seismic events have a relatively uniform distribution within intervals of depths of 80–150 km with a maximum credible magnitude of M = 7.5–7.8 (Gutenberg-Richter scale) (Drumea et al., 2003).

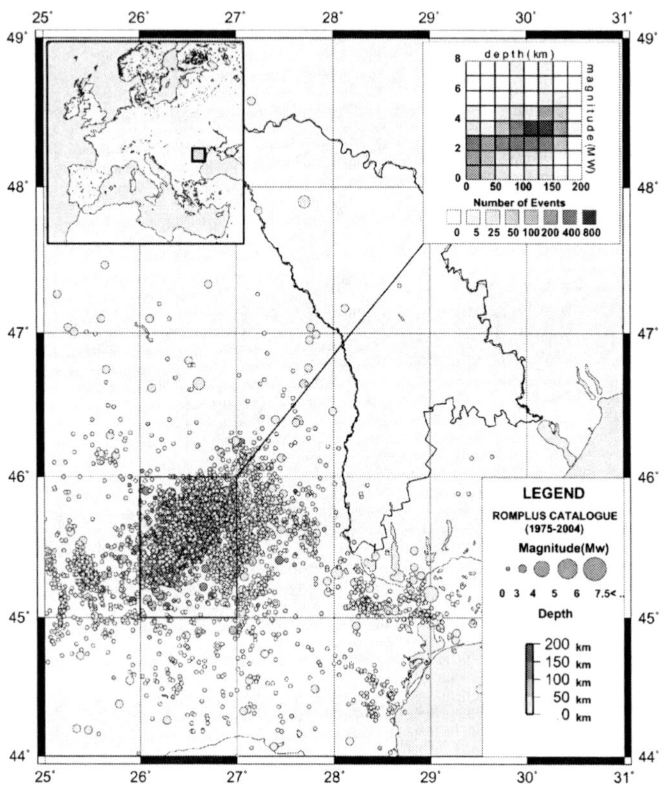

Figure 1: Vrancea zone

Regionally the Vrancea zone covers an area greater than 300,000 km^2 populated by more than 25 million people. The impact of seismic hazards on economic and social activities in the countries in the zone creates both a theoretical and a practical interest in a thorough analysis of seismic source parameters (Marza et al., 1991; Riznicenko et al., 1981). Recent investigations of strong earthquakes explain high values of seismic intensity in certain areas as the result of the directivity effect and of local soil conditions (Kramer, 1996; Lay and Wallance, 1995). Therefore, it is important to study the focal mechanisms and investigate their properties for both crustal and sub-crustal events. The catalogues of focal mechanisms in the Vrancea zone were compiled from rigorous analyses of the records of strong earthquakes (Trifu and Oncescu, 1987) using polarities of the P-waves, the method used by the authors in the current study and in their recent publications (Sandu and Zaicenco, 2007).

TABLE 1: The fault-plane solutions divergences from different source mechanism catalogues

Quake mm:dd:yyyy	M_w	Computed solution (1/2)			HVD, M&P*, ONC**		
		Str.	Dip	Rake	Str.	Dip	Rake
04:12:1969	4.9	347/191	57/36	−104/−70	329/232	73/69*	22/162*
08:01:1985	5.2	279/155	64/40	58/138	288/32	9/88	−14/−98
07:19:1987	4.8	189/80	36/76	−157/−56	178/248**	47/29**	248/292**
12:06:1989	4.4	158/60	26/86	−172/−65	34/232**	32/64**	72/101**
10:14:1990	4.6	264/160	19/85	15/109	203/89*	87/8*	−96/−23*
07:12:1992	4.1	205/309	29/82	−16/−118	126/29**	42/84**	188/311**
07:20:2001	4.9	241/62	15/75	90/90	119/17	44/48	17/162

TABLE 2: Magnitude divergences from different source mechanism catalogues

Quake mm:dd:yyyy	Magnitude, M_w					
	ISC	ROM+	ANSS	ONC	M&P	HVD
03:05:1967	4.1	5.0	–	–	–	–
01:06:1968	3.9	5.0	–	–	–	–
03:04:1977	6.5	7.4	7.3	–	6.5	7.5
11:05:1982	4.8	4.2	4.3	–	4.8	–
05:14:1984	4.2	3.4	4.2	4.6	–	–
04:27:1985	5.2	4.4	4.5	–	5.2	–
05:30:1990	7.3	6.9	7.1	–	7.3	6.9

Recently, the method of Centroid Moment Tensor (Ardeleanu and Radulian, 1998) has been used more frequently due to the advantages offered by digital records (Bala et al., 2003).

The objective of the current work is to consolidate available catalogues of focal mechanisms for the southeastern Carpathians (the Vrancea zone), to compute the parameters of the mechanisms from the available data and to compile a unique database for the time span 1967–2006.

A preliminary comparative analysis of solutions of focal mechanisms re-computed by the authors from ISC data and of solutions provided from Romanian local seismic networks (NIEP, INCERC) found inconsistencies for several events including those on July 20, 2001, July 12, 1992, October 14, 1990, December 6, 1989, July 19, 1987, August 1, 1985 and April 12, 1969. These inconsistencies were in the magnitude values and focal mechanism solutions (Tables 1–3).

TABLE 3: Magnitude scales used in catalogues of focal mechanisms

ISC	ANSS	HVD	M&P	ONC	ROM+	MED	FSU
M_S,m_B,M_w	m_B,M_L	M_w	m_B	M_S	M_w	MI,M_w	M_S

Based on these findings, we re-computed the focal mechanism solutions using the ISC catalogue and compared the results with already existing solutions (Radulian et al., 1996). A similar effort was undertaken recently in Italy (Pondrelli et al., 2006). In addition to being useful for statistical analysis, this database could demonstrate the complex process of mechanical energy accumulation and release by characterizing the stress-strain distribution within the zone.

2. Methodology

The assessment of magnitude values as measures of energy released during seismic events has an important role in seismic source study.

The use of a particular magnitude scale in these catalogues could be justified by the data available, by the recording equipment, by the methodology and by the objectives set by researchers (Engdahl and Villasenor, 2003). One of the primary goals of the current study was the application of a uniform magnitude scale (M_w) to the database of the focal mechanisms.

The body-wave magnitude m_B (used in ISC), which can be converted into M_w as well as into magnitude scales M_S and M_L (Utsu, 2002), was considered the primary energy parameter characterizing the source.

The application of a single magnitude scale and the compilation of a uniform database allowed a comparative analysis of the catalogues in chronological order (Sandu and Zaicenco, 2007). The final database of focal mechanisms is provided in Table 5.

The main objectives of the work are focal mechanism computations (slip, dip and strike angles) as well as axes P and T based on the initial data reported in the ISC catalogue and their comparison with solutions from other catalogues (Sandu and Zaicenco, 2007). Therefore, the methodology applied to derive these solutions was determined by the nature of the data and was based on the application of the double-couple method (Suetsugu, 1995a, Sandu and Zaicenco, 2007).

The main idea of the methodology is using the information about polarities (up, down) of body P waves that are recorded by the global network of seismic stations. In addition, the spatial solution (3-D focal sphere) and its planar projection as a diagram of the focal mechanism (Suetsugu, 1995b) require knowledge of the seismic station's coordinates and hypocentral location.

The orthogonal condition of the nodal planes allows the derivation of the parameters of both planes from the source characteristics (Suetsugu, 1995c). The final database reports both solutions (angles) of planes in order not to create confusion by selecting only one of them. The double-couple method cannot establish a single original solution of plane angles but only admits the equivalent existence of both planes for a point source (Suetsugu, 1995a).

A supplementary study that takes into account geomorphological peculiarities and GPS data on the crustal movements at the surface makes it possible to select a unique focal mechanism on the basis of the crustal stress-strain distribution (Yeats et al., 1997).

The applied computational procedure includes several stages (Sandu and Zaicenco, 2007; Sandu, 2005; Sandu, 2006) that are formulated as follows:

1. Conversion of the reported ISC data for each event into the focal sphere and graphical representation of the distribution of polarities;

   – Azimuth calculation for each station
   – Epicentral distance (curvilinear) calculation for each station
   – Computation of the takeoff angle or Ritsema curve considering the internal structure of the Earth, namely the model of P-wave velocity distribution with depth, along with the application of the Snell law of wave propagation
   – Projection of the 3-D focal sphere of signal polarities on a 2-D plane and the derivation of a focal mechanism diagram

2. Selection of nodal planes:

   – Automatic selection implies use of mathematical procedures for clustering and curvature optimization and separation in quadrants.
   – Manual selection (applied in the current study) implies visual selection of the nodal lines that separate quadrants.
   – Computation of source parameters on the basis of the derived diagram: strike, dip, axes T, P, and the direction of the displacements vector 'rake'.

In the following calculations, these two stages are described in detail. Assume the data about the location of the epicenter, north pole and seismic station on the globe are known. The azimuth angle and curvilinear epicentral distance are computed (Suetsugu and Yagi, 2005) as shown in Figure 2.

$$\Delta = \cos^{-1}(\cos L1 \cdot \cos L2 + \sin L1 \cdot \sin L2 \cdot \cos D2) \tag{1}$$

$$Azimuth = \cos^{-1}\left(\frac{\cos L2 - \cos \Delta \cdot \cos L1}{\sin \Delta \cdot \sin L2}\right) \tag{2}$$

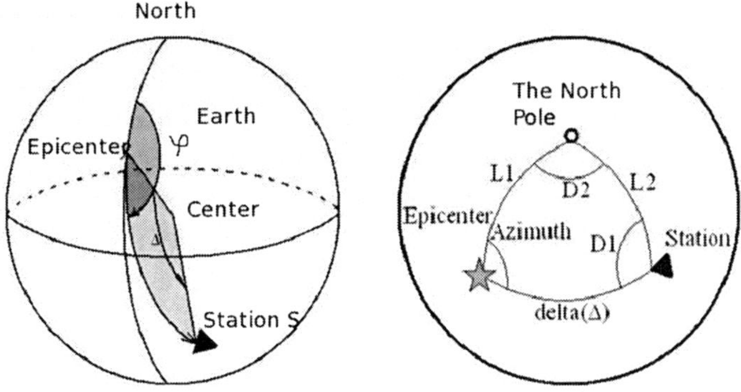

Figure 2: Determining the Station's Azimuth

where ϕ is the azimuth angle of the station with respect to the epicenter of the event, and Δ is the curvilinear distance from the epicenter to the recording station. L1, L2 and D1, D2 are the arc lengths and azimuths on a 3-D sphere.

The graphical presentations of the focal mechanisms include a list of 2-D projections of the solutions and a map of the Vrancea region indicating the corresponding tabulated events. The size of the solution plotted is proportional to the magnitude of the seismic event.

After positioning the stations in respect to the epicenter and establishing curvilinear epicentral distance, two major groups could be distinguished: (a) teleseismic ($\Delta > 18°$) and (b) local or regional ($\Delta < 18°$). For the teleseismic events, the global model of the Earth's structure applied is known as "Jeffrey's Model" (Suetsugu, 1995c). For the regional stations, the Earth's structure was updated from known regional studies to take into account local shallow-velocity structure (depths less than 50 km), which is particularly important for low-magnitude events in which the number of regional stations that report P-wave polarities dominates. Recent investigations performed by researchers from

TABLE 4: Regional velocity profile obtained from clustering and averaging the results of the Vrancea-99 project (Hauser et al., 2000)

Nr.	Depth interval (km)	Thickness (km)	V_p (km/s)
1	0–5	5	3.5
2	5–10	5	5.0
3	10–20	10	6.0
4	20–30	10	6.5
5	30–40	10	7.0
6	40–50	10	8.0

Karlsruhe University resulted in a new velocity profile for the Vrancea zone (Hauser et al., 2000). The project, Vrancea-99, interpreted tomography results using the refraction method and evidenced lateral heterogeneities. In order to incorporate these heterogeneities into the current study, geometric clustering and averaging were applied. In this way, a simplified model was obtained that provides the regional velocity profile shown in Table 4 and indicates separation between Mohorovich and Conrad surfaces.

Taking into account the spherical symmetry of the Earth, the Snell law is applicable for teleseismic and local waves in media with different properties:

$$p = \frac{r \cdot \sin i}{\upsilon} \tag{3}$$

where r is the position vector with its origin in the center of the Earth, v is the velocity of wave propagation and i is the incidence angle at the surface of media separation (Lay and Wallance, 1995). This is a general expression that allows the computation of the takeoff angle for each station as a function of the Earth's structure. Next, a 3-D projection is replaced with a 2-D one as is shown in Figure 3, (Suetsugu, 1995c).

Once the distribution of stations has been obtained, the nodal planes are selected. In this study, manual selection was used (Suetsugu, 1995c). The important parameter in establishing the optimal separation of quadrants is the ratio of consistent to inconsistent numbers of stations. In the current study, only focal solutions that maximize this ratio are provided.

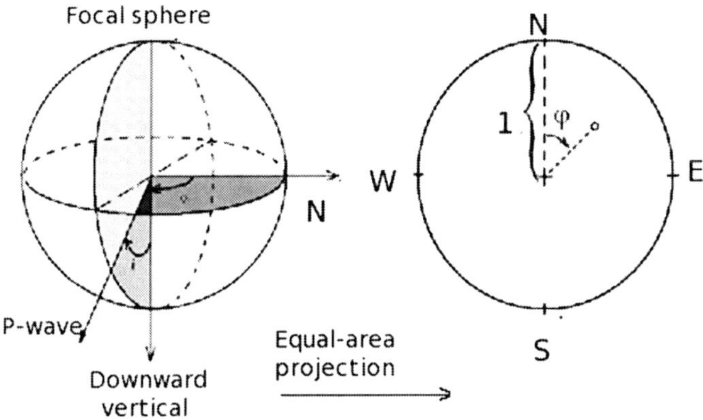

Figure 3: The focal mechanism diagram

Additional constraints are applied to specify the location of the nodal planes (Suetsugu, 1995a). They allow for obtaining the final values of the focal mechanism solution: strike, dip and rake angles and axes T and P.

3. Data Analysis

Data for the current study were taken from the following regional and global catalogues that refer to seismic events in the Vrancea zone:

- **ISC** – International Seismological Center (Catalogue/Bulletin since 1904) (ISC, 2002)
- **ANSS** – Advanced National Seismic System Catalogue (since 1898) (Allen, 1978; Dziewonski et al., 1981)
- **HVD** – Harvard Centroid Moment Tensor Catalogue (Huang et al., 1997; Giardini, 1984)
- **M&P** – Moscow CMT catalogue, by Mostreakov and Petrov (1964–1992) (Mostrioukov and Petrov, 1994)

- **ONC** – Oncescu, crustal CMT earthquake catalogue (1956–1994) (Radulian et al., 1996)
- **ROM+** – RomPlus catalogue: (1000–2004) (Oncescu and Bonjer, 1997)
- **MED** – European-Mediterranian Regional CMTs catalogue (1997–2004) (Pondrelli et al., 2002; Pondrelli et al., 2004)

The fault-plane solutions were obtained numerically according to the methodology described earlier using the polarities of the P-waves reported in ISC bulletin as the primary data. These solutions were compiled and compared with the known focal mechanisms from the catalogues HVD, M&P, ONC and MED. In the recent study (Sandu and Zaicenco, 2007) authors have added to the list of the analyzed catalogues the one reported by Shebalin (Shebalin and Leydecker, 1997) (referred to in the following as FSU).

ANSS and ROM+ catalogues were used to reduce uncertainties in values of magnitude and focal depth observed in data from ISC, HVD, ONC, MED and M&P.

The following are the main steps taken to perform the comparative analysis of the focal mechanism catalogues for Vrancea zone during the period 1967–2006 (Table 5, Figures 4–10):

1. Conducting search of catalogues and bulletins which reported seismic events for the selected area: Latitude 44–50°, Longitude 24–30°, and time span 1967–2006.
2. Establishing the computational methodology for the fault-plane solutions that is applicable for the available data sets. The reference catalogue has been assigned that provided data in the necessary format for the applied algorithm (Suetsugu, 1995c). In this study the ISC bulletin served as the reference catalogue.
3. Computing the focal mechanism solutions for the strong seismic events.
4. Applying a uniform magnitude scale for the analyzed seismic events (M_w).

TABLE 5: Focal mechanism parameters (1967–2006)

yyyy/mm/dd	Lat.	Long.	Depth, km	Strike 1/2	Dip 1/2	Rake 1/2	m_B	M_w, (ROM+)	Cons./ incons.
1967/02/27	26.69	44.86	32.0	188/310	50/58	−43/−130	4.7	4.7(5.0)	M&P
1967/03/05	26.78	45.80	140.0	*	*	*	4.0	4.1(4.9)	*
1967/04/04	26.32	45.72	161.0	*	*	*	4.4	4.5(5.0)	*
1967/05/26	26.17	45.43	163.0	*	*	*	4.1	4.2(4.8)	*
1967/07/25	26.55	45.68	165.0	*	*	*	3.8	4.0(4.7)	*
1967/10/27	26.70	45.80	99.0	*	*	*	3.7	3.9(4.5)	*
1968/01/06	26.46	45.76	173.0	171/305	57/43	−60/−127	3.7	3.9(5.0)	23/10
1968/02/09	26.42	45.61	128.0	221/47	60/30	87/95	4.7	4.7(5.0)	29/4
1968/02/24	26.55	45.74	142.0	348/93	52/71	−24/−140	4.3	4.4(3.6)	16/2
				351/185	48/42	81/101	4.3	4.4(3.6)	16/2
1968/08/14	26.41	45.68	129.0	*	*	*	4.3	4.4(4.8)	*
1968/09/21	26.31	45.55	147.0	*	*	*	4.2	4.3(4.7)	*
1968/10/20	26.59	45.81	120.0	240/63	29/61	88/91	4.6	4.6(5.0)	16/4
1968/11/20	26.60	45.65	124.0	*	*	*	4.0	4.1(4.7)	*
1968/11/26	27.85	45.71	46.0	*	*	*	4.4	4.5(4.7)	*
1969/01/15	26.55	45.62	129.0	343/133	55/39	108/66	4.6	4.6(5.0)	18/2
1969/04/12	25.12	45.31	23.0	347/191	57/36	−104/−70	4.9	4.9(5.2)	29/4
1969/07/27	26.40	45.72	165.0	*	*	*	4.3	4.4(4.8)	*
1969/12/21	26.91	45.68	37.0	*	*	*	4.6	4.6(4.6)	*
1970/01/02	26.35	45.50	133.0	*	*	*	4.7	4.7(3.6)	*
1970/05/30	26.70	45.90	104.0	*	*	*	3.8	4.0(2.4)	*
1970/06/05	26.57	45.66	131.0	*	*	*	4.4	4.5(3.6)	*
1970/07/09	26.58	45.80	138.0	322/53	17/90	−1/−107	4.4	4.5(3.8)	13/1
				319/98	61/37	−67/−125	4.4	4.5(3.8)	13/1
1970/07/10	25.83	47.95	33.0	267/72	69/22	96/76	4.7	4.7(4.7)	9/1
1970/08/04	25.40	47.85	33.0	*	*	*	3.8	4.0(4.0)	*
1971/07/18	26.50	45.72	145.3	287/191	73/73	−162/−17	4.5	4.5(3.8)	27/5
				162/30	42/59	50/120	4.5	4.5(3.8)	27/5
1971/07/22	26.59	45.43	152.9	*	*	*	4.2	4.3(3.3)	*
1972/04/16	26.38	45.52	130.5	262/117	19/74	56/101	4.4	4.5(3.8)	20/1
1972/08/23	26.70	45.79	90.3	157/327	60/30	95/82	4.9	4.9(4.0)	20/2
1972/10/01	26.56	45.77	145.0	270/66	22/70	113/81	4.4	4.5(3.8)	18/0
1973/01/05	26.43	45.50	135.4	132/327	37/54	−102/−81	4.4	4.5(3.6)	6/0
				128/294	31/60	103/82	4.4	4.5(3.6)	6/0
1973/03/12	26.08	45.60	129.4	*	*	*	3.8	4.0(2.6)	*
1973/03/31	26.73	45.77	147.2	*	*	*	4.3	4.4(3.1)	*
1973/08/20	26.52	45.73	70.4	263/168	19/88	−174/−71	5.3	5.4(6.0)	42/8
1973/09/07	26.64	45.81	126.7	191/46	70/24	76/123	4.6	4.6(3.8)	20/2
				170/284	23/80	−26/−111	4.6	4.6(3.8)	20/2
1973/10/23	26.48	45.72	171.1	310/103	46/47	109/72	4.9	4.9(4.3)	23/6
1973/12/21	26.60	45.65	154.8	274/112	76/14	−94/−73	3.9	4.0(3.0)	7/0
1974/02/22	26.49	45.66	149.1	287/53	27/74	141/69	4.2	4.3(3.3)	11/0
1974/06/10	26.55	45.71	147.0	216/124	70/83	8/160	4.0	4.1(3.3)	8/0
1974/07/17	26.61	45.76	134.9	154/345	69/21	86/100	5.0	5.0(4.6)	61/7
1975/02/08	26.13	45.19	23.0	241/22	30/66	125/72	4.5	4.5(3.3)	9/4
1975/03/07	26.72	45.93	21.0	251/15	57/50	−50/−134	4.9	4.9(4.5)	21/9
1975/03/08	26.67	45.67	146.0	304/138	47/44	−100/−79	4.3	4.4(3.3)	9/0
1975/03/31	26.46	45.56	140.0	113/206	89/14	104/3	4.6	4.6(4.0)	21/3

1976/10/01	26.54	45.71	142.1	350/228	23/77	34/110	5.0	5.0(6.0)	48/9
1977/03/04	26.71	45.82	85.8	279/52	75/21	105/46	6.1	6.5(7.4)	82/23
1977/03/05	27.09	45.54	9.6	267/21	71/40	−54/−149	4.7	4.7(4.2)	11/2
1977/04/20	26.12	44.18	11.0	318/17	16/82	32/104	3.9	4.4(3.0)	ONC
1977/06/16	26.63	45.77	148.0	287/148	72/23	−104/−50	*.*	*.*(4.7)	M&P
1978/01/01	26.51	45.69	141.8	285/95	41/50	97/84	5.0	5.0(5.1)	48/12
1978/01/15	26.75	45.77	141.0	311/213	73/63	29/161	4.7	4.7(4.8)	M&P
1978/09/05	26.47	45.66	154.2	293/121	84/6	89/98	4.5	4.5(4.8)	10/1
1978/10/02	26.54	45.71	160.8	124/324	39/53	74/103	4.9	4.9(5.2)	61/10
1978/09/30	26.60	45.67	150.3	220/7	42/53	115/69	4.1	4.2(4.4)	4/0
1979/05/31	26.29	45.57	120.8	249/73	67/23	88/94	5.1	5.2(5.3)	50/8
1979/09/11	26.31	45.59	153.6	262/29	28/73	140/68	5.0	5.0(5.3)	55/10
1980/01/14	26.59	45.78	141.3	229/76	12/79	64/95	5.0	5.1	42/7
1980/06/15	26.26	45.69	10.0	*	*	*	4.3	4.4(2.9)	*
1980/07/03	26.44	45.60	163.2	290/56	24/76	142/71	4.2	4.3(3.8)	9/0
1980/09/11	28.06	45.38	16.7	214/44	75/15	−93/−80	4.6	4.2(4.2)	17/4
				18/108	87/77	167/3	4.6	4.2(4.2)	17/4
1981/02/16	27.31	45.76	24.0	78/222	83/8	−85/−124	3.3	4.1(3.0)	ONC
1981/03/07	26.58	45.72	150.1	309/127	70/20	−88/−91	4.6	4.6(4.6)	M&P
1981/07/18	26.38	45.68	143.9	39/235	65/26	83/105	5.0	5.0(5.5)	55/19
1981/11/13	29.00	45.26	8.0	0/158	75/16	96/69	5.2	5.3(5.1)	48/18
				265/153	72/40	−127/−28	5.2	5.3(5.1)	45/21
1982/05/12	26.47	45.62	146.1	307/88	42/55	121/65	4.3	4.4(4.1)	10/0
				307/127	41/49	90/90	4.3	4.4(4.1)	10/0
1982/10/16	26.54	45.60	137.7	310/57	42/75	157/50	4.6	4.6(4.2)	11/0
1982/11/05	26.45	45.70	156.8	127/263	78/16	101/47	4.8	4.8(4.2)	M&P
1982/12/01	26.61	45.69	146.0	220/112	83/23	68/161	*.*	*.*(4.3)	M&P
1983/01/09	26.67	45.70	99.0	126/217	90/40	−49/−178	4.8	4.8(4.0)	M&P
1983/01/25	26.74	45.66	160.4	340/99	58/52	133/42	4.9	4.9(5.6)	60/14
1983/02/16	26.80	45.73	136.5	276/96	43/47	90/90	4.2	4.3(4.9)	32/9
1983/02/21	26.88	45.28	21.3	128/298	38/52	−82/−96	4.7	4.7(3.5)	26/8
1983/03/11	26.59	45.72	145.8	77/288	40/54	65/110	4.7	4.7(4.5)	38/17
1983/04/02	26.32	45.58	130.0	2/267	79/63	28/168	*.*	*.*(3.8)	M&P
1983/04/12	26.58	45.73	145.6	227/71	47/45	73/107	4.8	4.8(4.4)	37/9
1983/05/03	26.97	45.36	55.0	346/209	57/41	63/126	*.*	*.*(3.7)	M&P
1983/05/05	26.91	45.81	122.0	347/77	65/89	0/−154	*.*	*.*(4.6)	M&P
1983/05/21	26.50	45.63	152.3	87/188	65/69	156/27	4.5	4.5(4.1)	30/5
1983/06/06	26.79	45.72	135.8	121/333	55/40	70/115	4.3	4.4(4.1)	25/3
1983/08/17	26.30	45.55	165.7	212/105	68/54	40/153	4.1	4.2(4.1)	20/4
1983/08/25	26.51	45.62	132.5	63/238	41/49	94/87	4.5	4.5(3.8)	22/5
1983/08/30	26.43	45.57	146.0	247/158	82/43	48/168	*.*	*.*(3.7)	M&P
1983/09/20	26.47	45.67	170.0	350/260	89/88	2/179	*.*	*.*(4.3)	M&P
1983/12/26	26.46	45.65	156.7	207/11	83/7	92/74	4.5	4.5(3.9)	18/8
1984/01/20	26.33	45.51	135.9	231/90	48/49	62/118	4.7	4.7(4.4)	51/18
1984/02/07	26.35	45.49	135.9	157/310	55/38	106/69	4.8	4.8(4.1)	18/5
				74/192	51/60	140/47	4.8	4.8(4.1)	18/5
1984/02/11	28.81	45.39	133.0	*	*	*	3.7	3.9(3.2)	*
1984/02/12	26.71	45.73	132.5	340/95	28/78	153/65	3.7	3.9(3.2)	23/7
1984/05/14	26.86	45.20	47.6	124/23	29/84	−167/−62	4.1	4.2(3.4)	ONC
1984/06/05	26.82	45.07	37.0	252/52	32/59	107/80	3.3	4.1(3.2)	ONC
1984/08/07	26.44	45.66	160.5	205/97	72/47	46/155	5.0	5.0(4.2)	34/6

1985/01/28	26.48	45.65	145.4	194/100	77/75	16/167	4.6	4.6(4.2)	19/6
				300/190	32/78	24/119	4.6	4.6(4.2)	20/5
1985/02/04	28.12	45.29	11.0	197/76	38/69	−143/−58	3.3	4.1(3.2)	ONC
1985/03/05	26.38	45.64	164.0	67/218	51/43	109/68	*.*	*.*(3.6)	M&P
1985/03/17	26.45	45.39	133.0	19/209	22/69	81/94	*.*	*.*(4.7)	M&P
1985/03/26	26.60	45.69	144.8	204/323	71/35	120/34	4.5	4.5(4.9)	24/9
1985/04/27	26.67	45.69	84.6	20/118	76/60	−31/−164	5.1	5.2(4.4)	28/8
1985/06/21	26.52	45.63	122.3	14/227	47/48	66/113	4.5	4.5(4.5)	30/9
1985/06/26	26.50	45.73	124.0	16/257	57/54	45/138	*.*	*.*(3.8)	M&P
1985/06/26	26.90	45.41	10.0	20/150	86/6	95/40	*.*	*.*(3.1)	M&P
1985/07/08	27.42	46.07	27.0	212/87	67/36	61/139	*.*	*.*(3.3)	M&P
1985/07/17	26.44	45.69	166.0	297/46	82/24	113/20	*.*	*.*(4.4)	M&P
1985/08/01	26.70	45.79	119.9	235/29	39/54	111/74	4.7	4.7(5.2)	53/16
1985/08/01	26.51	45.73	105.3	279/155	64/40	58/138	5.1	5.2(5.8)	81/27
1986/03/04	26.46	45.64	166.0	318/194	62/44	55/137	*.*	*.*(3.8)	M&P
1986/03/16	26.81	45.72	137.0	310/197	64/50	45/146	*.*	*.*(3.8)	M&P
1986/04/27	26.94	45.48	41.3	209/28	39/51	91/89	5.0	5.0(3.7)	56/27
				66/188	38/68	142/58	5.0	5.0(3.7)	59/24
1986/04/27	26.95	45.45	71.4	29/259	76/22	74/22	4.3	4.4(3.1)	14/4
1986/08/16	26.34	45.58	154.3	280/106	67/23	88/96	4.8	4.8(4.7)	58/12
1986/08/17	26.49	45.70	128.0	120/28	82/76	14/172	*.*	*.*(4.4)	M&P
1986/08/30	26.31	45.53	137.0	248/28	70/26	106/52	6.3	7.0(7.1)	232/27
1986/09/02	26.40	45.53	150.6	272/71	38/54	107/77	4.3	4.4(4.5)	16/1
1986/09/17	26.26	45.53	159.0	10/100	88/90	0/−177	*.*	*.*(4.0)	M&P
1986/09/19	26.55	45.57	147.7	323/190	37/63	50/116	4.4	4.5(4.4)	22/6
1986/10/10	26.33	45.48	132.3	313/189	51/55	48/130	4.1	4.2(4.4)	23/2
				356/211	23/71	57/103	4.1	4.2(4.4)	23/2
1986/10/18	26.34	45.51	151.0	295/167	14/81	39/101	*.*	*.*(3.8)	M&P
1986/10/24	26.33	45.58	167.0	276/178	78/57	34/165	*.*	*.*(3.6)	M&P
1986/10/26	26.38	45.51	141.0	174/58	74/33	60/150	*.*	*.*(3.6)	M&P
1986/12/02	26.25	45.49	151.0	149/306	51/41	105/72	4.8	4.8(4.2)	18/4
1986/12/02	25.54	45.02	31.0	178/36	24/36	−125/−75	*.*	*.*(3.4)	M&P
1986/12/10	26.39	45.46	151.6	216/351	78/17	102/47	4.0	4.1(3.9)	14/3
				354/177	50/40	88/92	4.0	4.1(3.9)	14/3
1986/12/16	26.44	45.61	142.5	109/311	43/49	73/105	4.8	4.8(4.5)	63/21
1987/02/19	26.49	45.57	186.0	160/67	69/81	10/159	*.*	*.*(3.6)	M&P
1987/06/22	26.51	45.68	162.0	294/171	17/81	34/104	*.*	*.*(4.2)	M&P
1987/03/18	26.64	45.62	118.0	227/85	52/45	64/120	4.9	4.9(4.4)	34/4
1987/03/21	26.66	45.71	165.9	167/320	62/31	103/67	4.8	4.8(4.0)	22/0
1987/03/21	26.63	45.75	146.8	312/198	49/65	34/134	4.7	4.7(3.9)	17/1
1987/03/30	26.74	45.80	82.1	286/87	17/74	108/85	4.6	4.6(4.3)	27/7
1987/07/19	27.63	45.53	37.2	189/80	36/76	−157/−56	4.8	4.8(4.4)	25/7
1987/08/30	27.25	45.75	56.0	65/160	17/89	−4/−100	*.*	*.*(3.5)	M&P
1987/09/04	26.48	45.68	159.0	276/106	38/52	82/96	4.8	4.8(5.0)	37/13
				230/121	70/48	46/152	4.8	4.8(5.0)	37/13
1987/09/04	26.45	45.63	164.7	75/323	75/36	57/153	4.5	4.5(4.7)	24/5
1987/09/23	26.60	45.58	145.0	28/264	25/75	37/111	*.*	*.*(3.9)	M&P
1987/09/25	26.62	45.74	145.3	179/79	33/83	13/123	4.7	4.7(4.4)	40/8
1987/10/05	26.50	45.64	155.0	56/147	89/74	−15/−178	3.6	3.8(3.9)	M&P
1987/10/09	26.60	45.62	154.0	228/138	85/84	6/175	*.*	*.*(4.4)	M&P
1987/10/19	26.66	45.68	127.8	340/125	48/48	115/65	4.8	4.8(4.8)	31/2
1987/11/25	26.80	45.78	126.0	167/356	73/17	87/99	*.*	*.*(4.3)	M&P

1988/01/07	26.58	45.67	158.3	16/155	51/48	119/59	4.4	4.5(4.6)	15/1
				351/114	19/79	145/74	4.4	4.5(4.6)	15/1
1988/01/08	26.28	45.52	134.6	205/79	67/36	61/139	4.7	4.7(4.6)	51/13
1988/03/12	25.49	45.13	33.0	180/24	78/13	−94/−66	3.5	3.7(3.3)	M&P
1988/03/20	26.71	45.75	121.0	149/37	84/15	−103/−22	3.5	3.7(4.0)	M&P
1988/06/15	26.80	45.81	113.0	10/267	87/12	79/167	4.2	4.3(3.8)	M&P
1988/06/27	26.82	45.74	99.2	170/15	46/47	72/108	4.8	4.8(4.3)	18/3
1988/07/20	24.89	45.27	10.0	164/287	29/73	−36/−113	*.*	*.*(*.*)	M&P
1988/11/03	26.93	45.78	113.0	37/196	54/37	103/73	4.0	4.1(4.4)	M&P
1988/11/26	26.70	45.55	130.0	348/210	62/36	66/127	*.*	*.*(3.9)	M&P
1988/12/11	26.48	45.56	130.0	347/211	75/21	76/132	3.5	3.7(4.1)	M&P
1988/12/31	26.30	45.58	145.0	142/320	74/16	90/89	3.5	3.7(3.9)	M&P
1989/01/01	26.19	45.82	151.0	*	*	*	4.4	4.5(3.9)	*
1989/01/03	26.67	45.72	134.4	332/213	22/79	31/110	4.3	4.4(4.3)	17/0
1989/04/13	26.64	45.75	123.7	286/165	44/64	40/127	4.9	4.9(4.0)	20/5
1989/05/16	26.20	45.46	128.0	196/104	80/74	16/170	3.5	3.7(4.0)	M&P
1989/05/21	26.38	45.50	145.0	193/357	39/52	102/80	4.5	4.5(4.4)	18/3
1989/06/08	26.30	45.55	148.0	37/306	83/77	13/173	*.*	*.*(*.*)	M&P
1989/06/22	26.61	45.73	110.0	1/271	88/69	21/178	4.2	4.3(3.8)	M&P
1989/07/08	26.46	45.55	148.6	203/3	70/21	97/71	4.6	4.5(4.4)	26/6
1989/07/16	26.50	45.65	161.8	278/75	78/13	95/67	4.5	4.5(4.2)	19/2
1989/08/15	26.72	45.66	109.7	286/181	86/13	78/164	4.9	4.9(4.4)	26/4
1989/11/23	26.31	45.45	143.0	141/28	72/40	54/150	4.1	4.2(4.3)	M&P
1989/11/24	25.31	46.80	10.0	179/71	79/32	60/159	4.4	4.5(3.1)	M&P
1989/12/06	25.43	46.83	28.0	158/60	26/86	−172/−65	4.3	4.4(3.8)	20/5
1989/12/28	27.12	45.73	27.0	320/214	90/2	−91/−15	4.1	4.2(3.2)	M&P
1990/01/13	26.45	45.66	158.3	345/93	15/85	161/76	4.4	4.5(4.1)	17/2
1990/03/07	26.50	45.80	175.0	342/202	14/79	51/99	3.2	3.6(4.0)	M&P
1990/05/30	26.66	45.85	89.0	227/13	60/35	109/61	6.4	7.3(6.9)	237/49
1990/05/30	26.78	45.94	97.0	143/352	53/41	71/113	4.0	4.1(4.2)	14/6
				251/89	17/73	73/95	4.0	4.1(4.2)	14/6
1990/05/31	26.77	45.81	90.1	304/105	66/25	98/72	5.9	6.1(6.4)	173/44
1990/05/31	26.79	45.87	111.9	258/50	18/74	117/82	4.0	4.1(4.1)	15/1
1990/06/12	26.41	43.95	10.0	56/171	86/9	98/25	3.3	3.6(3.4)	M&P
1990/07/19	27.19	46.24	10.0	230/29	57/34	−78/−106	4.1	4.2(3.3)	M&P
1990/09/10	26.59	45.81	118.0	64/187	60/46	127/44	4.5	4.5(4.1)	M&P
1990/09/30	26.45	45.62	78.4	319/220	65/71	−159/−26	4.4	4.5(4.3)	33/13
				280/87	18/73	102/86	4.4	4.5(4.3)	35/11
1990/10/03	26.45	45.64	159.7	271/167	88/10	80/166	4.7	4.7(4.2)	27/5
1990/10/06	26.24	45.50	140.1	286/104	65/25	91/88	4.7	4.7(4.6)	49/10
1990/10/14	26.98	45.92	49.2	264/160	19/85	15/109	4.6	4.6(3.4)	17/5
1990/11/11	27.13	46.03	25.0	210/57	85/6	−92/−61	3.4	3.7(3.4)	M&P
1990/11/18	26.60	45.73	152.0	206/90	7/87	26/96	4.2	4.3(4.1)	M&P
1990/12/29	26.60	45.67	176.0	64/331	87/51	39/176	4.4	4.5(4.0)	M&P
1991/01/04	26.50	45.72	121.0	47/189	5/86	128/87	*.*	*.*(2.8)	M&P
1991/01/13	26.70	45.70	131.7	302/111	38/52	99/83	4.9	4.9(4.4)	80/13
1991/01/26	23.90	44.26	10.0	291/35	49/74	−22/−137	3.4	4.1(*.*)	ONC
1991/01/31	26.68	45.73	121.8	288/135	52/41	73/111	5.0	5.0(4.2)	66/15
1991/04/01	26.49	45.66	171.0	30/285	83/24	67/163	3.6	3.8(4.0)	M&P
1991/04/28	26.32	45.63	168.0	28/283	73/52	41/158	4.1	4.2(3.9)	M&P

1991/04/29	26.61	45.69	170.0	343/210	20/76	45/104	3.4	3.7(3.9)	M&P
1991/04/30	26.44	45.55	155.0	69/105	58/47	126/47	3.4	3.7(3.7)	M&P
1991/05/03	26.50	45.60	166.0	295/201	73/76	14/162	*.*	*.*(3.7)	M&P
1991/05/19	26.82	45.87	110.0	61/268	85/6	87/117	3.4	3.7(3.6)	M&P
1991/05/25	26.76	45.75	93.0	28/119	51/88	−1/−140	3.5	3.7(3.7)	M&P
1991/08/07	27.15	46.13	12.0	220/89	60/42	−120/49	2.8	3.7(3.2)	ONC
1991/08/31	26.97	45.48	24.0	40/275	76/24	71/143	3.2	4.0(3.1)	ONC
1991/09/01	26.92	45.48	33.0	225/31	66/25	−84/−102	4.9	4.9(3.5)	52/19
1991/09/01A	26.87	45.48	33.0	170/356	19/71	−96/−88	4.6	4.6(2.6)	15/2
				165/321	17/75	−66/−97	4.6	4.6(2.6)	15/2
1991/09/29	26.44	45.48	131.4	226/21	76/16	97/66	4.8	4.8(4.1)	19/2
				251/11	83/15	103/31	4.8	4.8(4.1)	19/2
1991/12/09	26.28	45.55	163.4	*	*	*	3.4	3.7(4.1)	*
1992/03/31	26.54	45.67	139.8	200/56	43/53	63/114	4.6	4.6(4.7)	55/11
1992/04/23	26.50	45.52	145.7	311/55	57/69	155/36	4.4	4.5(4.4)	29/8
1992/05/03	26.40	45.54	157.0	294/175	65/44	52/143	3.1	3.5(4.1)	18/0
				292/102	66/24	94/81	3.1	3.5(4.1)	18/0
1992/07/12	27.03	45.63	25.0	205/309	29/82	−16/−118	4.0	4.1(0.0)	14/4
1992/08/17	26.53	45.56	109.1	288/79	25/67	117/78	5.0	5.0(*.*)	40/8
1992/10/12	26.48	45.55	105.6	283/56	30/68	133/68	5.0	5.0(4.6)	42/16
1992/10/20	27.28	45.81	11.0	195/324	52/51	127/52	3.0	3.9(2.5)	ONC
1992/10/23	26.75	45.79	70.1	201/39	63/29	81/106	4.8	4.8(*.*)	29/17
1992/11/10	26.47	45.64	152.0	240/67	69/21	88/96	4.6	4.6(4.5)	34/9
1992/11/11	26.50	45.65	149.8	240/67	54/36	86/96	4.5	4.5(4.4)	38/7
1992/11/21	26.65	45.69	134.4	327/173	16/75	65/97	5.1	5.2(4.6)	86/16
1992/12/09	26.74	45.59	138.2	207/13	24/67	103/84	4.4	4.5(4.1)	22/3
1992/12/15	26.28	45.44	144.3	271/135	78/16	79/132	4.2	4.3(4.2)	30/3
1993/03/09	26.39	45.60	151.5	358/268	89/89	1/179	4.1	4.2(4.2)	18/2
				331/85	59/55	−42/−141	4.1	4.2(4.2)	18/2
1993/05/06	26.27	45.47	159.7	272/23	32/77	155/61	3.3	3.6(*.*)	11/0
1993/05/07	25.27	45.21	10.2	*	*	*	3.5	3.7(3.0)	*
1993/05/20	25.46	45.38	10.0	*	*	*	3.8	4.0(3.0)	*
1993/05/23	25.52	45.36	10.0	*	*	*	4.9	4.9(3.4)	*
1993/07/30	26.59	45.69	144.5	244/76	80/10	88/102	4.7	4.7(4.4)	36/14
1993/08/26	26.56	45.70	144.1	219/31	57/33	94/83	5.0	5.0(4.3)	67/17
1993/09/18	26.49	45.68	159.0	229/42	32/58	96/86	4.9	4.9(4.0)	9/0
1993/09/21	26.99	45.42	10.0	100/190	88/90	−180/2	3.7	4.3(2.4)	ONC
1993/09/29	27.11	44.47	10.0	301/196	57/50	50/135	3.4	4.1(3.1)	ONC
1993/11/04	26.62	45.69	131.2	185/21	65/26	83/105	4.7	4.7(4.1)	17/4
1994/02/19	26.38	45.59	163.0	311/111	64/27	99/72	4.3	4.4(4.2)	11/1
				2/250	57/61	36/141	4.3	4.4(4.2)	11/1
1994/02/27	26.77	45.74	130.1	134/257	55/52	131/46	3.7	3.9(4.1)	20/3
1994/03/04	26.75	45.70	152.4	343/183	46/46	76/104	4.3	4.4(4.1)	11/0
1994/03/05	28.70	44.97	50.3	*	*	*	3.0	3.4(3.1)	*
1994/06/06	26.64	45.68	131.7	217/311	88/22	112/5	4.8	4.8(4.3)	20/8
1994/08/10	23.96	47.16	39.0	58/311	46/73	−156/−46	3.9	4.4(3.3)	ONC
1994/09/06	26.43	45.46	118.3	283/17	84/49	139/7	4.9	4.9(4.2)	17/1
1994/10/18	26.44	45.55	160.5	272/13	78/83	173/4	5.3	5.4(4.3)	18/1
1995/01/28	25.07	45.62	10.0	292/99	50/40	−82/−100	3.7	3.9(2.9)	8/0
1995/03/09	26.47	45.69	148.8	204/48	37/55	70/104	3.9	4.0(4.0)	12/0
				72/238	73/17	94/77	3.9	4.0(4.0)	12/0

1995/03/14	26.72	45.74	91.5	*	*	*	4.0	4.1(3.4)	*
1995/03/30	26.45	45.53	132.2	*	*	*	4.2	4.3(3.6)	*
1995/06/25	26.83	45.71	96.8	*	*	*	3.6	3.8(3.5)	*
1995/08/20	27.62	45.05	43.4	*	*	*	3.4	3.7(*.*)	*
1995/09/06	26.40	45.52	130.7	300/136	41/50	78/100	4.2	4.3(4.1)	23/2
1995/10/16	25.01	47.52	10.0	*	*	*	3.7	3.9(3.2)	*
1996/01/30	26.61	45.69	100.6	135/234	76/59	147/16	4.0	4.1(3.9)	16/8
1996/02/03	26.97	45.79	113.7	78/295	36/60	59/110	3.0	3.4(3.7)	11/0
1996/03/02	26.35	45.54	144.1	*	*	*	3.3	3.6(3.8)	*
1996/04/19	26.45	45.70	138.2	37/210	23/67	96/88	3.9	4.0(3.9)	11/0
				84/174	87/89	179/3	3.9	4.0(3.9)	11/0
1996/04/24	26.45	45.62	117.9	67/240	35/55	95/86	3.5	3.7(3.7)	12/1
				205/111	76/74	16/165	3.5	3.7(3.7)	12/1
1996/04/26	26.90	45.71	4.0	*	*	*	3.5	3.7(2.8)	*
1996/05/09	26.66	45.69	60.8	*	*	*	3.9	4.0(3.2)	*
1996/05/12	27.38	45.70	26.1	*	*	*	3.7	3.9(3.0)	*
1996/05/25	26.83	45.59	64.3	*	*	*	3.3	3.6(3.1)	*
1996/06/07	26.39	45.46	133.7	90/238	59/36	108/63	4.4	4.5(4.6)	36/7
1996/06/11	27.38	45.61	42.9	*	*	*	4.1	4.2(3.1)	*
1996/06/26	26.39	45.52	122.2	200/19	78/12	90/89	3.7	3.9(3.9)	16/1
				189/99	90/84	6/180	3.7	3.9(3.9)	15/2
1996/08/14	26.71	45.77	149.1	*	*	*	4.3	4.4(4.1)	*
1996/09/01	26.51	45.68	163.0	*	*	*	3.2	3.6(3.8)	*
1996/09/27	26.78	45.74	128.3	88/187	70/67	155/22	3.8	4.0(3.9)	15/0
				186/359	61/29	94/83	3.8	4.0(3.9)	14/1
1997/03/01	26.65	45.71	118.2	278/110	60/30	84/101	3.9	4.0(4.2)	24/6
				284/142	60/36	68/123	3.9	4.0(4.2)	25/5
1997/03/13	28.80	45.15	*	*	*	*	3.6	3.8(2.8)	*
1997/03/27	26.21	45.40	143.2	95/244	52/42	109/67	3.5	3.7(3.8)	12/2
1997/04/02	27.63	45.67	5.0	*	*	*	3.0	3.4(2.9)	*
1997/04/28	27.16	45.23	25.0	*	*	*	3.0	3.4(2.7)	*
1997/05/17	28.39	45.17	5.0	*	*	*	3.3	3.6(2.8)	*
1997/05/31	26.50	45.61	3.2	*	*	*	3.2	3.6(3.8)	*
1997/07/14	26.67	45.83	138.3	*	*	*	3.8	4.0(4.2)	*
1997/07/24	26.53	45.60	135.9	*	*	*	3.9	4.0(*.*)	*
1997/07/27	27.42	45.96	97.0	*	*	*	3.9	4.0(*.*)	*
1997/10/01	26.38	45.60	149.7	*	*	*	3.8	4.0(3.8)	*
1997/10/11	26.73	45.72	115.4	57/306	72/42	51/153	4.4	4.5(4.5)	23/6
1997/10/22	26.34	45.53	138.0	*	*	*	3.8	4.0(3.9)	*
1997/11/06	28.22	46.75	10.0	*	*	*	3.9	4.0(3.1)	*
1997/11/18	26.68	45.68	124.6	237/327	90/82	−8/−180	4.4	4.5(4.7)	30/7
1997/11/25	27.89	45.80	59.1	*	*	*	3.4	3.7(3.2)	*
1997/12/06	27.01	45.64	40.7	46/261	52/44	67/116	3.9	4.0(3.2)	6/0
1997/12/18	26.14	45.58	135.3	*	*	*	3.9	4.0(3.9)	*
1997/12/30	26.34	45.56	139.1	221/74	15/77	57/98	4.5	4.5(4.6)	42/6
1998/01/14	26.59	45.68	154.2	*	*	*	3.9	4.0(4.0)	*
1998/01/19	26.66	45.61	106.0	*	*	*	3.9	4.0(4.0)	*
1998/01/31	26.34	45.45	141.7	*	*	*	3.5	3.7(3.6)	*
1998/03/13	26.26	45.57	156.2	49/269	83/9	84/130	4.7	4.7(4.7)	52/22
				238/108	21/76	42/106	4.7	4.7(4.7)	53/21
1998/04/14	26.56	45.70	150.0	*	*	*	3.6	3.8(3.8)	*

1998/04/23	26.68	45.77	112.3	*	*	*	3.1	3.5(3.8)	*
1998/05/04	26.51	45.69	148.0	*	*	*	4.0	4.1(4.0)	*
1998/07/03	26.80	45.67	147.5	*	*	*	3.8	4.0(4.2)	*
1998/07/27	26.47	45.63	135.5	*	*	*	4.3	4.4(4.4)	*
1998/08/14	27.32	45.53	14.3	*	*	*	4.1	4.2(2.7)	*
1998/12/29	26.54	45.53	116.5	*	*	*	3.7	3.9(3.8)	*
1999/01/23	26.51	45.63	141.3	*	*	*	3.9	4.0(4.1)	*
1999/03/17	26.50	45.69	144.5	*	*	*	4.1	4.2(4.1)	*
1999/03/22	26.30	45.47	144.2	*	*	*	4.0	4.1(4.4)	*
1999/03/23	26.51	45.72	164.4	*	*	*	3.2	3.6(4.0)	*
1999/04/04	26.54	45.66	154.0	*	*	*	3.0	3.4(3.7)	*
1999/04/28	26.22	45.48	150.2	126/279	41/52	111/73	5.1	5.2(5.3)	40/9
				213/2	59/35	107/63	5.1	5.2(5.3)	40.9
1999/04/29	26.41	45.59	145.3	*	*	*	4.1	4.2(4.0)	*
1999/05/25	26.43	45.52	130.1	*	*	*	4.0	4.1(3.9)	*
1999/06/29	26.47	45.60	128.0	*	*	*	4.2	4.3(4.2)	*
1999/07/13	26.49	45.55	140.3	*	*	*	3.7	3.9(4.0)	*
1999/11/08	26.39	45.51	134.5	286/62	43/57	125/62	4.2	4.3(4.6)	32/0
1999/11/14	26.27	45.48	130.4	*	*	*	4.6	4.6(4.6)	*
2000/01/19	26.83	48.05	10.0	*	*	*	3.9	4.0(3.6)	*
2000/03/08	26.80	45.87	27.4	*	*	*	4.6	4.6(4.4)	*
2000/04/06	26.58	45.73	136.8	234/57	53/37	88/93	4.9	4.9(5.0)	48/4
2000/05/10	26.49	45.55	134.2	*	*	*	3.9	4.0(4.1)	*
2000/07/01	26.77	45.76	65.0	*	*	*	3.6	3.8(3.4)	*
2000/11/30	26.43	45.61	139.7	*	*	*	3.8	4.0(4.1)	*
2001/03/04	26.26	45.49	146.6	302/107	63/28	97/76	4.3	4.4(4.8)	31/8
2001/03/18	26.30	45.46	146.4	337/78	67/68	156/25	3.9	4.0(4.1)	10/0
2001/03/28	26.70	45.72	131.7	*	*	*	3.7	3.9(4.3)	*
2001/05/20	26.48	45.59	148.1	300/201	68/68	24/156	3.7	3.9(4.2)	10/0
2001/05/24	26.42	45.69	142.4	215/49	60/31	83/102	4.8	4.8(4.9)	78/9
2001/07/20	26.71	45.74	127.1	241/62	15/75	90/90	4.9	4.9(4.8)	40/12
2001/08/01	26.60	45.68	132.2	*	*	*	3.7	3.9(4.1)	*
2001/10/17	26.50	45.62	89.8	268/127	30/66	55/108	4.7	4.7(4.2)	27/4
2002/01/25	26.76	45.64	138.1	*	*	*	3.9	4.0(4.0)	*
2002/03/11	25.81	44.93	9.8	*	*	*	3.8	4.0(3.1)	*
2002/03/16	26.54	45.56	146.5	*	*	*	3.7	3.9(4.3)	*
2002/05/03	26.33	45.63	158.0	100/273	77/14	91/84	4.4	4.5(4.6)	21/6
2002/06/14	26.60	45.67	139.7	*	*	*	3.9	4.0(4.0)	*
2002/08/03	26.61	45.69	145.6	*	*	*	4.0	4.1(4.4)	*
2002/08/27	26.41	45.55	151.2	*	*	*	2.9	3.4(4.0)	*
2002/09/06	26.40	45.59	102.9	*	*	*	4.1	4.2(4.1)	*
2002/09/10	26.85	45.73	129.5	*	*	*	3.4	3.7(4.0)	*
2002/11/03	26.91	45.78	112.7	*	*	*	4.0	4.1(4.0)	*
2002/11/30A	26.50	45.69	172.1	*	*	*	5.0	5.0(4.7)	*
2002/11/30B	26.40	45.72	138.1	*	*	*	3.9	4.0(2.5)	*
2002/12/15	26.75	45.78	115.7	*	*	*	3.5	3.7(3.6)	*
2002/12/30	26.56	45.71	156.5	222/30	77/13	93/79	4.1	4.2(4.1)	14/1
2003/04/13	26.28	45.46	141.3	*	*	*	3.5	3.7(4.0)	*
2003/08/02	26.44	45.62	156.5	*	*	*	3.3	3.6(4.1)	*
2003/10/05	26.32	45.65	151.6	69/235	84/7	92/76	4.5	4.5(4.6)	27/7
2003/11/02	26.87	45.75	103.0	*	*	*	3.4	3.7(3.7)	*

2004/01/21	26.55	45.61	117.4	*	*	*	4.0	4.1(4.1)	*
2004/02/07	26.62	45.74	138.0	206/44	51/40	78/104	4.5	4.5(4.4)	23/2
2004/03/17	26.53	45.79	88.0	*	*	*	4.0	4.1(4.1)	*
2004/04/04	26.50	45.71	150.0	200/102	42/83	11/131	4.8	4.8(4.3)	15/0
2004/04/30	26.86	45.58	33.0	*	*	*	4.5	4.5(3.2)	*
2004/07/22	26.34	45.62	154.0	295/34	71/64	153/21	3.9	4.0(4.2)	15/0
2004/09/27	29.73	46.27	0.0	*	*	*	5.1	5.2(3.2)	*
2004/09/27	26.45	45.70	147.0	226/52	56/34	87/95	5.0	5.0(4.6)	32/6
				172/348	48/42	92/88	5.0	5.0(4.6)	33/5
2004/10/03	28.95	45.21	0.0	*	*	*	4.7	4.7(3.1)	*
2004/10/24	26.14	45.59	153.3	*	*	*	3.7	3.9(4.4)	*
2004/10/27	26.60	45.79	93.0	337/181	24/67	68/100	5.5	5.6(6.0)	44/5
2004/11/17	26.69	45.73	129.0	*	*	*	4.0	4.1(4.4)	*
2004/11/23	26.75	45.68	115.4	*	*	*	3.7	3.9(4.0)	*
2005/01/10	26.50	45.82	102.0	207/34	62/29	87/96	4.6	4.6(3.8)	9/0
				62/332	90/89	−179/0	4.6	4.6(3.8)	9.0
2005/02/17	26.60	45.65	116.0	*	*	*	4.0	4.1(3.8)	*
2005/03/06	26.45	45.78	132.0	285/30	63/64	151/30	4.0	4.1(4.1)	13/0
2005/05/09	26.32	45.50	145.0	*	*	*	3.9	4.0(4.1)	*
2005/05/14	29.64	45.64	0.0	*	*	*	5.0	5.0(*.*)	*
2005/05/14	26.51	45.71	143.0	207/17	72/18	93/80	5.1	5.2(5.1)	50/6
2005/05/14	26.55	45.70	150.0	*	*	*	4.0	4.1(4.2)	*
2005/06/18	26.68	45.72	143.0	281/105	17/73	86/91	4.9	4.9(4.9)	25/4
2005/07/16	26.67	45.81	95.0	*	*	*	3.9	4.0(3.9)	*
2005/08/17	26.39	45.62	150.0	*	*	*	4.4	4.5(4.1)	*
2005/08/21	27.89	45.53	10.0	162/322	55/37	102/73	4.0	4.1(3.0)	8/0
				295/205	90/90	0/180	4.0	4.1(3.0)	8/0
2005/08/23	26.36	45.58	138.0	*	*	*	4.0	4.1(4.0)	*
2005/09/05	26.74	45.76	80.0	257/166	65/87	−177/−25	4.4	4.5(4.3)	18/1
2005/09/08	26.37	45.53	140.0	59/316	73/54	38/159	4.3	4.4(4.3)	19/0
				276/95	54/36	91/89	4.3	4.4(4.3)	17/2
2005/09/10	27.15	45.27	32.0	*	*	*	4.0	4.1(3.2)	*
2005/09/17	26.38	45.60	152.0	*	*	*	4.0	4.1(4.2)	*
2005/12/13	26.79	45.78	144.0	*	*	*	5.0	5.0(4.6)	*
2006/01/04	29.32	45.27	113.0	*	*	*	4.0	4.1(3.0)	*
2006/01/30	27.91	45.99	10.0	*	*	*	4.0	4.1(3.3)	*
2006/02/16	26.71	45.72	135.0	270/25	64/50	134/36	4.0	4.1(3.9)	13/0
2006/02/19	25.94	44.03	33.0	*	*	*	3.3	3.6(3.1)	*
2006/03/06	26.55	45.72	145.0	200/43	36/56	71/103	4.6	4.6(4.6)	17/1
2006/03/19	26.49	45.70	153.0	204/75	43/60	48/122	4.3	4.4(4.4)	9/0
2006/03/27	26.45	45.56	125.0	228/50	77/13	89/92	3.7	3.9(4.0)	14/0
2006/03/31	26.93	45.88	98.0	236/56	30/60	89/90	3.9	4.0(3.8)	13/0

4. Aspects of Uncertainty Analysis in the Input Data

The main input data for determining focal mechanism solutions are (a) event location parameters, i.e., origin time (OT) and hypocenter coordinates; (b) polarities of the first P-waves and (c) velocity structure of the Earth.

First, the velocity model is used by ISC to localize the seismic event and report it in the bulletin. The second model used in the numerical computations to build the focal mechanism diagram and provide source parameters is Jeffrey's Model. An adjustment of the regional crustal structure (depths 0–50 km) was possible due to recent results obtained in the Vrancea-99 project.

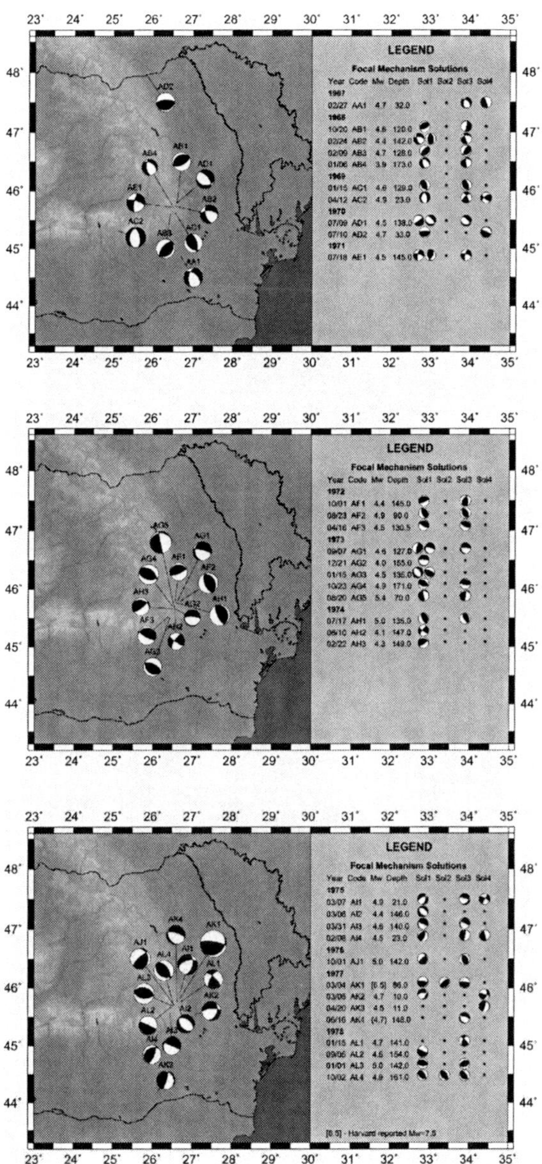

Figure 4: Focal mechanisms for Vrancea zone (1967–1978)

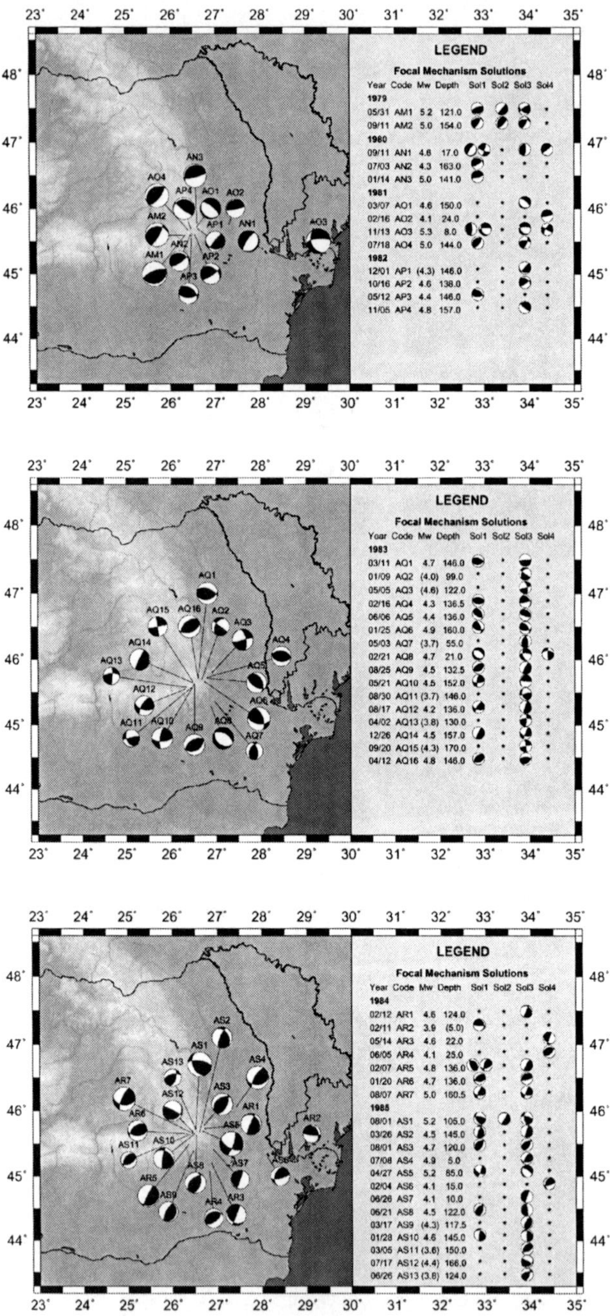

Figure 5: Focal mechanisms for Vrancea zone (1979–1985)

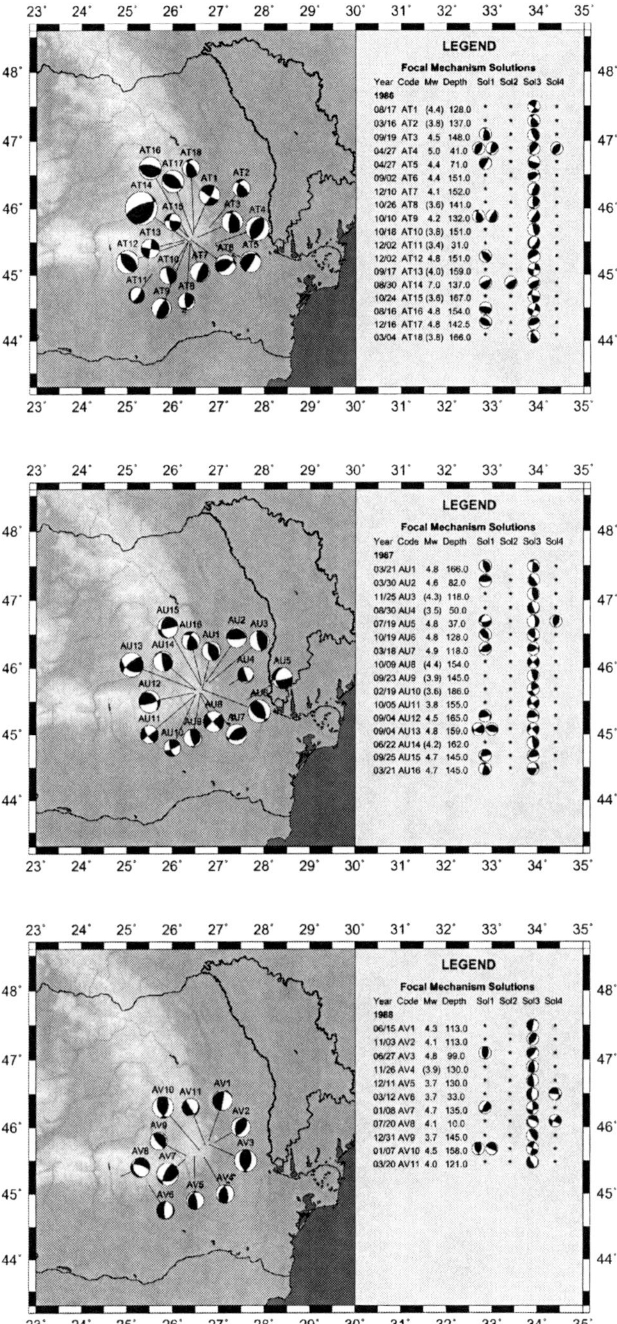

Figure 6: Focal mechanisms for Vrancea zone (1986–1988)

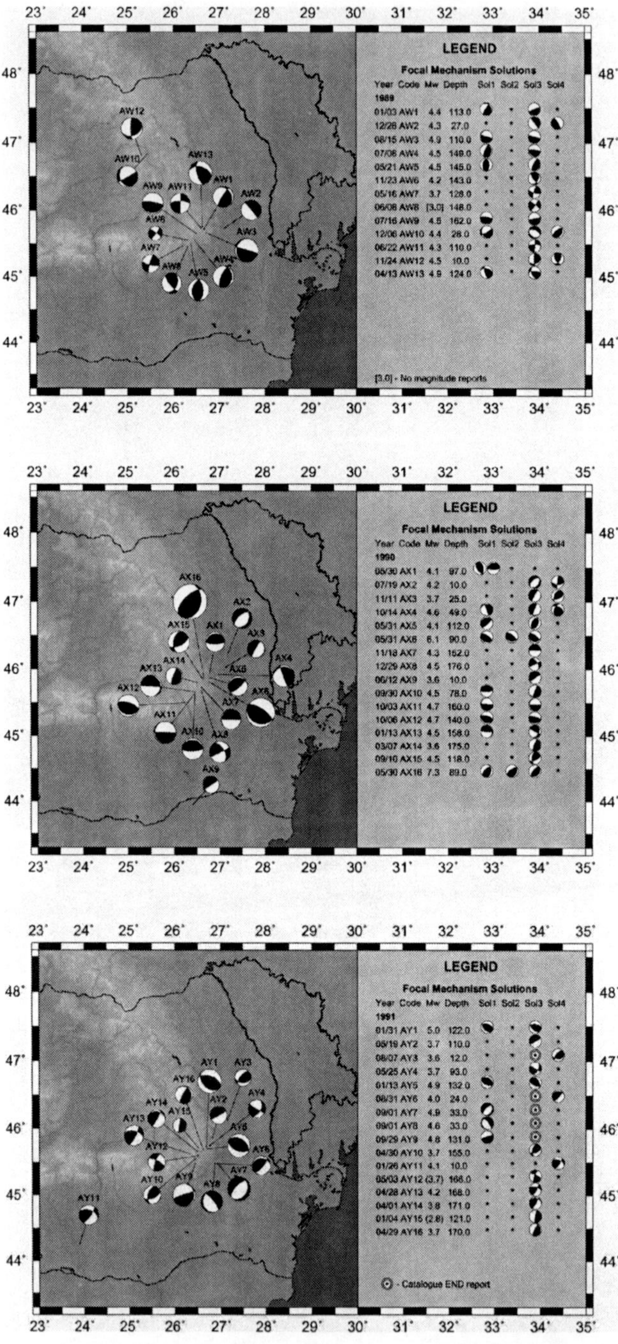

Figure 7: Focal mechanisms for Vrancea zone (1989–1991)

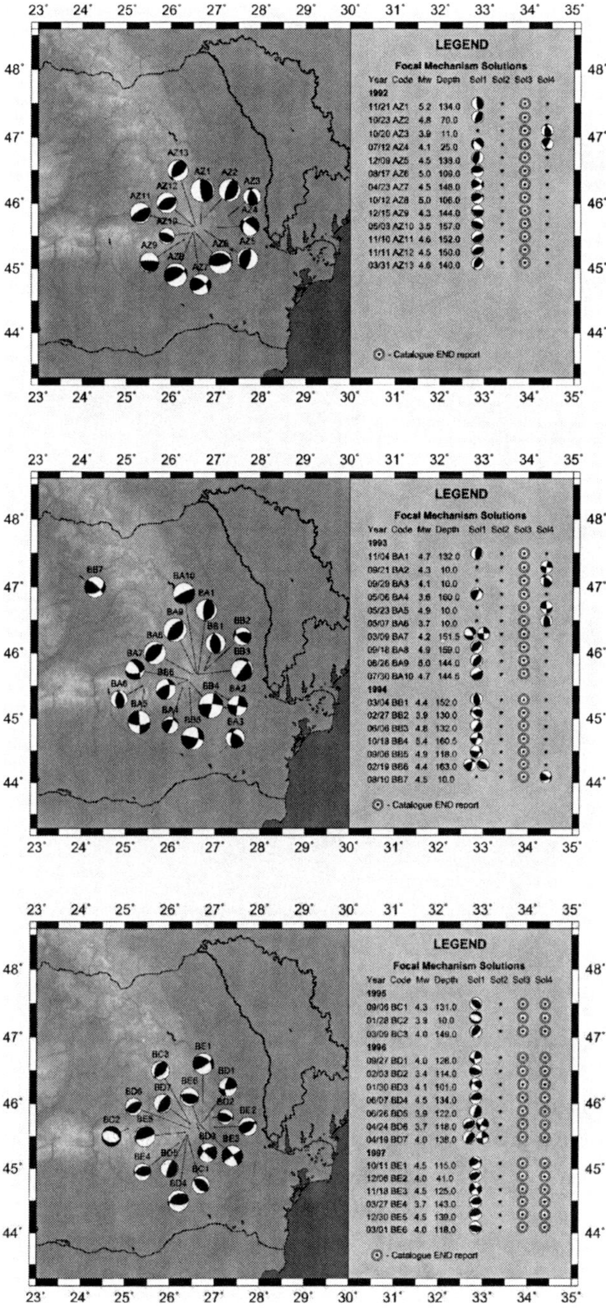

Figure 8: Focal mechanisms for Vrancea zone (1992–1997)

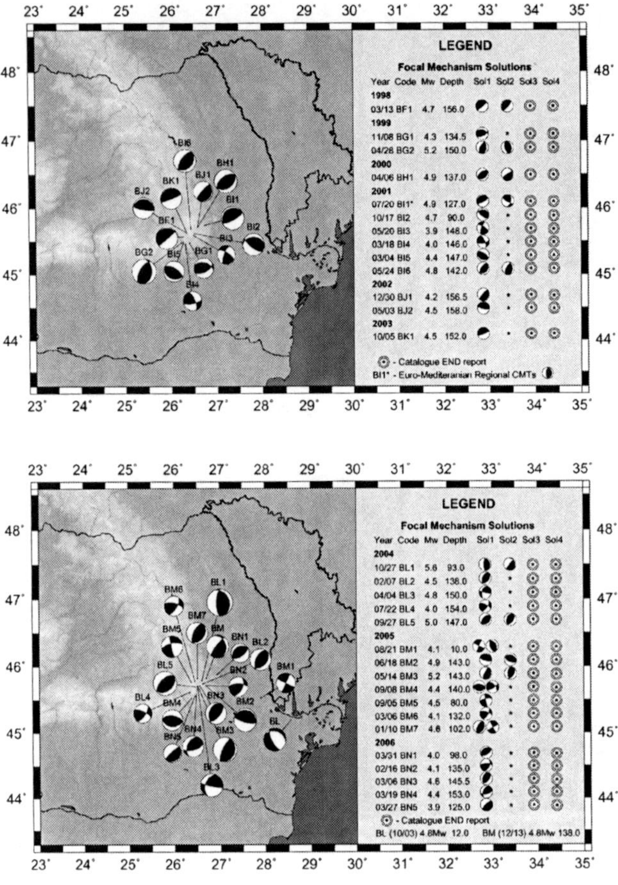

Figure 9: Focal mechanisms for Vrancea zone (1998–2006)

Uncertainties in the velocity model directly influence the take-off angle, giving rise to variations in the values of the epicentral distances of the stations (radial errors in the focal diagram) and azimuth (for lateral heterogeneities). In this way, nodal planes and dip, rake and strike angles are influenced directly by these uncertainties.

The current study did not consider lateral heterogeneities, and computations were done with a 1-D model for the longitudinal wave velocity profile which was clustered and averaged to build a horizontally layered model. Uncertainties in the polarities of P-waves could be caused only by the non-sampling errors. Therefore, the values of the polarities might be considered as deterministic variables. These errors directly influence the determination of the OT values through the Wadati diagram and earthquake localization values.

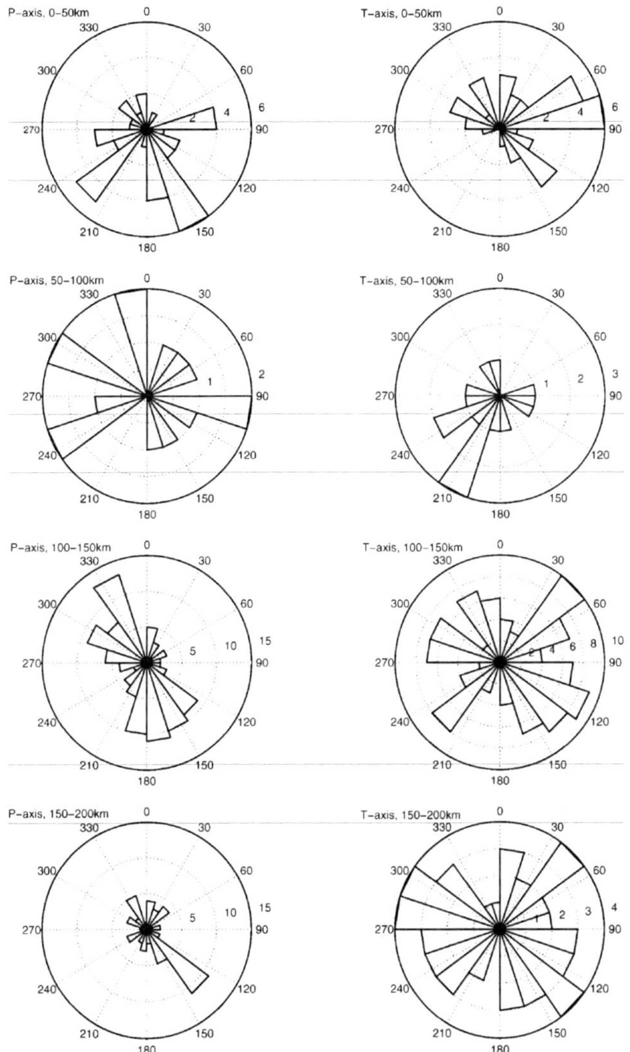

Figure 10: Angle histograms of P- and T-axis of mechanisms for Vrancea zone: depth intervals 0–50, 50–100, 100–150 and 150–200 km

The level of confidence for the focal mechanism solution is proportional to the number of stations that report P-wave polarities. Their number should be greater than six with a homogeneous distribution in the focal diagram, i.e., four stations distributed over four quadrants and two stations for control. The study of 292 seismic events (ISC) produced 173 (34 double and 139 unique) focal mechanism solutions that entirely meet this condition. Two thirds of the solutions were derived on the basis of more then 20 stations, Figure 11, which inspired a high degree of confidence in the results obtained.

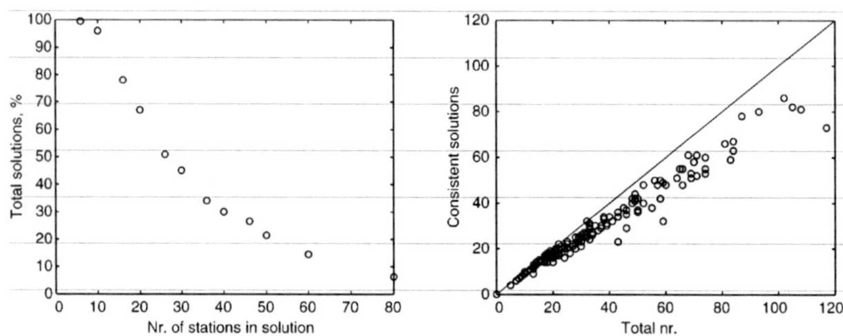

Figure 11: Percentage of number of stations in solutions, and total number of stations in solutions vs. consistent results

The degree of confidence in the results obtained in this study is affected by the presence of errors in measurements and computational results provided by ISC. Parameters of the localization of seismic events in space-time coordinates – OT, epicenter, focal depth – are the main sources of errors in the catalogues. For the entire set of 292 events, these parameters are:

- OT errors are distributed over the interval [0 3.6] s, with 3% of them exceeding 1 s.
- Epicentral location errors are represented by the axis of ellipse (S_{max} and S_{min}); 99% of them do not exceed 25 km.
- Focal depth errors are distributed over the interval [0 30.1] km, with less then 4% of them exceeding 10 km.

The fluctuation of errors in localizing an earthquake with a magnitude $3 \leq M_w < 8$ reported by ISC is almost constant for the entire period of observation. This is explained by the precision of the recording equipment and the density of the seismic network at the regional and global scales.

Out of the entire set of ISC seismic events (292) for the period 1967–2006, 15% (45 events) are reported without any information about errors in the values of focal depths. To these "dark horses", another nine events with OT values greater than 1 s could be added. In this way, 238 seismic events form the "working" data set. Error analysis of the available input data demonstrates a direct link between the degree of precision of random variables input and the amount of energy released during the seismic event (magnitude).

The "working" data set of 238 events was divided into two magnitude classes – $M_w \leq 5$, and $M_w > 5$ – that were split in temporal sub-intervals of 5 years each. An analysis of mean and standard deviation values suggests a large dispersion of OT errors for events with $M_w \leq 5$ with mean values within interval [0.25 1.0] s. For events with $M_w > 5$, mean values of OT errors do not exceed

0.3 s, and standard deviations are smaller (Figure 12). OT errors directly influence the parameters for localizing seismic events: epicentral coordinates and depth. Figure 12 plots errors in OT values vs errors in epicenter location.

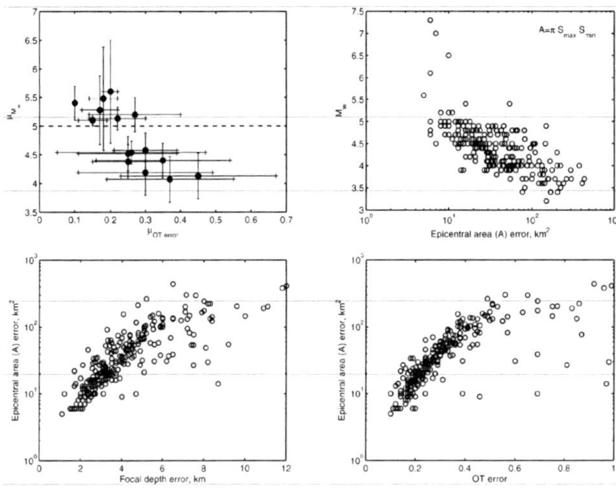

Figure 12: Input data errors

5. Conclusions

The paper presented a complex study of the source parameters of seismic events of magnitude $M_w \geq 3.0$ in SE Carpathians that took place from 1967 to 2006. It is based on ten independent catalogues taking as a reference the instrumental data (P-wave polarities) given by ISC for a period of 40 years. The source parameters of interest are epicentral coordinates, magnitude and focal mechanism. Conversion to the unique magnitude scale (M_w) was done for the entire data set. The scope of the study was not to give any priority to a particular catalogue during comparison, but rather to identify the differences and similarities of the seismic source parameters. The contribution of each catalogue is different, and two major groups could be defined: ANSS, ROM+, FSU, USGS, SBL which report epicentral coordinates, focal depths and magnitudes and ISC, HVD, MED, ONC, M&P which also contain information about the focal mechanisms.

The primary goal was to represent the focal mechanism data to which supplementary information-hypocentral location and magnitude was added. The values of source parameters from catalogues ISC, M&P and ONC were compared and merged into a new uniform database. Applying the first P-wave polarity method to the entire data set of 292 events reported by ISC, 139 unique and 34 double (equally probable) solutions were obtained, forming a total of 173 fault plane solutions (ISC). For the rest of the 119 ISC seismic events, the methodology applied is inefficient due to scarce data on polarities. Another set of 62 events from the M&P catalogue was added to these 173 focal mechanism solutions in addition to 12 events reported by ONC, resulting in a final number of 247 known solutions and 115 ISC events without focal mechanism, as 362 seismic events tabulated chronologically. We need to remark also that 20 events reported by SBL were not included into the events table, in order to save space but were used at comparison for focal parameters between catalogues.

Comparison criteria for the source parameters of the same events were set as follows:

– Epicentral coordinates not exceeding the interval 25 km
– Hypocentral depths not exceeding the interval ±10 km
– Magnitude (M_w) values not differing more than ±0.1

The results of the comparative analysis are presented in Table 6–9. Zero values indicate the cases when no common events were found. It should be mentioned that the use of different magnitude scales also contributed negatively to the precision of conversion to the uniform magnitude scale M_w. The proposed intervals for the comparative analysis have a physical background, namely the geometric shape of the source and the finite length of the rupture area as approximated by a point source.

To determine an "acceptable" maximum value of the fault size, a corresponding magnitude should be selected. As Vrancea earthquakes with M_w ≥ 6.5 have been observed five times during the last century, this range was selected as the candidate to define the maximum fault geometry. If a rupture area is approximated by a circular plane, the corresponding radius is R = 25 km for M_w = 7.5 and R = 10 km for M_w = 6.5 (Engdahl and Villasenor, 2003). In this way the epicentral location as a point can be determined within ±25 km. Focal depth requires more precision due to focal mechanism and seismicity aspects. In order to distinguish crustal and sub-crustal events (d > 40 km), the focal depths should be located with a precision of ±20 km. On the other hand, statistics on focal mechanism solutions show that ≈90% of the events have dips (angles between horizontal and fault planes) below 60°, and over 50% of them are less than 30°. Therefore, the horizontal projection of the approximation of the fault plane for angles of a dip less then 30° would suggest a precision of ±10 km.

The range of ±0.1 for the magnitudes comes from the recurrence relations for European earthquakes given for the interval $3.0 < M_w < 8.0$ according to ISC procedures. In this way the differences are within the interval 0.2 units (Bormann, 2002).

A comparative analysis of focal mechanism solutions was done for 173 events computed by the authors on the basis of ISC data against solutions provided in major catalogues: M&P, ONC, MED, HVD. Compatibility is considered to be achieved if solutions lie within the limits: $\pm45°$ (strike) $\pm12°$ (dip) $\pm45°$ (rake), which is dictated by the necessity to keep solutions within a single quadrant ($\pm45°$) and smaller variations of the dip angle ($0–90°$). The results of the comparative analysis are summarized in Table 9 which demonstrates 54% compatibility of the computed solutions with the HVD catalogue.

The results of the computations are presented in the form of GMT maps (Figures 4–9) and Table 5. An immediate by-product of these data is T- and P-axes distribution, Figure 10. A recent study of the depth interval 100–150 km (Sandu and Zaicenco, 2007; Heidbach et al., 2007), which corresponds to the high seismic intensity segment of the Vrancea zone emphasized the well-known directivity of these intermediate-depth earthquakes.

TABLE 6: Seismic catalogues: convergence in epicenter location (%)

Catal.	ISC	M&P	ONC	HVD	MED	SBL	ROM+	ANSS	USGS	FSU
ISC	100	100	90	63	92	100	82	95	96	100
M&P	100	100	91	34	0	100	97	99	96	100
ONC	90	91	100	0	0	83	90	92	83	100
HVD	63	34	0	100	100	33	68	67	67	50
MED	92	0	0	100	100	0	100	100	100	0
SBL	100	100	83	33	0	100	89	86	89	100
ROM+	82	97	90	68	100	89	100	93	93	100
ANSS	95	99	92	67	100	86	93	100	98	100
USGS	96	96	83	67	100	89	93	98	100	100
FSU	100	100	100	50	0	100	100	100	100	100

TABLE 7: Seismic catalogues: convergence in focal depth location (%)

Catal.	ISC	M&P	ONC	HVD	MED	SBL	ROM+	ANSS	USGS	FSU
ISC	100	99	90	74	69	91	70	82	83	56
M&P	99	100	83	75	0	87	64	93	94	57
ONC	90	83	100	0	0	67	66	43	47	33
HVD	74	75	0	100	80	67	63	67	72	50
MED	69	0	0	80	100	0	67	92	92	0
SBL	91	87	67	67	0	100	77	75	71	70
ROM+	70	64	66	63	67	77	100	63	63	81
ANSS	82	93	43	67	92	75	63	100	98	75
USGS	83	94	47	72	92	71	63	98	100	64
FSU	56	57	33	50	0	70	81	75	64	100

TABLE 8: Seismic catalogues: convergence in magnitude values (%)

Catal.	ISC	M&P	ONC	HVD	MED	SBL	ROM+	ANSS	USGS	FSU
ISC	100	98	85	53	54	83	30	51	47	75
M&P	98	100	79	25	0	82	25	51	51	7
ONC	85	79	100	0	0	33	14	30	28	33
HVD	53	25	0	100	40	50	26	100	71	0
MED	54	0	0	40	100	0	58	100	100	0
SBL	83	82	33	50	0	100	21	54	68	0
ROM+	30	25	14	26	58	21	100	27	30	75
ANSS	51	51	30	100	100	54	27	100	100	44
USGS	47	51	28	71	100	68	30	100	100	31
FSU	75	7	33	0	0	0	75	44	31	100

TABLE 9: Seismic catalogues: convergence in focal mechanism parameters (%)

Catal.	ISC	M&P	ONC	HVD	MED	SBL	ROM+	ANSS	USGS	FSU
ISC	100	41	15	54	22	-	-	-	-	-

Acknowledgements

This research is sponsored by NATO's Scientific Affairs Division in the framework of the Science for Peace Programme, project SfP-980468.

References

Allen, R. (1978). Automatic earthquake recognition and timing from single traces, *Bull. Seis. Soc. Am.*, **68**, 1521–1532.

Ambraseys, N. N., Jackson, J. A., Melville, C. P. (2003). Historical Seismicity and Tectonics. In: *International Handbook of Earthquake and Engineering Seismology*, part A, vol. 81A. Academic Press, New York/London, pp. 747–763.

Ardeleanu, L., Radulian, M. (1998). 21–22 February, 1983 seismic sequences of Ramnicu Sarat: focal mechanism and source time function inferred from short period waveform inversion, *Rev. Roum. Geophys*, **47**, Bucharest, 27–38.

Bala, A., Radulian, M., Popescu, E., Benetatos, C. (2003). Earthquake distribution and their focal mechanisms in correlation with the active tectonic zones of Romania, *J. Geodynamics*, **36**, 129–145.

Bormann, P. (2002). Seismic Sources and Source Parameters: Magnitude of seismic events, *Volume 1 – NMSOP, IASPEI*, Potsdam, Germany, p. 49

Drumea, A., Ginsari, V., Zaicenco, A. (2003). Moldova (Country Report), *International Handbook of Earthquake and Engineering Seismology*, vol. 81B. Academic Press, New York/London.

Dziewonski, A. M., Chou, T.-A., Woodhouse, J. H. (1981) Determination of earthquake source parameters from waveform data for studies of global and regional seismicity, *J. Geophys. Res.*, **86**, 2825–2952.

Engdahl, E. R., Villasenor, A. (2003). Global Seismicity: 1900–1999, In: *International Handbook of Earthquake and Engineering Seismology*, vol. 81A. Academic Press, New York/London, pp. 665–690.

Giardini, D. (1984). Systematic analysis of deep seismicity: 200 centroid-moment tensor solutions for earthquakes between 1977 and 1980, *Geophys. J. R. Astron. Soc.*, **77**, 883–914.

Hauser, F., Raileanu, V., Prodehl, C., Bala, A., Schulze, A., Denton, P. (2000). The Seismic-Refraction Project VRANCEA-99, *Open-File Report*, Geophysical Institute, University of Karlsruhe.

Heidbach, O., P. Ledermann, D. Kurfess, G. Peters, T. Buchmann, B. Sperner, B. Mueller, A. Nuckelt, G. Schmitt (2007). Attached or not attached: Slab dynamics beneath Vrancea, Romania *International Symposium on Strong Vrancea Earthquakes and Risk Mitigation*, October 4–6, Bucharest, Romania.

Huang, W.-C., Okal, E. A., Ekstrom, G., Salganik, M. P. (1997). Centroid-moment tensor solutions for deep earthquakes predating the digital era: the WWSSN dataset (1962–1976), *Phys. Earth Planet. Inter.*, **99**, 121–129.

ISC (1964–2002). Bulletins of the International Seismological Centre. **6–13**, CD, Pipers Lane, Thatcham, Berkshire, U.K. RG 19 4NS. http://www.isc.ac.uk

Kramer, S. L. (1996). *Geotechnical Earthquake Engineering*, Prentice Hall, Inc., Upper Saddle River, New Jersey, 653pp.

Lay, T., Wallance, T. C. (1995). *Modern Global Seismology*, Academic Press, New York, 517 pp.

Marza, V., Kijko, A., Mantyniemi, P. (1991). Estimate of Earthquake Hazard in the Vrancea (Romania) Region, *PAGEOPH 136*, 143–154.

Mostrioukov, A. O., Petrov, V. A. (1994). Catalogue of Focal Mechanisms of Earthquakes 1964–1990. *Materials of the World Data Center*, Moscow, 87 p.

Oncescu, M.-C. Bonjer, K.-P. (1997). A note on the depth recurrence and strain release of large Vrancea earthquakes, *Tectonophysics*, Elsevier, **272**:2, 291–302.

Pondrelli, S., Morelli, A., Ekstrom, G, Mazza, S., Boschi, E., Dziewonski, A. M. (2002), European-Mediterranean regional centroid-moment tensors: 1997–2000, *Phys. Earth Planet. Int.*, **130**, 71–101.

Pondrelli, S., Morelli, A., Ekstrom, G. (2004). European-Mediterranean Regional Centroid Moment Tensor catalog: solutions for years 2001 and 2002, *Phys. Earth Planet. Int.*, **145**: 1–4, 127–147.

Pondrelli, S., Salimbeni, S., Ekstrom, G., Morelli, A., Gasperini, P., Vannucci, G. (2006). The Italian CMT dataset from 1977 to the present, *Phys. Earth Planet. Int.*, **159**: 3–4, 286–303.

Popescu, E., Enescu, M., Radulian, M., Bazacliu, O. (2003). Clustering properties in time and space for Vrancea (Romania) earthquakes, *Rev. Roum. Geophys*, **47**, Bucharest, 89–108.

Purcaru, G. (1979). The Vrancea, Romania, earthquake of March 4, 1977 – a quite successful prediction. *Physics of the Earth and Planetary Interiors*, **18**, 274–287.

Radu, C. (2003). Catalogul istoric al cutremurelor din Vrancea n perioada 984–1900. In: *Constructii amplasate in zone cu miscari seismice puternice*. Orizonturi Universitare, Timisoara, p. 25.

Radulian, M., Mindrescu, N., Popescu, E., Utale, A., Panza, G. F. (1996). Seismic Activity and Stress Field Characteristics for the Seismogenic Zones of Romania, *International Center for Theoretical Physics*, Miramare-Trieste, December, IC/96/256

Riznicenko, Yu., Drumea, V., Stepanenko, N. (1981). Seismicity and Seismic Hazard of Carpathian Region, In: *Carpathian Earthquake of 4 March 1977, and its consequences*, Nauka, Moscow, pp. 46–85, (in Russian).

Sandu, I. (2005). Relocation of Hypocenters and Composite Fault Plane Solutions for Earthquakes in Republic of Moldova, *Individual Studies by Participants at the International Institute of Seismology and Earthquake Engineering*, Tsukuba, Japan, vol. **41**, pp. 61–70.

Sandu, I. (2006). Composite fault plane Solutions for Earthquakes in the Republic of Moldova, *First European Conference on Earthquake Engineering and Seismology*, Geneva, Switzerland, p. 266.

Sandu, I., Zaicenco, A. (2007). Focal Mechanism Solution Catalogue for SE Carpathian Region, *Thirty Years from the Romania Earthquake of March 4, 1977*, Bucharest, Romania.

Shebalin, N. V., Leydecker, G. (1997). Earthquake Catalogue for the Former Soviet Union and Borders up to 1988. – 135 pp., 13 fig.; *European Commission, Report No. EUR 17245 EN, Nuclear Science and Technology Series*, ISSN 1018–5593 – Office for Official Publications of the European Communities, Luxembourg.

Suetsugu, D. (1995a). Shear Faulting and Double Couple Model. In: *Source Mechanism Practice*, IISEE, Tsukuba, Japan, 10–16.

Suetsugu, D. (1995b). Mathematical Description of Double Couple Model. In: *Source Mechanism Practice*, IISEE, Tsukuba, Japan, 17–19.

Suetsugu, D, (1995c). Focal Mechanism Diagram. In: *Source Mechanism Practice*, IISEE, Tsukuba, Japan, 20–57.

Suetsugu, D., Yagi, Y. (2005). Part II. Source Mechanism. In: *Source Mechanism Practice*, IISEE, Tsukuba, Japan, p. 15.

Utsu, T. (2002). Relationships between Magnitude Scales, in: Lee, W.H.K., Kanamori, H., Jennings, P.C., and Kisslinger, C., editors, *International Handbook of Earthquake and Engineering Seismology*: Academic Press, a division of Elsevier, two volumes, International Geophysics, vol. **81-A**, pp. 733–746.

Trifu, C., Oncescu, M. (1987). Fault geometry of August 30, 1986 Vrancea earthquake, *Ann. Geophys.*, **5B**, 727–729.

Wenzel, F., Lorentz, F. P., Sperner, B., Oncescu, M. C. (1999). Seismotectonics of the Romanian Vrancea Area. In: *Vrancea Earthquakes: Tectonics, Hazard and Risk Mitigation*, 15–25, Wenzel et al. (eds.), Kluwer, Dordrecht.

Yeats, R. S., Sieh, K., Allen, C. R. (1997). *Geology of Earthquakes*. Oxford University Press, New York/Oxford.

http://www.ncedc.org/anss/
http://www.seismology.harvard.edu/
http://wwwbrk.adm.yar.ru/russian/1_512/1_512_3e.htm
http://streaming.ictp.trieste.it/preprints/P/96/256.pdf
http://www.infp.ro/catal.php
http://legacy.ingv.it/seismoglo/RCMT/index.html

GROUND MOTION PATTERNS OF INTERMEDIATE-DEPTH VRANCEA EARTHQUAKES: THE OCTOBER 27, 2004 EVENT

K.-P. BONJER[*,1], C. IONESCU[2], V. SOKOLOV[1],
M. RADULIAN[2], B. GRECU[2], M. POPA[2], E. POPESCU[2]

[1]*Geophysical Institute, University of Karlsruhe, Germany*
[2]*National Institute for Earth Physics, Bucharest-Magurele, Romania*

Abstract. Weak and strong intermediate-depth Vrancea earthquakes occur in a small zone beneath the southeastern Carpathian Mountains. Despite their small seismogenic dimension, the seismic moment release is unusually high. During the last century, four major earthquakes occurred in this source zone which is confined to an area of only 20 x 60 km^2. The focal depths range from 60 to 170 km, but a few events were found in depths to about 220 km. In this paper we analyze the October 27, 2004 $M_w = 5.8$ event and the aftershock pattern in the context of the background seismicity. We further present ground motion maps for the mainshock using the recordings of the accelerometer networks installed in Romania and the macroseismic intensities as observed by the Internet community provided by the United States Geological Survey. The combination of these instrument and Internet-based observations enabled the construction of ground motion maps covering the whole territory of Romania and its bordering countries. The geographical pattern of the peak ground acceleration as well as the seismic intensity of the moderate October 2004 earthquake show the well-known northeast-southwest trending of the maximum values as displayed by the big events of 1940, 1977, 1986 and 1990. The growing community of Internet users will provide denser, extensive and even cross-border information on the effects of future strong Vrancea earthquakes than instrument observations alone can do.

Keywords: Vrancea earthquakes, double Wadati-Benioff Zone, ground motion pattern

[*]Geophysical Institute, University of Karlsruhe, Germany, e-mail: kpbonjer@online.de

A. Zaicenco et al. (eds.), *Harmonization of Seismic Hazard in Vrancea Zone,*   47
© Springer Science + Business Media B.V. 2008

1. Introduction

The Vrancea intermediate-depth earthquake zone is a small area centered in the southeastern Carpathian Mountains. The majority of the earthquakes occur at depths between 60 and 170 km in an almost vertical stripe (e.g. Radu, 1974; Fuchs et al., 1978; Jung, 1983; Koch, 1982, 1985; Oncescu, 1984; Trifu and Oncescu, 1987; Trifu et al., 1992; Oncescu and Bonjer, 1997; Radulian et al., 2007a, b). Only two earthquakes with greater depths are well documented: the events of November 30, 2002 at 179 km (Bonjer et al., 2005; Radulian et al., 2007a) and of May 16, 1982 at 220 km (e.g. Oncescu and Bonjer, 1997).

Joint Hypocenter Determinations (JHD) based on the high-quality recordings of the K2 network in Romania (Figure 1) show that the intermediate-depth background seismicity of the Vrancea region is confined to an area of 20 x 60 km^2 only. Weak and moderate earthquakes occur at depths between 60 and 170 km on two parallel, nearly vertical planes. They are separated by less than 10 km (Bonjer et al., 2005; Radulian et al., 2007a). The discovery of such a double Wadati-Benioff Zone (WBZ) in the Vrancea earthquake zone is a strong argument for a subduction type process and therefore will influence the refinement of existing geodynamic models. A convincing documentation of a double-seismic zone was done first by Hasegawa et al. (1978) beneath northern Honshu, Japan. Since then, the increasing number of worldwide high-quality observations lead to the discovery of WBZs in different subduction zones (e.g. Rietbrock and Waldhauser, 2004). Brudzinski et al. (2007) showed recently that they are an inherent feature in subducting plates and could be explained by dehydration processes of different hydrous minerals in the oceanic lithosphere. In the case of Vrancea, the mechanism of the seismogenesis of a **vertical** double WBZ is not yet understood and might be more complicated than in most subduction zones where the plates have much smaller dip angles.

2. The Collaborative Research Center 461 – National Institute for Earth Physics Seismic Networks in Romania (Status: October 2004)

A new strong-motion seismic network (Figure 1) was installed in recent years jointly by Karlsruhe University's Collaborative Research Center (CRC) 461 "Strong Earthquakes" (http://www.gpi.physik-uni.karlsruhe.de/) and the National Institute for Earth Physics, Bucharest (http://www.infp.ro/). The network (Bonjer et al., 2000; Sokolov et al., 2005) is centered in the Vrancea epicentral zone and covers an area with a diameter of up to 500 km. It consists of 29 free field stations that are equipped with Kinemetrics K2 data loggers including Global Positioning System (GPS) timing units and three-component accelerometers (Kinemetrics Episensors).

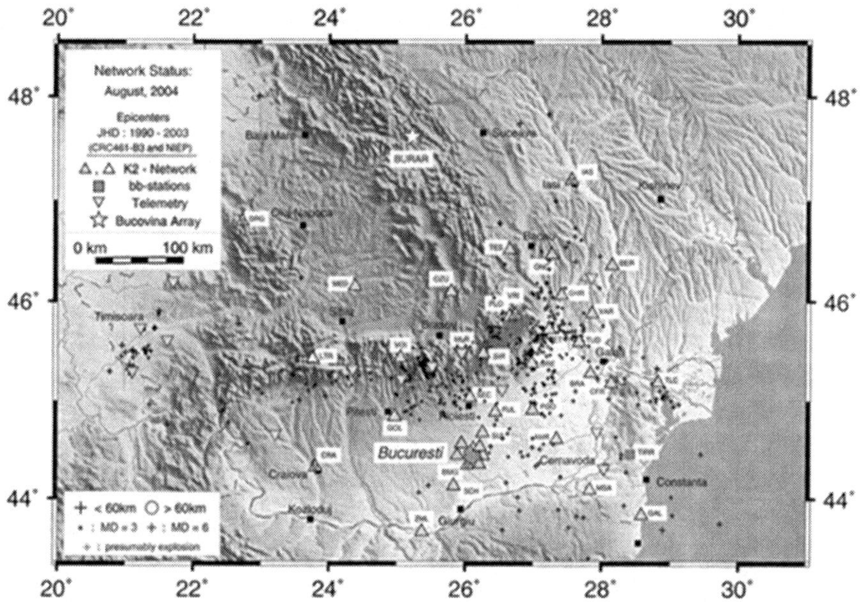

Figure 1: Location of the K2 accelerometers network (yellow triangles), installed within the German-Romanian collaborative program CRC461 (Bonjer et al., 2000) and the telemetry network (grey inverted triangles) of the National Institute for Earth Physics, Bucharest-Magurele. The K2-stations BMG, CRA, DRG, IAS, TLC, VRI, ZML, the Bucovina array, the NIEP-JICA station MLR, station Buzias, and the GEOFON station TIRR together with the fm-based telemetry stations, form the real-time network of NIEP (status August 2004)

Most stations also have three-component velocity transducers. Station Plostina (PLO) in the northeastern Vrancea epicenter zone has a three-component Kinemetrics borehole accelerometer at a depth of 50 m as well. Fifteen more stations are deployed in the Romanian capital Bucharest in nearly free field conditions, and one is in a 100 m deep borehole. Furthermore, two 12-channel K2s with 3 three-component accelerometers each are installed on the tenth floor of a typical multistory building in Bucharest and in a test building in the yard of the Building Research Institute Bucharest (INCERC). The joint K2 network of the CRC461 and NIEP consists of a total of 46 data loggers. The sample rates are set to 200 samples per channel and the pre-event memory is adjusted to about 35 s. This ensures the recording of the complete P-wave onset in case of S-wave triggering for hypocenter distances of up to about 350 km from the Vrancea source location. The accumulated database (Bonjer and Rizescu, 2000; Bonjer et al., 2002; Bonjer and Grecu, 2004, 2006) consists of hundreds of records obtained by scores of stations during many small and moderate earthquakes. It allows high-quality analysis of P- and S-wave arrival times and features of ground motion excitation and propagation in the region, including site response analysis (e.g. Bonjer et al., 2000; Wirth et al., 2003; Sokolov et al., 2004a, b, 2005, 2008).

3. Results

3.1. THE MAINSHOCK AND AFTERSHOCK PATTERN OF THE OCTOBER 27, 2004 SEQUENCE

With a moment magnitude of $M_w = 5.8$, the October 27, 2004 earthquake is the strongest intermediate-depth Vrancea event since the May 1990 earthquake sequence when two major shocks of $M_w = 6.9$ and $M_w = 6.3$, respectively, occurred (Global CMT Catalog). In this paper we determine the position of the October event within the fine-scale, double-layered WBZ by jointly inverting P- and S-wave arrival times of this event and the background seismicity from January 1996 to December 2004.

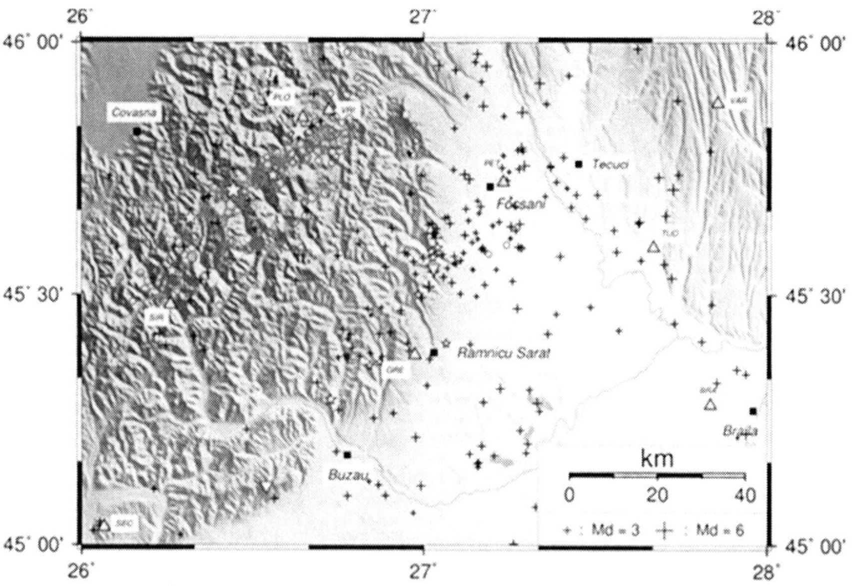

Figure 2: Joint hypocenter determinations (JHD) of the seismicity in the time period 1996–2004. Colored circles represent epicenters of Vrancea intermediate-depth earthquakes. Black crosses represent epicenters of crustal earthquakes. The epicenter of the event of October 27, 2004 is indicated by the large yellow star, the immediate aftershocks (first four days) by smaller yellow stars. The white star is the epicenter of the largest earthquake (September 27, $M_w = 4.7$) in 2004 before October 27, but in the deeper segment of the Vrancea seismic zone

For the inversion we apply the JHD method. Intermediate-depth earthquakes with a minimum number of 12 P-arrivals and at least 3 S-arrivals (average numbers are 30 P-arrivals and 25 S-arrivals) were used. Events with azimuthal gaps of more than 170° were omitted. A subset of 240 intermediate depth events including the October 2004 sequence met the criteria and is used for further analysis. In addition, arrival times of 540 crustal earthquakes less strongly selected were separately inverted with the JHD as well. In Figures 2 and 3, the spatial patterns of the background seismicity and the October main-shock and aftershock sequence are shown. Colored circles represent epicenters of Vrancea intermediate-depth earthquakes and black crosses represent epicenters of crustal earthquakes. The epicenter of the October 27, 2004 main-shock is indicated by the large yellow star, the immediate aftershocks (first 4 days) by smaller yellow stars.

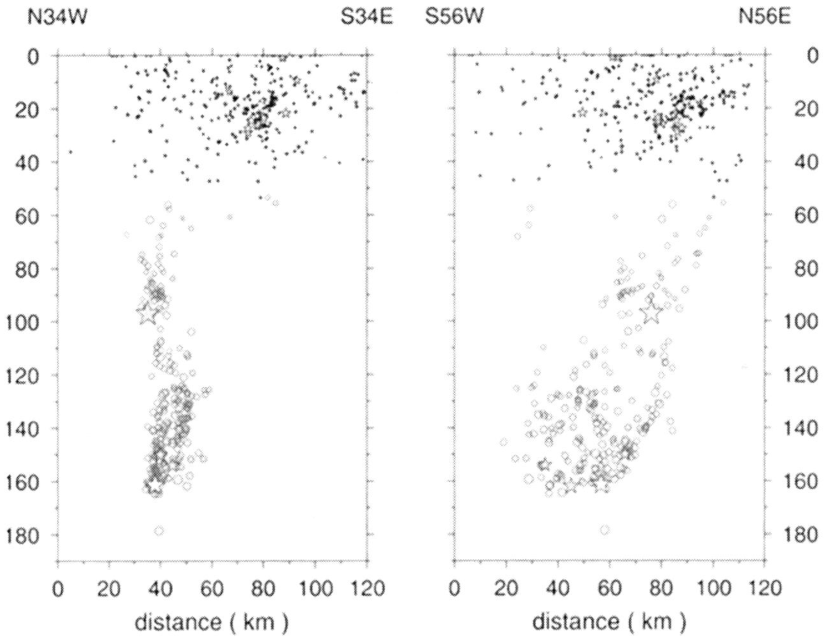

Figure 3: Depth distribution of the earthquakes of Figure 2 (background seismicity and the October 2004 sequence) projected on two perpendicular, vertical cross sections striking N34°W and N56°E. Symbols and colors are as those in Figure 2. Note the vertical plunge of the seismic activity on two parallel planes (*left figure*). The focus of the mainshock lies in an area of reduced activity. Aftershocks were very rare and not found near the mainshock. However, a few aftershocks are found in the crust and in the deepest floor of the seismic zone

The main event occurred in the northeastern part of the Vrancea source zone at a depth of 97 km in the more northwestern strand of the two seismogenic planes (Figures 2 and 3). Both planes dip at nearly 90°, and the main activity extends vertically to depths of about 170 km. The northeastern part is an area of several remarkable features. It is the zone where the rupture of the big events with hypocenter depths less than 100 km initiate as was found for the 1977 (Müller et al., 1978) and 1990 (Trifu et al., 1990) earthquakes. Furthermore, all JHDs regardless of the time windows used or the arrival time accuracy show that the intermediate and the crustal seismogenic zones seem to be connected at this edge. Starting from there, seismic activity deepens toward the southwest (e.g. Jung, 1983, Oncescu, 1984; Oncescu and Bonjer, 1997; Bonjer et al., 2005). This feature could be associated with a partial break-off that has started in the southwestern edge of the seismogenic part of the slab and which has still a weak coupling (e.g. Sperner et al., 2001; Heidbach et al., 2007) left to the seismically active parts of the overlaying crust at the northeastern edge of the seismic zone in Vrancea.

The distribution of seismic activity is not uniform in the double-seismic zone. It differs laterally and with depth. Roughly speaking, the activity in one layer is enhanced in those areas where the other layer shows less activity. In the depth range between 100 and 120 km, further structure in the pattern of seismic activity seems to exist: the background events are less numerous in comparison to the depth ranges above and below, and particular areas are nearly free of seismic activity. The October 27, 2004 mainshock occurred in the transition between the upper active zone and this intermediate zone of lower activity. As already mentioned, no aftershocks occurred in the close vicinity of the main shock. Only a few of them happened in the deepest floor of the seismogenic zone and in the crust above the location of the mainshock. Interestingly, the aftershocks of the 1986 event occurred in the 130–150 km depth range of the more southeastern seismogenic layer (red circles in Figures 2 and 3) and the aftershocks of the 1990 sequence in the 80–100 km depth range of the more northwestern layer (blue circles in Figures 2 and 3).

Concerning the separation of the two layers by about 10 km, this is best developed in the greater depth range, namely between 110 and 170 km (Figure 3). According to Hacker et al. (2003) and Brudzinski et al. (2007), the amount of separation is proportional to the time span since the subduction started, i.e. a distance of about 10 km would be indicative of a subducting plate age of about 30–35 million years. This is in coarse agreement with the development of the subduction process as discussed, e.g., by Sperner et al. (2001).

3.2. THE RADIATION PATTERN OF THE MAINSHOCK

The fault plane solution is a nearly pure dip-slip mechanism as deduced from teleseismic observations with the **C**entroid **M**oment **T**ensor (CMT) method (e.g. Harvard CMT Catalog: strike = 219°, dip = 81°, rake = 107°), as well as from local waveform inversion and first motion investigations (strike = 239°, dip = 88°, rake = 96°). Such a mechanism with one nodal plane dipping steeply and striking at about 230° is typical for the strong Vrancea intermediate-depth earthquakes (e.g. Müller et al., 1979; Oncescu, 1989; Trifu et al., 1990; Harvard CMT Catalog).

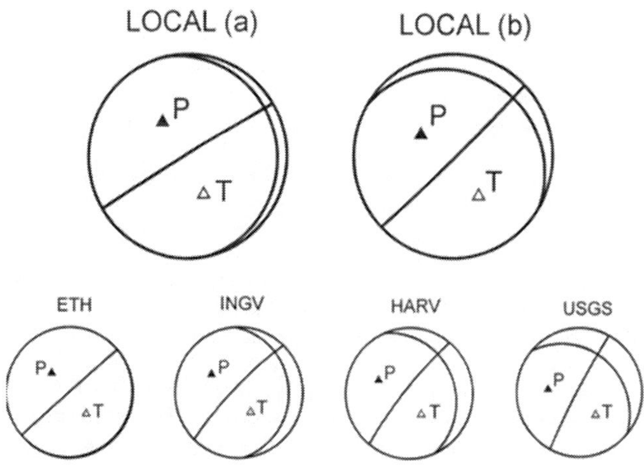

Figure 4: The fault plane solutions of the main shock obtained using local or regional/ global recordings: (a) from polarities and (b) from spectral amplitudes (Radulian et al., 2004). ETH: Swiss Seismological Service, Institute of Geophysics; ETH: Hoenggerberg, Zurich, Switzerland; INGV-MEDNET network: Istituto Nazionale di Geofisica e Vulcanologia and Mediterranean Very Broadband Seismological Network; HARV: Department of Earth and Planetary Sciences, Harvard University (Harvard Seismology); USGS: United States Department of the Interior Geological Survey

Different studies (for references see Radulian et al., 2007a) gave very similar results and confirmed the very steeply dipping nodal plane with values between 80° to 90° (Figure 4). With such a great dip, the October event differs from the bigger events where the maximum dip is not more than about 70°. Their steeper nodal planes do not fit as well into the vertical patterns of the background seismicity (see Figure 3) as the one of the October 2004 event does. On the other hand, the aftershocks of the main shocks of 1986 and 1990 do mimic perfectly the steeper nodal planes as shown by Trifu and Oncescu (1987) and

Trifu et al. (1992) with their JHDs. Such a congruency of the rupture plane and the planar area of immediate aftershocks could imply complexity in the rupture process leaving behind a heterogeneous stress-strength distribution that enables aftershock generation. Fast healing rates could be an alternate explanation. In the case of the October 2004 earthquake, however, no aftershocks were observed in the rupture area indicating complete stress release. This is supported by the high-stress drop of 100–200 Mpa, determined through deconvolution of the mainshock with empirical Green's functions (Radulian et al., 2007a). Such a high value is a common feature of big Vrancea events but not for a moderate $M_w = 5.8$ event. As already mentioned, the rupture and the aftershock zones of the big events do not match the course of either of the two seismogenic layers. Because the layers have a thickness of approximately 5 km only, the ruptures and aftershocks of the big events would have migrated into an area free of background activity, i.e. into unbroken material. If so, one would have expected some small to moderate events to have occurred since then in the vicinity of the main ruptures of the past century. However, neither after the 1986 nor the 1990 sequences nor even after the 1977 multi-shock sequence seismic activity could be observed outside the area defined by the background seismicity in Figures 2 and 3. An alternative explanation for the accommodation of the "big event scenario" within the area of background seismicity could be as follows: the rupture and the aftershock migration developed obliquely from one of the two thin, parallel seismogenic layers towards the second one, passing the apparently aseismic range in between. Such speculations need confirmation by a future event of about magnitude 7 or greater and its aftershock pattern, recorded with the dense local and teleseismic high-resolution strong- and weak-motion networks now existing.

3.3. THE GROUND MOTION PARAMETERS: PEAK GROUND ACCELERATION AND INTENSITY

Figure 5 shows the distribution of the peak ground acceleration (PGA) (maximum of the two horizontal components) recorded by the October 2004 earthquake. White triangles show the positions of those 40 K2 stations of the strong-motion network of CRC 461 of Karlsruhe University and the National Institute for Earth Physics (NIEP, Bucharest) that recorded the event. Three stations failed to record the earthquake because of power or maintenance problems, and three stations were switched off at the time of the event.

At several locations, peak accelerations are observed above 100 cm/s^2. The greatest horizontal acceleration recorded was 265 cm/s^2 at Seciu (SEC), a small village near the city of Ploiesti, about 100 km away from the epicenter. This peak acceleration comes close to the largest values found for the strong and damaging earthquakes of 1977, 1986 and 1990.

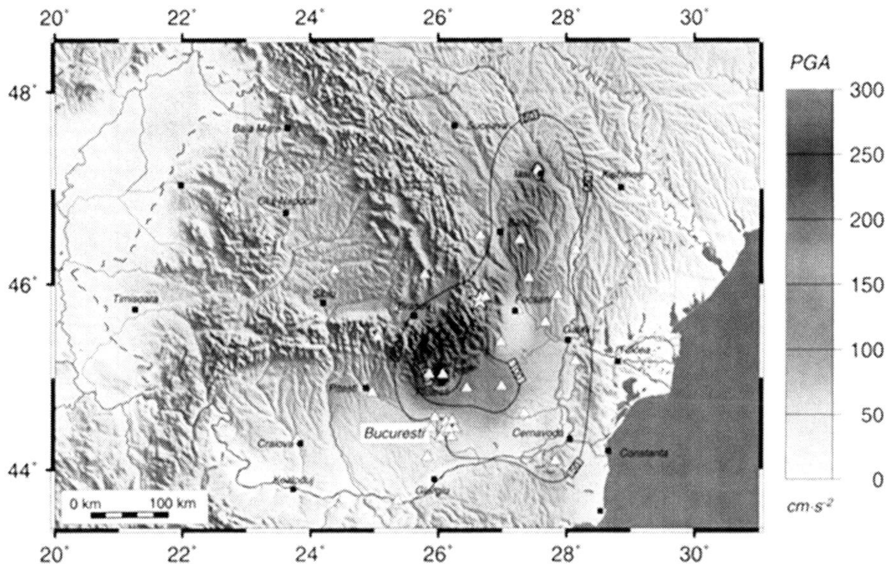

Figure 5: The distribution of the peak ground acceleration (maximum of the two horizontal components) recorded by the earthquake of 27 October 2004. White triangles show the positions of the 40 K2 stations of the CRC 461-NIEP network that recorded the event

Although the October 2004 earthquake was felt countrywide, very few reliable macroseismic data were collected. We therefore calculated Instrumental Intensities (II) using the Fourier amplitude spectra of the horizontal accelerograms of the K2 recordings (Sokolov, 2002). These intensities are approximately equivalent to the MSK and MMI scales. A maximum intensity of I = 6.5 was found at the location with the PGA of 265 cm/s^2. In the epicentral area, observed and calculated intensities reached values between 5 and 6.0 only. No intensities above these values were found (Figure 6).

Shakemaps for PGA and II were constructed by combining the following data sets: on-scale accelerograms of the CRC461-NIEP K2 network, peak amplitudes of analog and digital data of the National Institute for Building Research (INCERC) networks, the K2 recordings of the National Center for Seismic Risk Reduction (NCSRR), and the macroseismic observations compiled in the USGS Community Internet Intensity Map (CIIM).

The macroseismic observations were sent by the community of interested Internet users to the USGS "Did You Feel It" project (Wald et al., 1999a, b) from countries up to epicenter distances of more than 1,000 km (data are from the following countries: Romania, Bulgaria, Moldova, Ukraine, Serbia, and

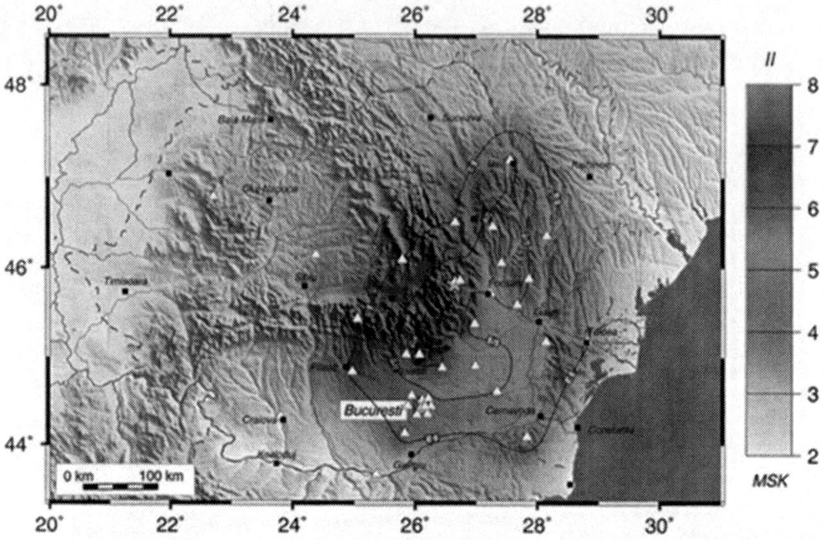

Figure 6: The instrumental intensity (II) map of the October 27, 2004 Vrancea event. The intensities were calculated from the accelerograms of the K2 stations of the CRC 461-NIEP network (white triangles). The singular, high intensity at the city of Iasi is generated by spurious oscillations of an antenna tower near the recording site

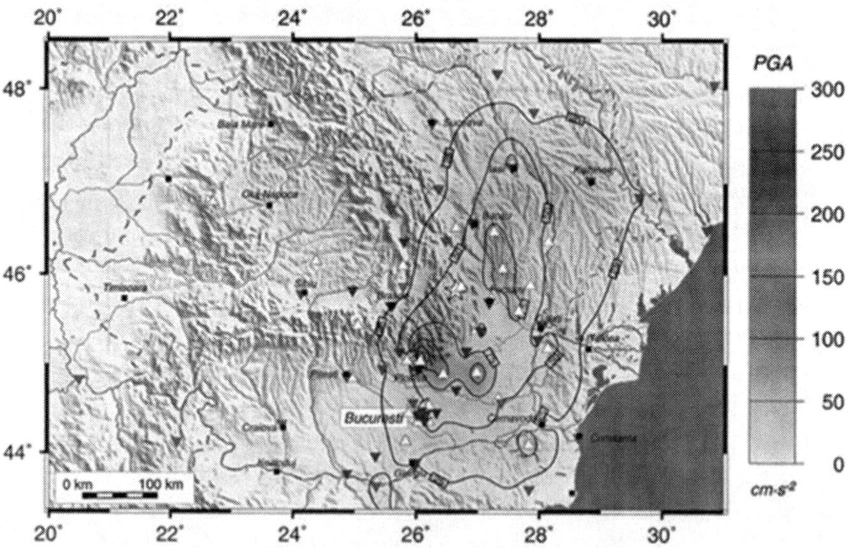

Figure 7: Composite PGA map: CRC461-NIEP-data (white triangles), the INCERC- and NCSRR-data (blue inverted triangles) and the CII-data (red inverted triangles), converted to acceleration with the following relation: II = 3.25 * log (PGA) − 0.18 (Bonjer et al., 2001)

Turkey). Macroseismic information or data that were available only as amplitude values like PGA were converted with the following relation: II = 3.25 * log (PGA) − 0.18 (Bonjer et al., 2001), valid only for intermediate-depth Vrancea earthquakes. The application of this relation enabled the joint use of the different parameters (see above) for the construction of linear-interpolated PGA and intensity shakemaps.

The PGA-pattern of the 2004 event (Figure 7) displays similar, symmetric features as they were reported in particular by Atanasiu (1959) for the intensity pattern of historical events. The intensity pattern of the October 2004 event (Figure 8), however, is more comparable to the generalized intensity and isoseismal patterns of the big events of the 20th century as presented by Demetrescu (1941) with the macroseismic map of the 1940 Vrancea earthquake (Figure 9) or by Corneliu Radu and his co-workers for the 1977, 1986 and 1990 Vrancea events (Radu et al., 1977, 1987; Radu and Utale, 1990).

4. Conclusions

A moderate earthquake (M_w = 5.8) occurred in the Vrancea seismic region on October 27, 2004 in a depth range that was affected by the last major shocks, i.e. the earthquakes of March 4, 1977 (M_w = 7.5), May 30, 1990 (M_w = 6.9), and May 31, 1990 (M_w = 6.3). The event did not create noticeable damage but was felt over a large area from Bulgaria to the Republic of Moldova.

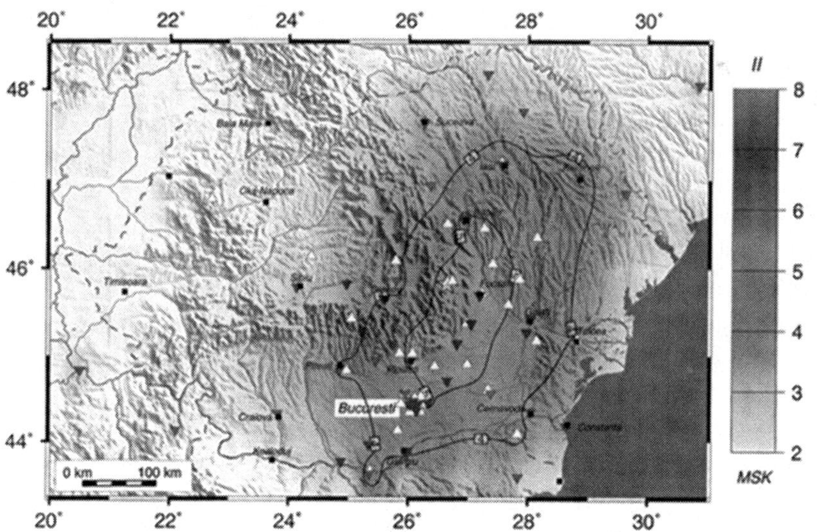

Figure 8: Composite intensity map: CRC461-NIEP-data (white triangles), the CII-data (red inverted triangles), the NCSRR-data (blue inverted triangles), and the INCERC PGA-values (blue inverted triangles) converted to intensity with the same relation as in Figure 7

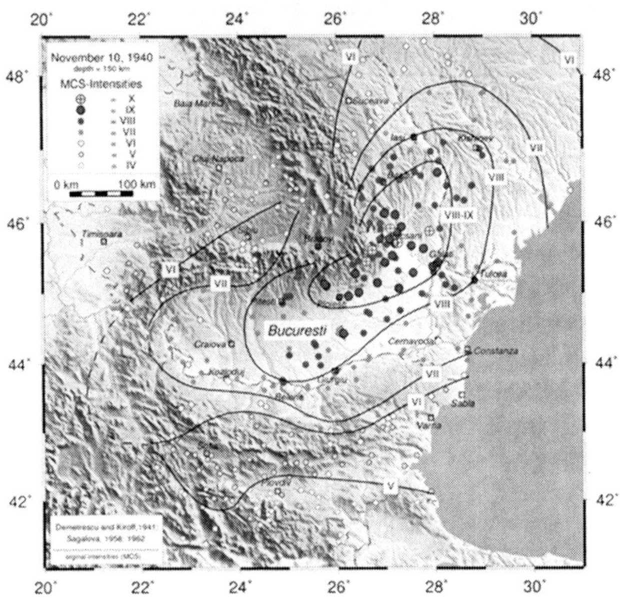

Figure 9: Macroseismic map of the M_w = 7.7 Vrancea intermediate-depth earthquake of November 10, 1940. The intensity values in Romania and Bulgaria were digitized from the map of Demetrescu (1941). The data from the Republic of Moldova and from Ukraine were taken from Sagalova (1958, 1962). The intensities are scaled according to the European macroseismic scale (Mercalli-Cancani-Sieberg) as used by Demetrescu, Kiroff and Sagalova for their original maps

We present first evidence for a double-layered WBZ in the Vrancea subduction scenario using high-precision JHD locations. Such zones are an inherent feature in subducting plates as Brudzinski et al. (2007) showed recently. In general, they are dipping at less than 70°. In the case of Vrancea, however, the WBZ is dipping vertically. This indicates a complicated time history from the onset of the steepening of the northwest dipping plate followed by a separation from the overlaying crust that began in the southwestern edge and lead to a sinking into the mantle.

The focal mechanism of the October 2004 earthquake is a nearly pure dip-slip event with one nodal plane dipping close to 90°. This fits perfectly with the strike and dip into the northern Benioff plane.

The greatest horizontal acceleration recorded was 265 cm/s^2 at Seciu (SEC), a small village near the city of Ploiesti, about 100 km away from the epicenter. This peak acceleration comes close to the largest values found for the strong and damaging earthquakes of 1977, 1986 and 1990. However, the dominating signal-frequencies are significantly different. For the 1977 and 1986 earthquakes they are about 1 Hz, but for the 2004 event they are only about 4

Hz, resulting in the small instrument intensity of only $I = 6.5$. This explains why no noticeable damage was registered, as confirmed by USGS CIIM data.

The combination of the on-scale recordings of the CRC461-NIEP K2 network, the analogue and digital data of the INCERC networks, the K2 recordings of the NCSRR and Moldovan network with the macroseismic observations of the CIIM enabled the construction of the most complete shakemaps for a Vrancea earthquake so far. These maps enlarge the database for the quantitative modeling of Vrancea scenario events needed to improve the seismic hazard assessment in Romania and in neighboring countries.

Acknowledgements

This study is based on data recorded by the K2 network installed in the framework of the Karlsruhe University Collaborative Research Center 461 "Strong Earthquakes" in collaboration with the National Institute for Earth Physics, Bucharest-Magurele. These data are also a contribution of the National Institute for Earth Physics to the EC-project MEREDIAN-2 (Mediterranean-European Rapid Earthquake Data Information and Archiving Network contract EVR1-CT-2000-40007) as well as to the NATO research project SfP-980468 (Harmonization of Seismic Hazard and Seismic Risk Reduction in Countries influenced by Vrancea Earthquakes).

We thank Iolanda Craifaleanu for providing the INCERC data, Natalia Poiata for providing the K2 records of the NCSRR network, and A.V. Drumea for providing the Chisinau K2-record of the October 2004 Vrancea mainshock. The macroseismic map of Demetrescu (1941) was made available by Traian Moldoveanu, GEOTEC Consultant Bucharest, and was digitized and geo-referenced by W. Weisbrich, Institute of Photogrammetry and Remote Sensing, Karlsruhe University.

We thank George Chircea, Viorel Pirvu and Werner Scherer as well as Olivia Bazacliu, Mioara Pompilian and Rainer Plokarz for their efforts in collecting and pre-processing the K2 data. The maps and depth profiles were plotted with the Generic Mapping Tool (Wessel and Smith, 1991).

References

Atanasiu, I., 1959. Cutremurele de Pamint din Romania, editura academmiei republicii populare Romine, p. 194.

Bonjer, K.-P., Grecu, B., 2004. Data Release **January 1996 – June 2004** of the Vrancea K2 Seismic Network. One DVD with evt-files. Bucharest – Karlsruhe, December 6.

Bonjer, K.-P., Grecu, B., 2006. Data Release **January 1996 – December 2004** of the Vrancea K2 Seismic Network. One DVD with evt-files. Karlsruhe – Bucharest, March 1.

Bonjer, K.-P., Rizescu, M., 2000. Data Release **1996 – 1999** of the Vrancea K2 Seismic Network. Six CD's with evt-files and KMI VOL1-, 2-, v-files. Karlsruhe-Bucharest, July 15.

Bonjer, K.-P., Oncescu, M.C., Rizescu, M., Enescu, D., Driad, L., Radulian, M, Ionescu, C., Moldoveanu, T., 2000. Source- and site-parameters of the April 28, 1999 intermediate depth Vrancea earthquake: First results from the new K2 network in Romania, XXVII General Assembly of the European Seismological Commission, Lisbon, Portugal, Book of Abstracts and Papers, SSA-2-13-O, p. 53.

Bonjer K.-P., Sokolov, V., Wirth, W., Rizescu, M., Driad, L., Treml, M., 2001. The seismogenic potential of the Vrancea subduction zone – quantification of source- and site-effects of great events. Collaborative Research Center (CRC) 461 "Strong Earthquakes: A Challenge for Geosciences and Civil Engineering", Report of 1999–2001, University of Karlsruhe, pp. 259–293 (in German).

Bonjer, K.-P., Rizescu, M., Grecu, B., 2002. Data Release **2000 – 2001** of the Vrancea K2 Seismic Network. Bucharest – Karlsruhe, October 24.

Bonjer, K.-P., Ionescu, C., Sokolov, V., Radulian, M., Grecu, B., Popa, M., Popescu, E., 2005. Source parameters and ground motion pattern of the October 27, 2004 intermediate depth Vrancea earthquake, EGU General Assembly 2005, Vienna.

Brudzinski, M.R., Thurber, C.H., Hacker, B.R., Engdahl, E.R., 2007. Global Prevalence of Double Benioff Zones, Science, 316, 5830, 1472–1474.

Demetrescu, G., 1941. Détermination provisoire de l'épicentre du tremblement de terre de Roumanie du 10 Novembre 1940. Observatoire de Bucuresti, Station seismic, p. 10.

Fuchs, K., Bonjer, K.-P., Bock, G., Cornea, I., Radu, C., Enescu, D., Jianu, D., Nourescu, A., Merkler, G., Moldoveanu, T., G. Tudorache, 1978. The Romanian earthquake of 4 March 1977, II. Aftershocks and migration of seismic activity, Tectonophysics, 53, 225–247.

Hacker, B.R., Abers, G.A., Peacock, S.M., 2003. Subduction factory 1. Theoretical mineralogy, densities, seismic wave speeds, and H_2O contents, J. Geophys. Res., 108, 2029, doi:10.1029/2001JB001127.

Harvard Seismology: CMT-Project (http://www.seismology.harvard.edu/projects/CMT/)

Hasegawa, A., Umino, N., Takagi, A., 1978. Double-planed structure of the deep seismic zone in the northeastern Japan arc, Tectonophysics, 47, 43–58.

Heidbach, O., Ledermann, P., Kurfeß, D., Peters, G., Buchmann, T., Matenco, L., Negut, M., Sperner, B., Müller, B., Nuckelt, A., Schmitt, G., 2007. Attached or not attached: Slab Dynamics beneath Vrancea, Romania. In: Proceedings of the International Symposium on Strong Vrancea Earthquakes and Risk Mitigation, October 4–6, Bucharest, Romania, 3–20.

Jung, P., 1983. Herdparameter und Ausbreitungseffekte krustaler und mitteltiefer Erdbeben in den Rumänischen Karpaten unter besonderer Berücksichtigung der Vrancea – Region, Diplomarbeit, Universität Karlsruhe, p. 133.

Koch, M., 1982. Seismicity and structural investigations of the Romanian Vrancea region: Evidence for azimuthal variations of P-wave velocity and Poisson's ratio, Tectonophysics, 90, 91–115.

Koch, M., 1985. Nonlinear inversion of local seismic travel times for the simultaneous determination of the 3D-velocity structure and hypocenters – application to the seismic zone Vrancea, J. Geophys., 56, 160–173.

Müller, G., Bonjer, K.-P., Stöckl, H., Enescu, D., 1978. The Romanian earthquake of March 4, 1977. I. Rupture process inferred from fault-plane solution and multiple event analysis, J. Geophys., 44, 203–208.

Oncescu, M.C., 1984. Deep structure of Vrancea Region, Romania, inferred from simultaneous inversion for hypocenters and 3-D velocity structure, Ann. Geophys., 2, 23–28.

Oncescu, M.C., 1989. Investigation of a high stress drop earthquake on August 30, 1986 in the Vrancea region, Tectonophysics, 163, 35–43.

Oncescu, M.-C., Bonjer, K.-P., 1997. A note on the depth recurrence and strain release of large Vrancea earthquakes, Tectonophysics, 272, 291–302.

Radu, C., 1974. Contribution à l'étude de la sismicité de la Roumanie et comparison avec la sismicité de sud-est de la France, Thèse Dr. Sci., Université Strasbourg, France, p. 404.

Radu. C., Utale, A., 1990. Vrancea earthquake of 30 May, Intensity distribution, Contract CFP/IFA 30.86.3/1990 (in Romanian), IV, A2.

Radu, C., Polonic, G., Apopei, I., 1977. Macroseismic map of March 4, Vrancea Earthquake (in Romanian), Report CSEN, Bucharest.

Radu, C., Polonic, G., Apopei, I., 1979. Macroseismic field of the March 4, Vrancea Earthquake, Rev. Roum. Géol., Géophys, et Géogr. – GÉOPHYSIQUE, 23, Bucarest, 19–26.

Radu. C., Utale, A., Winter, V., 1987. The August 30, 1986 Vrancea earthquake. Seismic intensity distribution, National Institute for Earth Physics report, II, A-3.

Radulian, M., Bălă, A., Popescu, E., 2004. Fault plane solutions as indicators of specific stress field characteristics in Vrancea and adjacent seismogenic zones, in "Earthquake Loss Estimation and Risk Reduction" – Proceedings of the International Conference Earthquake Loss Estimation and Risk Reduction, 24–26 October 2002, Bucharest, Romania (eds. D. Lungu, F. Wenzel, P. Mouroux, I. Tojo), vol. 1, 151–160, 2004.

Radulian, M., Bonjer, K.-P., Popa, M., Popescu, E., 2007a. Seismicity patterns in SE Carpathians at crustal and subcrustal domains: Tectonic and geodynamic implications. In: International Symposium on Strong Vrancea Earthquakes and Risk Mitigation, October 4–6, Bucharest, Romania, 93–102.

Radulian, M., Bonjer, K.-P., Popescu, E., Popa, M., Ionescu, C., Grecu, B., 2007b. The October 27th, 2004 Vrancea (Romania) earthquake, Orfeus Newsletter, Vol. 7, No. 1, p. 1.

Rietbrock, A., Waldhauser, F., 2004. A narrowly spaced double-seismic zone in the subducting Nazca plate, Geophys. Res. Lett., 31, L10608, 1–4.

Sagalova, E.A., 1958. On the development of Carpathian Earthquakes during 1908–1953. Katalog Karpatskikh Zemlettryasenii No. 1(4), Kiev.

Sagalova, E.A., 1962. The seismic regime of the Vrancea region. Geofiz. Sbornik No. 1(3), Izdatel'stvo Akad. Nauk. UkrSSR.

Sokolov, V., 2002. Seismic intensity and Fourier acceleration spectra: Revised relationship, Earthquake Spectra, 18, 161–187.

Sokolov, V., Bonjer, K.-P., Wenzel, F., 2004a. Accounting for site effect in probabilistic assessment of seismic hazard for Romania and Bucharest: A case of deep seismicity in Vrancea zone. Soil Dyn. Earthq. Eng., 24, 12, 927–947.

Sokolov, V., Bonjer, K.-P., Rizescu, M., 2004b. Assessment of site effect in Romania during intermediate depth Vrancea earthquakes using different techniques. In: IUGG Special Volume Earthquake Hazard, Risk, and Strong Ground Motion. Seismological Press, Beijing, 295–322.

Sokolov V, Bonjer, K.-P., Oncescu, M., Rizescu, M., 2005. Hard rock spectral models for intermediate-depth Vrancea (Romania) earthquakes. Bull. Seism. Soc. Am., 95, 1749–1765.

Sokolov, V., Bonjer, K.-P., Wenzel, F., Grecu, B., Radulian, M., 2008. Ground-motion prediction equations for the intermediate depth Vrancea (Romania) earthquakes, Bull. Earthq. Eng., doi:10.1007/s10518-008-9065-6 (2008), Vol. 6, pp. 367–388.

Sperner, B., Lorenz, F., Bonjer, K., Hettel, S., Müller, B., Wenzel, F., 2001. Slab break-off – abrupt cut or gradual detachment? New insights from the Vrancea Region (SE Carpathians, Romania), Terra Nova, 13, 172–179.

Trifu, C.I., Oncescu, M.C., 1987. Fault geometry of 1986, August 30 Vrancea earthquake, Ann. Geophys., 5B, 727–730.

Trifu, C.I., Radulian, M., Popescu, E., 1990. Characteristics of the intermediate depth microseismicity in Vrancea region, Rev. de Geofisica 46, 75–82.

Trifu, C.I., Deschamps, A., Radulian, M., Lyon-Caen, H., 1992. The Vrancea earthquake of May 30, 1990: An estimate of the source parameters. In: Proceedings of the XXIInd ESC General Assembly, Barcelona, 1990, Vol. 1, 449–454.

Wald, D.J., Quitoriano, V., Heaton, V.T.H., Kanamori, H., 1999a. Relationships between peak ground acceleration, peak ground velocity and modified Mercalli intensity in California, Earthquake Spectra, 15, 557–564.

Wald, D.J., Quitoriano, V., Heaton, V.T.H., Kanamori, H., Scrivner, C.W., Worden, C.B., 1999b. TriNet ShakeMaps: Rapid generation of instrumental ground motion and intensity maps for earthquakes in Southern California, Earthquake Spectra, 15, 537–556.

Wessel, P., Smith, W.H.F., 1991. Free software helps map and display data, EOS Trans. AGU, 72, 441, 445–446.

Wirth, W., Wenzel, F., Sokolov, V., Bonjer, K.-P., 2003. A uniform approach to urban seismic effect analysis, Soil Dyn. Earthq. Eng. 23, 735–758.

A PARAMETRIC MODEL COMBINING GABOR WAVELET AND STOCHASTIC COMPONENT FOR THE AUGUST 30, 1986 VRANCEA EARTHQUAKE

A. ZAICENCO[*,1], H.P. GAVIN[1], B.W. DICKINSON[2]

[1]*Institute of Geology and Seismology, Academy of Sciences, Str. Acedemiei 3, Republic of Moldova, e-mail:anton.az@gmail.com; henri.gavin@duke.edu*

[2]*Department of Civil and Environmental Engineering, Pratt School of Engineering, Duke University, Box 90287, 119 Hudson Hall, Durham, NC 27708–0287, USA, e-mail: bryce.dickinson@duke.edu*

Abstract. An analytical model for the representation of strong ground motions is proposed for the August 30, 1986 Vrancea earthquake. The earthquake simulation model is represented by a short-duration, long-period pulse-like function based on the Gabor wavelet, and a long-duration stochastic record that has a frequency content higher than that of the long-period pulse. The simple physical meaning of the input parameters of the model adequately represents the impulsive character of the records, and successful simulation of the entire data set proves the potential of the method for use in ground-motion simulation. The modified Gabor wavelet is capable of capturing the time-history and response spectra characteristics of the coherent component of the records. The incoherent component of ground motion is simulated with the stochastic approach, providing good compatibility of the resulting linear response spectra.

Keywords: Strong ground motion, Vrancea earthquakes

1. Introduction

The displacement signature associated with a shear dislocation has two distinct terms: near-field (decaying with distance as R^{-2}) and far-field (decaying as R^{-1}). In the simplest case, if a fault is described by a single point with a ramp particle

---
[*]Institute of Geology and Seismology, Academy of Sciences, Str. Acedemiei 3, Republic of Moldova, e-mail: anton.az@gmail.com

A. Zaicenco et al. (eds.), *Harmonization of Seismic Hazard in Vrancea Zone,*
© Springer Science + Business Media B.V. 2008

time history, the far-field displacements would be shaped like boxcars. For a simple line source, or Haskell fault model (Haskell, 1964), the far-field S-wave displacements should be trapezoidally shaped. Near-field ground motions are frequently characterized by intense velocity and displacement pulses of relatively long periods. This observation was first made during the 1966 Parkfield, California earthquake at a distance of 80 km from the fault (Housner and Trifunac, 1967).

Since then, the existence of pulses in near-field records has been confirmed by numerous publications. Their destructive character on structures has been emphasized. The work of Bertero et al. (1978) recognized the special character of near-source ground motions from an earthquake engineering point of view.

Since the 1994 Northridge, California earthquake the engineering community accepted the idea that these pulses represent a serious impact on long-period structures and that near-source effects have to be incorporated in design procedures. Since then experimental and analytical studies of elastic and inelastic response of structures subjected to near-field records have been performed (Makris, 1997; Chopra and Chintanapakdee, 1998; Alavi and Krawinkler, 2001).

However, there is still a need for simple and robust models that could provide simulated ground motions with an impulsive character. Such models should have clear physical meaning inputs and capture the time-history and the response spectrum characteristics of the actual near-field records (Mavroeidis and Papageorgiou, 2003).

Analytical models incorporating near-source effects would allow engineers to study the response of common and non-conventional structures (base-isolated, or those equipped with control elements) to seismic ground motion as a function of input parameters of the analytical model describing the earthquake (Gavin and Zaicenco, 2007).

The current paper presents an application of a robust model proposed by Mavroeidis and Papageorgiou (2003) to a dataset of strong-motion records from the August 30, 1986 Vrancea earthquake. The model simulates the entire set of strong-motion data in their peak phase fairly well, as well as the corresponding response spectra.

2. Stochastic Strong Ground-Motion Models

A review of the history of strong ground motion characterization and simulation has been recently offered by Dickinson (2008) in connection with a statistical generalization of SAC Steel Project suites of strong ground motions.

The stochastic behavior in acceleration records was emphasized by (Housner, 1947) who used a series of impulses arriving randomly in time to model a random earthquake record. The Kanai-Tajimi model (Kanai, 1957) was a simple and effective way to generate seismic waves propagating through the ground which was proposed to be modeled as a linear elastic spring mass damper system excited by broadband input. The variation of acceleration in amplitude over time was pioneered by Iyengar (2001) who proposed to multiply stationary random processes by a deterministic envelope function. This idea was further developed in (Liu, 1970) considering frequency content variation over time described by evolutionary power spectral density function. A review of stochastic process models was provided in (Shinozuka and Deodatis, 1988), while Kozin (Kozin, 1988) discussed autoregressive moving average (ARMA) models to match target ground motions. A simple procedure for ground motion simulation matching a power spectrum and modulating envelope function was given by Grigoriu (1995). A review of stationary and nonstationary models is provided in Conte and Peng (1997), who also propose a method to generate evolutionary PSD that match target functions from two recorded events. Near-fault ground motions generated by a source model were addressed in the study by Shinozuka et al. (1999). The influence of input parameters of the model on the response spectra was studied as well. A frequency content evolution model was suggested by Wen and Gu (2004) who used a Hilbert transform. Averaged Hilbert spectra were used as target spectra. The procedure was further developed by Gu and Wen (2007) to accommodate tri-directional response spectra. It was shown that a Hilbert-Huang transform is capable of reproducing a near-fault pulse. All the models mentioned are efficient in reproducing existing earthquake ground motions.

Combined seismological models of ground motion in a frequency domain with the engineering interpretation of the random nature of high-frequency motions (Vanmarcke et al., 1997) were proposed by Hanks and Kanamori (1979) and McGuire and Hanks (1980). The simulated motions are S-waves, which are the most significant seismic hazard. The model describes the spectrum of ground motion assuming that accelerations on an elastic half-space are band-limited, finite-duration, white Gaussian noise. The total spectrum of the motion at a site breaks into contributions from the earthquake source (E), path (P), site (G), and instrument (I) (Boore, 2003):

$$Y(M_0, R, f) = E(M_0, f) \, P(R, f) \, G(f) \, I(f) \tag{1}$$

where M_0 is the seismic moment (Aki, 1966).

The most commonly used model of an earthquake source spectrum is the ω-square model (Aki, 1967). The effects of the path might be represented, generally, by simple functions that account for geometrical spreading, attenuation, and the general increase of duration with distance due to wave propagation and scattering (Boore, 2003). The ground-motion duration is the sum of the source duration related to the inverse of a corner frequency $1/f_0$ and a path-dependent duration (Atkinson and Boore, 1995). Site effect is usually separated from path effects and accounts for modifications due to local site geology.

3. Records of the August 30, 1986 Vrancea Earthquake

It was emphasized by Bolt (Bolt, 1999) that "Unfortunately for ground motion estimates in Romania, near-source recordings from intermediate depth sources and $M_w > 7.0$ earthquakes around the world remain sparse." The presence of pulses in near-source records of Vrancea earthquakes was evidenced by the same author:

> In the large Vrancea earthquakes of 1977 and 1986 (M_w = 7.4 and 7.1, respectively), strong motion recordings in the free field in Bucharest show a main pulse from S wave energy with period between 1 and 2 seconds. These pulses, particularly in the velocity time history of ground motion, resemble those recorded in the 1971 and 1994 California earthquakes which also had thrust mechanisms.

The coordinates of the August 30, 1986, 21:28 GMT Vrancea earthquake were: 45.53N, 26.47E, focal depth h = 133 km and moment magnitude M_w = 7.1 (Radu, 1994). The following properties of the media are assumed, as an approximation, in the hypocentral region: $\beta_p \approx 8$, $\beta_s \approx 4$ km/s, density $\rho \approx 3.4$ g/cm^3, shear modulus $G \approx 7.27 \cdot 10^{11}$ dyn/cm^2 (Raileanu et al., 2005; Benetatos and Kiratzi, 2004).

These media properties suggest the corner frequency of the source spectrum (Brune, 1971):

$$f_0 = 2.34 \cdot \beta_s / (\frac{7M_0}{16\Delta\sigma})^{1/3} \approx 0.57 \, rad/s \tag{2}$$

where: seismic moment $M_0 \approx 10^{20}\,\mathrm{N} \cdot \mathrm{m}$, and stress drop $\Delta\sigma \approx 40 \cdot 10^5$ Pa.

The observed corner frequencies of the source spectrum were: f_{01} = 0.12 Hz and f_{02} = 0.6 Hz. The second corner frequency, f_{02}, was attributed to a sub-event with higher compression stress inside a bigger event in Kondorskaya et al. (1990). Their study was based on broad-band records in the frequency range 0.04–2.5 Hz.

Several stations which provided strong-motion ground records that exhibit "distinct" pulses for August 30, 1986 event are given in Figure 1. The largest hypocentral distance of a station in the database is 250 km, which is still comparable with a focal depth of 133 km of the intermediate-depth Vrancea earthquake. These data were used for testing the analytical model presented hereafter.

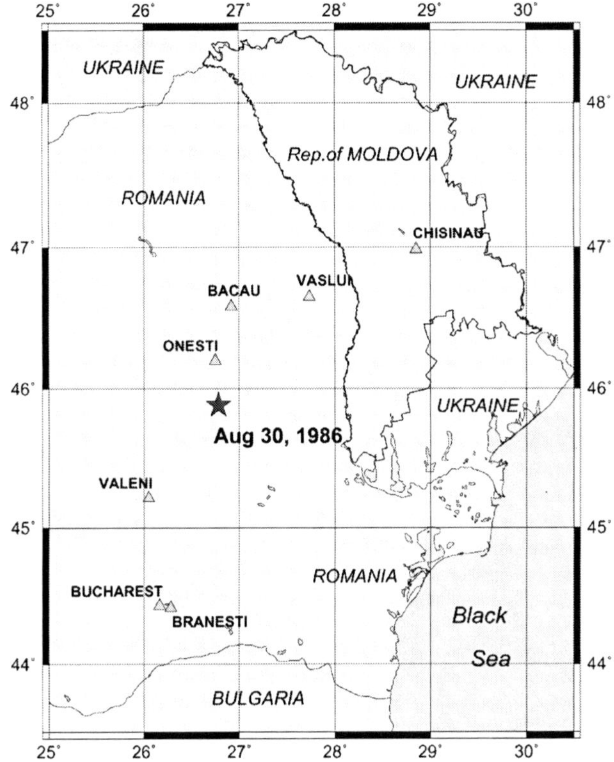

Figure 1: The August 30, 1986 earthquake: selected near-field seismic stations

4. Velocity Pulse Model

The basic features of the near-field earthquake ground motions are short duration, strong directivity, and low-frequency impulsive motion in the velocity time-history. To adequately represent a coherent signal, which is localized in time, wavelets are employed.

The *pulse duration, pulse amplitude, number* and *phase of half cycles* are the key parameters describing near-field velocity pulses (Mavroeidis and Papageorgiou, 2003). The Gabor wavelet meets the requirements necessary for use in the analytical modeling of seismological signals: it has a simple mathematical expression, it is capable of representing all records in the current study, and it allows the derivation of closed-form expressions of its spectral characteristics (Fourier and response spectra). This wavelet is a simple product of the Gauss-shaped taper and harmonic function:

$$\upsilon_{p}(t) = V_{p} \cdot e^{-(2\pi/(T_{p}\gamma))^2 t^2} cos(2\pi t/T_{p} + \phi) \tag{3}$$

where: V_p is amplitude, T_p is dominant period, ϕ is phase angle, and γ is oscillatory character. The presence of the Gauss-shaped taper in this format introduces difficulties in deriving closed-form solutions for the response of an SDOF system (Mavroeidis and Papageorgiou, 2003). Therefore, the Gaussian envelope is replaced by a similar one – a shifted haversed sine function:

$$\upsilon_p(t) = V_p \frac{1}{2}[1 + cos(2\pi t/(T_p\gamma))] \cdot cos(2\pi t/T_p + \phi) \qquad (4)$$

with the restriction that $\gamma > 1$, which means that the period of the harmonic function should be less than the period of the envelope.

The parameters of the wavelet are obtained in the following way: first the Fourier transform of the portion of the velocity record containing a pulse is taken which provides the pulse period, T_p (Figure 2).

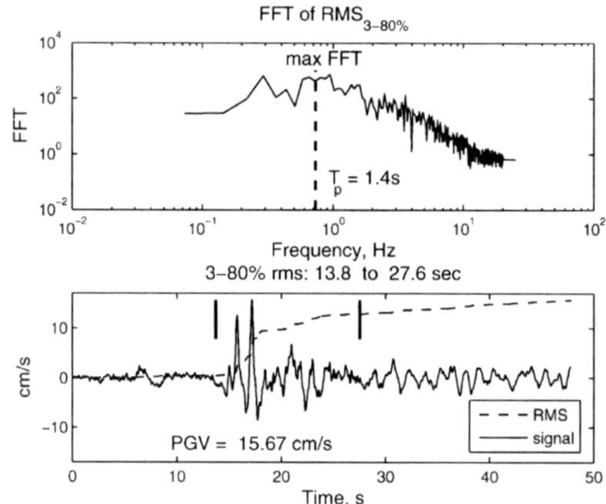

Figure 2: Search for the pulse period, T_p: velocity record of BRN N107W component

The lower and upper limits of this portion of the entire time-history could be set by the specified values of RMS in the stochastic process.

The amplitude of the pulse V_p should satisfy the condition that the velocity response spectrum of the synthetic and recorded ground motions have approximately the same peak value. It is noted that it is quite difficult to set the value of the oscillatory characteristic γ which is related to zero crossings, without visual control. The parameter γ is adjusted keeping in mind the physical characteristics of the earthquake. The phase angle ϕ controls the property of the wavelet related to its symmetry.

All parameters, except T_p and γ, are computed using the iterative method based on the Levenberg-Marquardt algorithm (LMA) (Levenberg, 1944; Marquardt, 1963). The minimization problem is formulated as follows:

$$\text{minimize} \sum_t [\hat{\upsilon}_p(t;\mathbf{p}) - \upsilon_p(t)]^2 \text{ such that:} \tag{5}$$

$$T = T_p, \gamma = \gamma_0$$

where: $\hat{\upsilon}_p(t;\mathbf{p})$ is the pulse function given by Equation (4) parameterized by the parameters in the vector ($\mathbf{p} = [V_p \; \phi]^T$), $\upsilon_p(t)$ is the velocity record, V_p is the peak velocity of the record, and $t_0 = t(V_p)$ is equal to the time instance when peak ground velocity occurs. T_p is provided by FFT analyses.

The LMA estimates the increment $\mathbf{q}$ of the vector $\mathbf{p}$ iteratively. The initial guess of parameters is assumed and a chi-square error χ is computed:

$$\mathbf{p_0} = [p_1 \; p_2 \cdots p_n]^T$$
$$\hat{\mathbf{v}}_p = \hat{\upsilon}_p(t;\mathbf{p_0}) \tag{6}$$
$$\chi_0 = \Sigma(\hat{\mathbf{v}}_p - \upsilon_p(t))^2$$

The iterations are performed:

$$\mathbf{J} = [\frac{d\hat{\upsilon}_p}{dp_1} \; \frac{d\hat{\upsilon}_p}{dp_2} \cdots \frac{d\hat{\upsilon}_p}{dp_n}]$$

$$\mathbf{q} = -(\mathbf{J}^T\mathbf{J} + \lambda\mathbf{I})^{-1}\mathbf{J}^T(\hat{\mathbf{v}}_p - \upsilon_p(t))$$

$$\mathbf{p_1} = \mathbf{p_0} + \mathbf{q}^T$$
$$\hat{\mathbf{v}}_p = \hat{\upsilon}_p(t;\mathbf{p_1})$$
$$\chi_1 = \Sigma(\hat{\mathbf{v}}_p - \upsilon_p(t))^2 \tag{7}$$

$$\lambda = \begin{cases} \lambda \cdot 10 & \text{if } \chi_1 > \chi_0 \\ \lambda/10 & \text{if } \chi_1 < \chi_0 \end{cases}$$

$$\mathbf{p_0} = \begin{cases} \mathbf{p_0} & \text{if } \chi_1 > \chi_0 \\ \mathbf{p_1} & \text{if } \chi_1 < \chi_0 \end{cases}$$

$$\chi_0 = \begin{cases} \chi_0 & \text{if } \chi_1 > \chi_0 \\ \chi_1 & \text{if } \chi_1 < \chi_0 \end{cases}$$

until the following condition is satisfied:

$$\frac{\|\mathbf{q}\|}{\|\mathbf{p_0}\|} < 0.01 \tag{8}$$

where $\mathbf{J}$ is the Jacobian of $\hat{u}_p$ at $\mathbf{p}$, and λ is the damping factor, which is adjusted at each iteration.

The additional requirement $SV_P(T_p) = SV_R(T_p)$, where SV_P and SV_R are spectral velocity values of the analytical pulse and the record, is satisfied after the vector $\mathbf{p}$ is computed and, consequently, V_p is found. The value of V_p simply is allowed to variate within the limits $\pm 30\%$ until the requirement is met. This is accomplished within a for-loop.

To demonstrate the proposed modified Gabor wavelet, the BRN N107W component velocity pulse was simulated. The synthetic signal reproduced the long-period part of the ground motion well. The harmonic oscillation, taper and fitted pulse are provided in Figure 3.

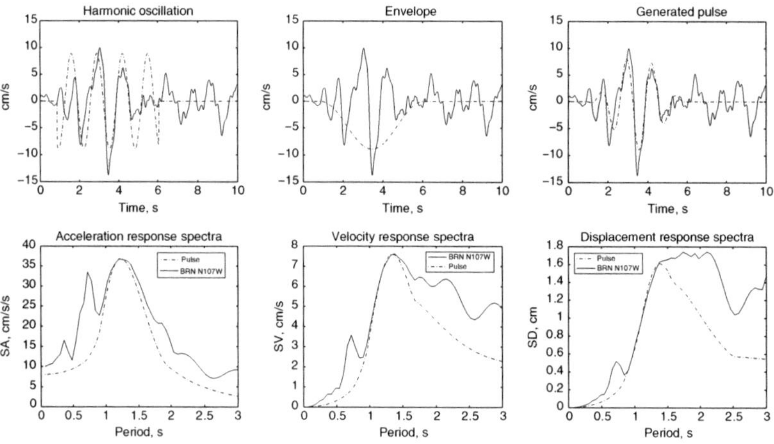

Figure 3: Procedure of fitting Gabor wavelet on BRN N107W component: harmonic oscillation, envelope, fitted pulse, and resulting response spectra

5. Model-Fitting Results

The results of the application of the pulse model suggested in Equation (4) to the databank of the strong-motion records of the August 30, 1986 Vrancea earthquake are summarized in Table 1. The table provides the parameters of a modified Gabor wavelet (V_p, T_p, γ, ϕ) where ϕ is optimized according to Equation (5).

It could be observed that analytical models for all records possess similar properties: $\gamma = 4.0$, $T_p \approx 1.5$ s, and have symmetric shapes ($\phi = 5.0$ rad). Moreover, these properties are preserved for both horizontal components of the records. The prevailing period of the pulse T_p for the records analyzed is the second-corner period of the source spectrum for the August 30, 1986 Vrancea earthquake. Therefore, it reflects the source mechanism which is very helpful in the study of the properties of propagation media and the influence of local soil conditions. It is necessary to investigate the scaling laws for the parameters of the model in order to use synthetic time histories in engineering designs. In this way, it is emphasized that the period of the pulse T_p is related to moment magnitude. Figure 4 displays some of the results of fitting the synthetic signal to strong ground-motions from Table 1. The analytical model was suggested in order to simulate the long-period properties of ground motion that are manifested in the velocity domain. The high-frequency components of ground motion which are stochastic in nature, are mostly presented in the acceleration records which could not be adequately simulated with the proposed pulse-generation method. It is the incoherent (high-frequency) component of ground motion that should be included in the model. The stochastic (engineering) approach described in the previous chapters is applicable here.

It is observed that all records in the current study have amplitudes of the analytic velocity pulses, V_p, smaller than the peak ground velocities (PGV) by an order of 70%. The PGV of a record is the superposition of a long period pulse and the velocity of a stochastic component that has shorter periods than the period of the pulse. The PGV can be higher than the amplitude of the pulse function, which passes through the higher-frequency oscillations of the full record. See, for example, the CHS records in Figure 4.

Polar plots of PGA and V_p clearly demonstrate the directivity effect of the earthquake studied along the NE-SW line which agrees with a general trend of the stress field in the Vrancea region (compression caused by the Black Sea sub-plate movement from SE to NW) (Figure 5).

The dataset of strong-motion records in the current study is characterized by a single multi-cycle velocity pulse that is a subset of a more general case when more than one pulse is generated by the seismic fault. In this case, a sequence of pulses could be simulated, describing the contribution of each individual fault segment (Mavroeidis and Papageorgiou, 2003). However, the fitting achieved with several pulses might be not unique. The idea is to not over-parameterize the data, keeping only several of them that have physical sense.

6. Synthetic Ground Motions with Incoherent Components

A simplified method to generate realistic ground motion is proposed in the current section. As mentioned previously, the modified Gabor wavelet with physically simple parameters (V_p, T_p, γ, ϕ) is capable of capturing the coherent component of the velocity pulse in ground records. Nevertheless, the incoherent (high-frequency) part of the ground motion containing a significant amount of energy remains unaddressed by the proposed methodology.

A. ZAICENCO ET AL.

TABLE 1: August 30, 1986 earthquake: Parameters of the analytical wavelet model

No	Location	Site code	Comp.	PGA (cm/s^2)	V_p (cm/s)	T_p (s)	γ	ϕ
1	Bolintin Vale	BLV	N65E	87.75	11.8	1.7	4	6.2
2	Branesti	BRN	N107W	98.86	13.9	1.5	4	6.1
		BRN	N163E	70.50	8.0	1.5	4	2.9
3	Campulung	CMP	NS	72.04	9.5	1.5	4	1.9
		CMP	EW	76.65	10.1	1.5	4	4.8
4	Giurgiu	GRG	NS	33.18	2.9	1.5	4	6.0
		GRG	EW	43.80	3.2	1.5	4	4.3
5	Ploiesti	PLS	N190E	216.70	25.5	1.5	4	6.7
		PLS	N100E	213.80	20.8	1.4	4	6.5
6	Peris	PRS	N80E	152.70	11.9	1.5	4	6.3
		PRS	N10W	114.80	9.7	1.5	4	6.4
7	Rimnicu Sarat	RMS	N55E	165.20	9.4	1.5	4	5.8
		RMS	N145E	93.15	3.8	1.5	4	5.4
8	Valenii de Munte	VLM	N174W	160.50	9.1	1.5	4	5.7
		VLM	N84W	191.40	14.1	1.3	4	2.1
9	Bucharest-EREN	ERE	N10W	156.00	10.4	1.4	4	6.7
		ERE	N10S	105.80	8.1	1.4	4	6.2
10	Bucharest-INCERC	INC	NS	88.69	11.4	1.5	4	6.0
		INC	EW	95.26	7.4	1.5	4	6.5
11	Bucharest-Metalurgiei	MET	W32S	69.78	10.4	1.5	4	5.9
		MET	N32W	43.68	4.4	1.5	4	2.8
12	Bucharest-Militari	MLT	NS	92.19	6.7	1.5	4	6.1
		MLT	EW	79.57	6.7	1.5	4	6.1
13	Bucharest-Metrou	MTR	N120W	72.21	11.3	1.5	4	3.6
		MTR	N30W	57.75	5.7	1.5	4	5.9
14	Bucharest-Otopeni	OTO	NS	187.30	10.5	1.5	4	2.4
		OTO	EW	140.00	5.7	1.5	4	4.2
15	Bucharest-Panduri	PND	N131E	90.61	5.2	1.5	4	3.7
		PND	N139W	101.20	8.3	1.5	4	3.6
16	Bucharest-Titulescu	TIT	N145W	89.57	11.6	1.6	4	2.7
		TIT	N55W	79.84	5.0	1.4	4	7.3
17	Bucharest-ISPH	ISP	N15E	86.75	10.6	1.5	4	6.0
		ISP	E15S	40.06	2.6	1.5	4	7.0
18	Bucharest-Magurele	MAG	NS	135.45	11.9	1.5	4	6.7
		MAG	EW	114.65	10.8	1.5	4	6.2
19	Dochia	DOC	NS	50.86	1.5	1.5	4	6.2
		DOC	EW	42.35	1.3	1.5	4	7.2
20	Focsani-UCA	FO1	N97W	224.20	11.7	1.5	4	5.3
		FO1	N07W	269.41	7.4	1.5	4	6.9
21	Focsani-	FOC	NS	273.29	14.3	1.5	4	3.2
		FOC	EW	297.14	8.1	1.5	4	4.2
22	Iasi	IAS	NS	66.88	3.0	1.3	4	3.9
		IAS	EW	99.62	3.8	1.3	4	2.3
23	Istrita	ISR	NS	109.07	10.8	1.3	4	3.1
		ISR	EW	71.63	5.5	1.3	4	3.0
24	Muntele Rosu	MLR	NS	79.12	11.5	1.5	4	2.4
		MLR	EW	33.74	5.0	1.5	4	6.5

25	Rimnicu	RMS	N55E	156.43	9.4	1.5	4	3.3
	Sarat	RMS	N145E	98.86	4.0	1.5	4	2.7
26	Vrincioaia	VRI	NS	82.33	7.3	1.5	4	6.5
		VRI	EW	140.78	-	-	-	-
27	Bacau	BAC	NS	105.00	5.6	1.5	4	4.4
		BAC	EW	68.48	3.1	1.5	4	6.9
28	Focsani	FO2	N97W	288.10	9.9	1.6	4	7.6
		FO2	N07W	207.40	13.4	1.5	4	6.5
29	Onesti	ONS	N200E	156.10	5.4	1.5	4	5.8
30	Vaslui	VSL	NS	152.90	3.9	1.5	4	3.6
		VSL	EW	179.50	5.8	1.5	4	5.1
31	Chisinau	CHS	NS	191.79	6.4	1.2	4	4.2
		CHS	EW	212.75	13.3	1.2	4	2.8
32	Krasnogorka	KRA	NS	81.96	1.6	1.5	4	2.5
		KRA	EW	69.24	1.7	1.5	4	2.5

The engineering (stochastic) approach presented previously could fill this gap and add effects generated by the propagation media and dynamic response of local soil conditions Equation 1.

Artificial earthquake ground motion is generated through the super-position of a random ground velocity record with a coherent long-period velocity pulse. The random ground motion acceleration record $a_u(t)$ is generated by first simulating an unscaled random ground acceleration record by multiplying noise generated from the 2-pole transfer function characterizing 2-degree-of-freedom (2-DOF) system:

$$|G(\omega)|^2 = \frac{(b_0 + b_1\omega + b_2\omega^2)^2}{[(\omega_1^2 - \omega^2)^2 + (2\xi\omega_1\omega)^2]^2[(\omega_2^2 - \omega^2)^2 + (2\xi\omega_2\omega)^2]} \tag{9}$$

with an envelope function $L_e(t)$ (Figure 6):

$$L_e(t,\tau_1,\tau_2,\tau_3) = \begin{cases} (t/\tau_1)^3 & \text{for } 0 \le t \le \tau_1 \\ 1 & \text{for } \tau_1 \le t \le \tau_1 + \tau_2 \\ exp[-(t-\tau_1-\tau_2)/\tau_3] & \text{for } \tau_1 + \tau_2 \le t \le \tau_1 + \tau_2 + \tau_3 \end{cases} \tag{10}$$

An unscaled random ground velocity record $v_u(t)$ is calculated from $a_u(t)$ using trapezoidal integration, and the scaled random ground velocity record is obtained by scaling $v_u(t)$ with the desired peak of the random ground velocity record V_r. The artificial ground velocity record is then found by combining the scaled random ground velocity with a smooth velocity pulse, with period T_p and amplitude V_p (Figure 7).

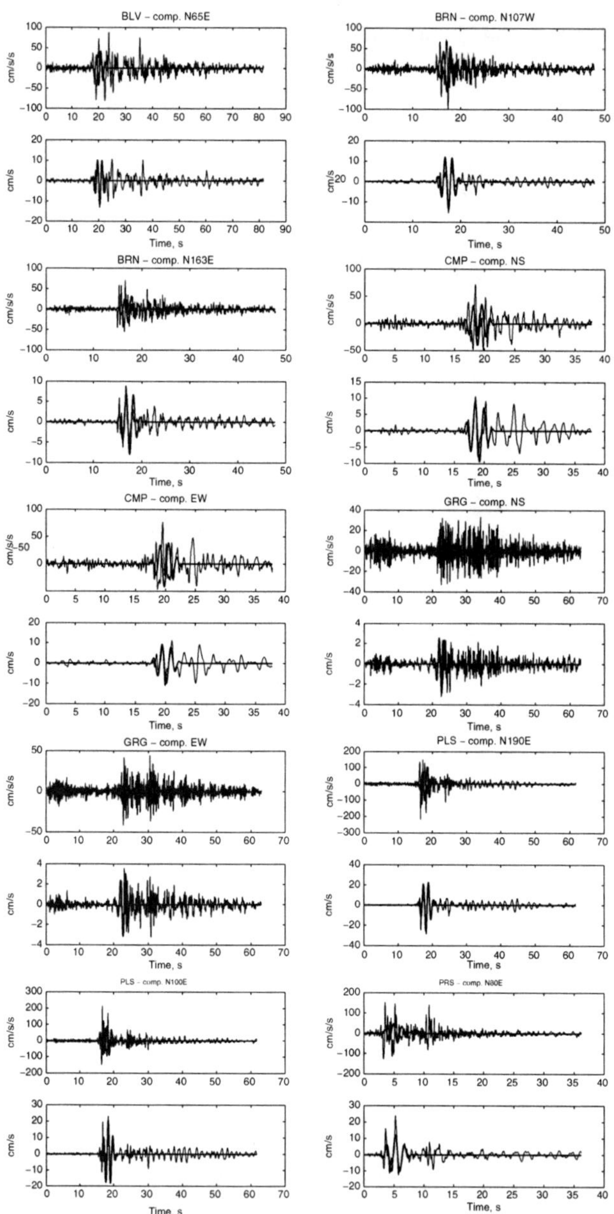

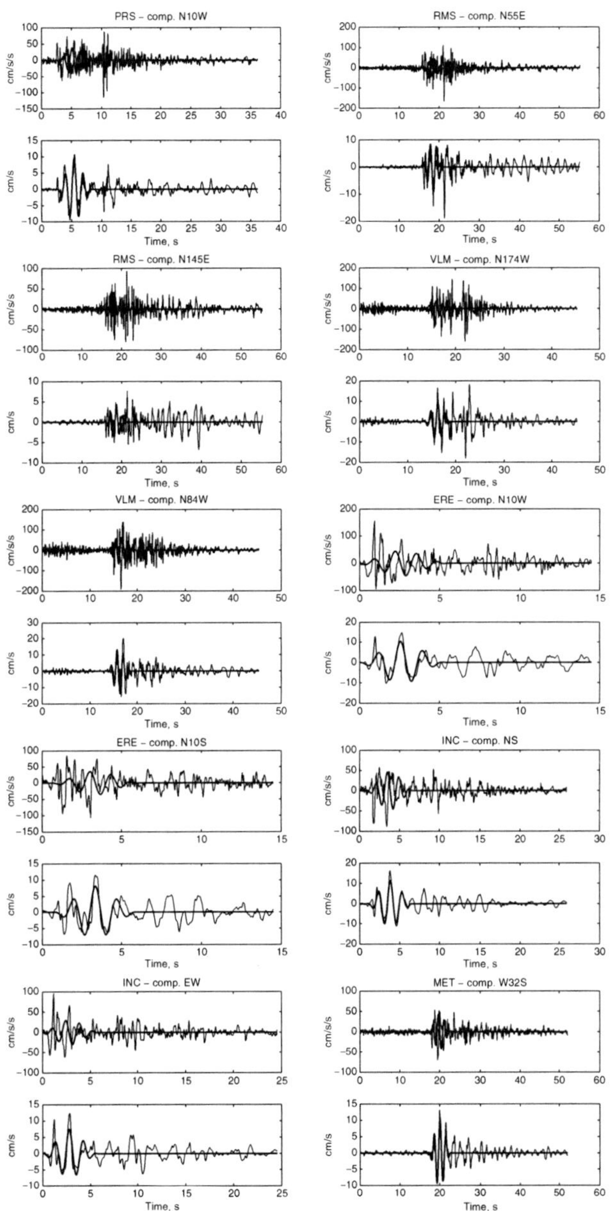

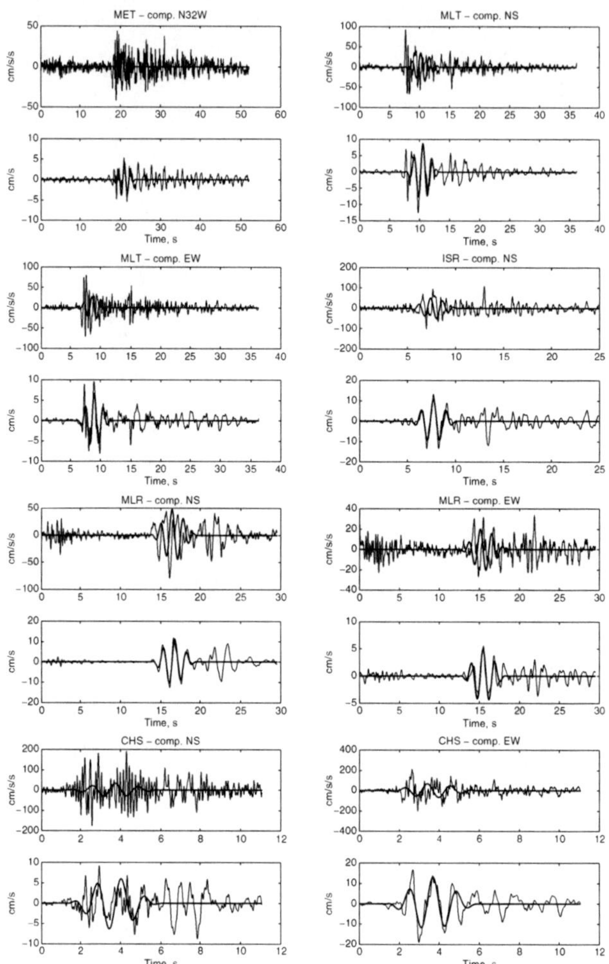

Figure 4: Results of fitting Gabor wavelet on several selected strong-motion records

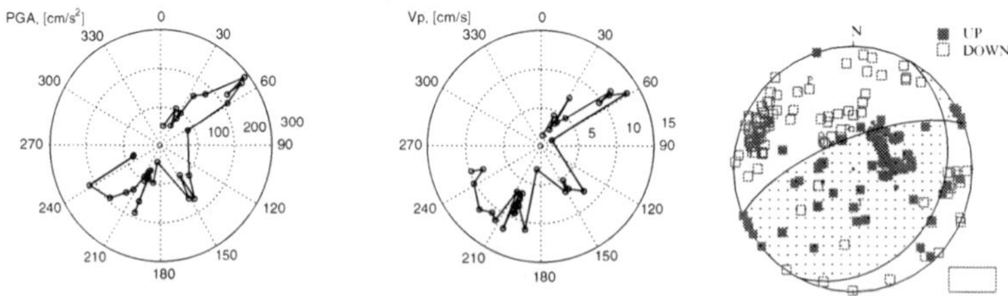

Figure 5: Polar plots of PGA and V_p together with focal mechanism solution indicating the directivity effect of the August 30, 1986 earthquake

Computed parameters of 2-DOF filter from Equation (9) matching the incoherent component of the selected strong-motion records using the Levenberg-Marquardt algorithm are provided in Table 2.

Using this procedure, the records of CHS station were simulated and results compared in terms of time-history shapes and linear response spectra with $\xi = 0.05$ (Figure 8).

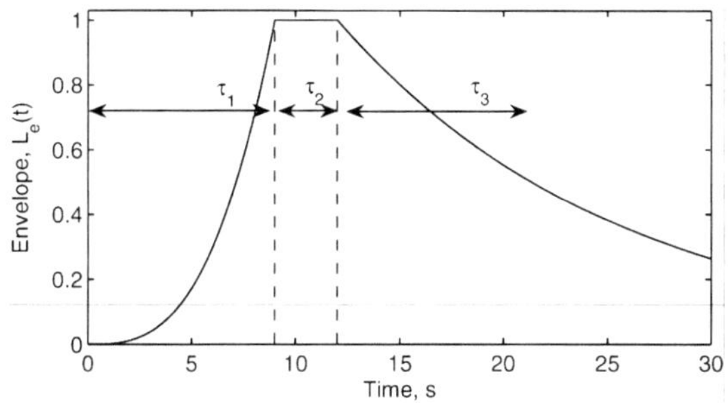

Figure 6: Envelope function

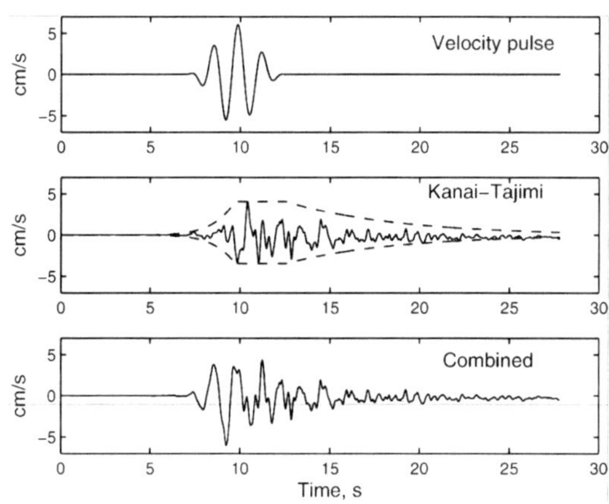

Figure 7: Simulation procedure of velocity time-history with incoherent component

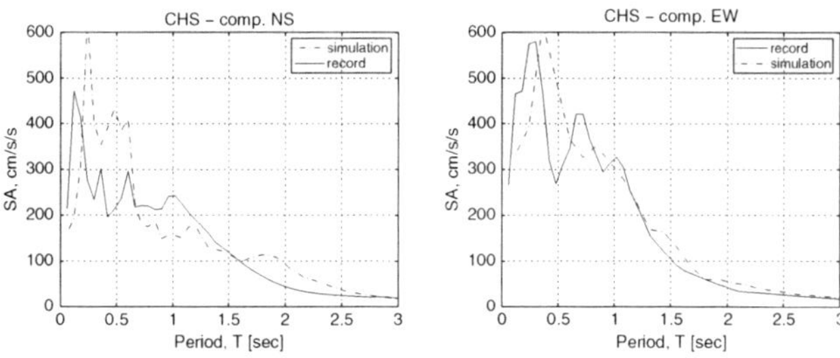

Figure 8: Recorded and simulated ground motions of CHS site: linear response spectra, $\xi = 0.05$

A. ZAICENCO ET AL.

TABLE 2: August 30, 1986 earthquake: parameters of the incoherent component of the analytical wavelet model

No	Site code	Comp.	$T_1(T_2)$ s	$\xi_1(\xi_2)$
1	BLV	N65E	1.1(0.4)	0.07(0.7)
	–	–	–	–
2	BRN	N107W	1.6(0.6)	0.09(0.39)
	BRN	N163E	1.7(0.5)	0.01(0.11)
3	CMP	NS	0.95(0.4)	0.04(0.25)
	CMP	EW	1.6(0.6)	0.13(0.28)
4	GRG	NS	0.9(0.4)	0.5(0.01)
	GRG	EW	1.1(0.4)	0.45(0.06)
5	PLS	N190E	1.6(0.5)	0.04(0.24)
	PLS	N100E	1.6(0.5)	0.08(0.29)
6	PRS	N80E	0.8(0.3)	0.1(0.25)
	PRS	N10W	1.6(0.3)	0.01(0.4)
7	RMS	N55E	1.3(0.5)	0.5(0.02)
	RMS	N145E	1.5(0.5)	0.5(0.02)
8	VLM	N174W	1.8(0.6)	0.5(0.02)
	VLM	N84W	0.7(0.3)	0.08(0.3)
9	ERE	N10W	1.6(0.4)	0.02(0.24)
	ERE	N10S	0.9(0.3)	0.13(0.15)
10	INC	NS	1.4(0.6)	0.04(0.16)
	INC	EW	2.1(0.8)	0.11(0.1)
11	MET	W32S	1.6(0.7)	0.1(0.13)
	MET	N32W	1.2(0.4)	0.12(0.02)
12	MLT	NS	1.3(0.4)	0.5(0.04)
	MLT	EW	1.3(0.5)	0.6(0.05)
13	MTR	N120W	1.7(0.7)	0.11(0.04)
	MTR	N30W	1.1(0.3)	0.06(0.27)
14	OTO	NS	1.4(0.5)	0.04(0.05)
	OTO	EW	1.2(0.4)	0.6(0.03)
15	PND	N131E	0.8(0.3)	0.04(0.17)
	PND	N139W	1.3(0.4)	0.01(0.03)
16	TIT	N145W	1.4(0.5)	0.1(0.02)
	TIT	N55W	0.9(0.4)	0.15(0.19)
17	ISP	N15E	1.0(0.2)	0.23(0.6)
	ISP	E15S	1.0(0.2)	0.6(0.1)
18	MAG	NS	0.8(0.3)	0.17(0.6)
	MAG	EW	1.7(0.7)	0.22(0.11)
19	DOC	NS	0.9(0.4)	0.6(0.05)
	DOC	EW	0.9(0.4)	0.6(0.07)
20	FO1	N97W	0.9(0.3)	0.05(0.27)
	FO1	N07W	1.1(0.4)	0.6(0.03)
21	FOC	NS	1.0(0.4)	0.12(0.15)
	FOC	EW	1.0(0.3)	0.6(0.01)
22	IAS	NS	0.9(0.4)	0.14(0.03)
	IAS	EW	0.8(0.3)	0.08(0.47)
23	ISR	NS	1.5(0.6)	0.45(0.01)
	ISR	EW	0.9(0.4)	0.22(0.38)
24	MLR	NS	1.6(0.7)	0.06(0.08)
	MLR	EW	1.3(0.5)	0.11(0.14)
25	RMS	N55E	1.6(0.5)	0.6(0.02)

	RMS	N145E	1.0(0.3)	0.37(0.03)
26	VRI	NS	0.8(0.3)	0.08(0.13)
	VRI	EW	-	-
27	BAC	NS	1.0(0.3)	0.04(0.22)
	BAC	EW	0.8(0.3)	0.02(0.02)
28	FO2	N97W	0.9(0.2)	0.6(0.01)
	FO2	N07W	1.0(0.3)	0.03(0.1)
29	ONS	N200E	1.4(0.6)	0.7(0.09)
30	VSL	NS	1.0(0.3)	0.7(0.03)
	VSL	EW	0.8(0.3)	0.06(0.08)
31	CHS	NS	0.4(0.1)	0.9(0.03)
	CHS	EW	0.65(0.2)	0.03(0.38)
32	KRA	NS	0.52(0.21)	0.9(0.06)
	KRA	EW	0.42(0.17)	0.42(0.07)

The results of the incoherent component contribution analysis for site CHS provided values of ground velocity $V_r = 12.7$ cm/s (comp. NS), $V_r = 17.7$ cm/s (comp. EW), frequency for the 2-DOF filter $T_1 = 0.4$ s (comp. NS), $T_1 = 0.65$ s (comp. EW) and damping $\xi_1 = 0.9$ (comp. NS) and $\xi_1 = 0.03$ (comp. EW). The natural period of the CHS site soft soil is $T_0 = 0.6$ s, which was obtained from a soil profile share-wave velocity distribution. It provides a good match with the frequency of the 2-DOF filter.

7. Correlation Spectra

The current study examines the wave-form characteristics of the August 30, 1986 Vrancea earthquake. Only horizontal ground motions are considered. Each of the 61 components was analyzed via the Levenberg-Marquardt algorithm in order to determine estimates for seven ground motion parameters: amplitude (V_p), period (T_p), phase (ϕ) and number of half-cycles (γ) of velocity pulse, peak value of random velocity component (V_r), period (f_g) and damping (ξ) of the Kanai-Tajimi filter. The probability density functions of these parameters, except γ which was found to be constant and ϕ which was irrelevant to structural response in the work of Dickinson (2008), were represented with a Weibull distribution (Table 3 and Figure 9). It has been demonstrated also in Dickinson (2008) that the parameters of the envelope function have a minor influence on the response spectra ordinates.

Assuming zero correlation between the random variables, a set of synthetic records was generated according to the methodology described in the previous section.

In order to evaluate the sensitivity of the response spectra to variability in the ground motion parameters, correlation spectra suggested by Dickinson

TABLE 3: Weibull distribution parameters for strong-motion data set

Parameter	Shape κ	Scale λ
V_p	3.0	8.0
T_p	20.0	1.5
V_r	2.0	4.0
f_g	2.0	2.0
ξ_g	1.7	0.7

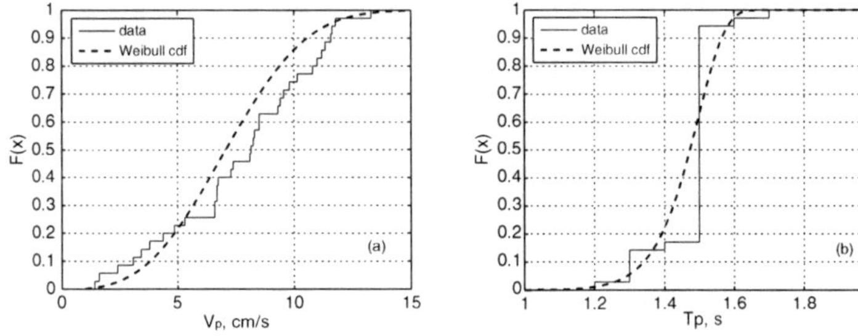

Figure 9: Cumulative distributions of ground motion parameters V_p and T_p

(2008) are computed. These spectra indicate a statistical correlation between a ground motion parameter and the peak response of the structure of a given natural period (Figure 10). Response spectra are computed for linear structures with 5% damping.

Correlation spectra are computed by generating several (a thousand or more) ground acceleration records using parameters drawn from the statistical distributions. Correlation spectra for acceleration response and displacement response are almost identical. This may not be the case for responses of inelastic structures, or heavily damped structures, which will be the topic of a future study. Correlations near zero indicate low sensitivity of the structural response to the particular ground motion parameter. Parameters with correlation spectral values near zero for the entire range of natural periods do not significantly affect response variability.

In order of decreasing correlation with the peak response statistics, the ground motion parameters are: V_p, T_p, V_r, f_g, and ξ_g. The peak response of flexible structures ($T > 1.0$ s) is better correlated with V_p and less correlated with V_r. Stiffer structures ($T < 1.0$ s) demonstrate the opposite. Kanai-Tajimi filter parameters, f_g and ξ_g, have some influence on the peak response of stiff structures ($T < 0.8$ s). Yet, this observation could be biased because for $T_p \approx 1/f_g$ it might be difficult to decouple coherent from incoherent component parameters.

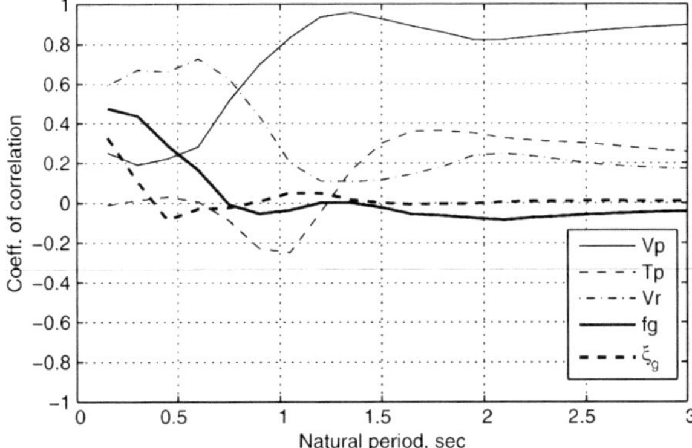

Figure 10: Correlation spectra between ground motion parameters and linear response spectra coordinates (5% damping)

8. Conclusions

The records of the August 30, 1986 intermediate-depth Vrancea earthquake clearly demonstrate the "fling", or velocity pulse, from the radiation mechanism and the directivity of the thrust source.

The parameters of the modified Gabor wavelet applied to the records selected show similar features of the low-frequency pulse on all horizontal components that correspond to the second corner frequency of the source spectrum reported from the broad-band records.

The incoherent components that are not addressed by the wavelet model, could be considered by the stochastic engineering method. Further research is required to derive the scaling laws for the model parameters, i.e. to investigate the influence of earthquake size on them. As a result, it might be possible to come up with a practical simulation model suitable for engineering aseismic designs from intermediate-depth earthquakes for which the near-source recordings remain sparse around the world.

The correlation spectra of the peak response statistics of linear SDOF systems with the parameters of the proposed ground-motion model are developed. These spectra allow the degree of influence of the selected model parameters on structural response to be established.

Acknowledgements

This research is sponsored by NATO's Scientific Affairs Division in the framework of the Science for Peace Programme, project SfP-980468. The data were kindly provided by Professor Lungu from the National Institute for

Building Research (INCERC), Romania, and Dr. Alkaz from the Institute of Geophysics and Geology, Academy of Sciences of Moldova.

References

Aki, K. (1966). Generation and Propagation of G Waves from the Niigata Earthquake of June 16, 1964. Part 2. Estimation of Earthquake Moment, Released Energy, and Stress-Strain Drop from the G-Wave Spectrum, *Bull. Earthq. Res. Inst.*, **44**, 73–88.

Aki, K. (1967). Scaling Law of Seismic Spectrum, *J. Geophys. Res.*, **72**, 1217–1231.

Alavi, P., and Krawinkler, H. (2001). Effects of near-fault ground motion on building structures, *CUREE-Kajima Joint Res. Program Rep.*, Richmond, CA.

Atkinson, G.M., and Boore, D.M. (1995). Ground Motion Relations for Eastern North America, *Bull. Seismol. Soc. Am.*, **85**, 17–30.

Benetatos C.A., and Kiratzi A.A. (2004). Stochastic strong ground motion simulation of intermediate depth earthquakes: the cases of the 30 May 1990 Vrancea (Romania) and of the 22 January 2002 Karpathos island (Greece) earthquakes, *Soil Dyn. Earthq. Eng..*, **24**, 1–9.

Bertero, V.V., Mahin, S.A., and Herrera, R.A. (1978). Aseismic Design Implications of Near-Fault San Fernando Earthquake Records, *Earthquake Eng. Struct. Dyn.*, **6**, 31–42.

Bolt, B.A. (1999). Modern recording of seismic strong motion for hazard reduction, in F. Wenzel et al. (eds.), *Vrancea Earthquakes: Tectonics, Hazard and Risk Mitigation*, pp. 1–14, Kluwer, Dordrecht.

Boore, D.M. (2003). Simulation of Ground Motion Using the Stochastic Method, *Pure Appl. Geophys.*, **160**, 635–676.

Brune, J.N. (1971). Correction, *J. Geophys. Res.*, **76**, 5002.

Chopra, A.K., and Chintanapakdee, C. (1998). Accuracy of response spectrum estimates of structural response too near-field earthquake ground motions: preliminary results, in *Proceedings of the Structural Engineers World Congress*, San Francisco, CA, 18–23 July.

Conte, J.P., and Peng, B.F. (1997). Fully Nonstationary Analytical Earthquake Ground-motion Model, *J. Eng. Mech.*, 15–24.

Dickinson, B.W. (2008). A Parametric Statistical Generalization and Simulation of Uniform Hazard Earthquake Ground Motions, *MS thesis, Duke University*, 117 pp.

Gavin, H.P., and Zaicenco, A. (2007). Performance and reliability of semi-active equipment isolation, *J. Sound Vibration*, **306**, 1–2, 74–90.

Grigoriu, M. (1995). Probabilistic Models and Simulation Methods for Seismic Ground Acceleration, *Meccanica*, **30**, 105–124.

Gu, P., and Wen, Y.K. (2007). A Record-based Method for the Generation of Tridirectional Uniform Hazard-response Spectra and Ground Motions Using the Hilbert-Huang Transform, *Bull. of Seism. Soc. of Am.*, **97**, 1539–1556.

Hanks, T.C., and Kanamori, H. (1979). A Moment Magnitude Scale, *J. Geophys. Res.*, **84**, 2348–2350.

Haskell, N.A. (1964). Total Energy an Energy Spectral Density of Elastic Wave Radiation from Propagating Faults, *Bull. Seism. Soc. Am.*, **54**, 1811–1841.

Housner, G.W. (1947). Characteristics of Strong-motion Earthquakes, *Bull. Seism. Soc. Am.*, **37**, 19–31.

Housner, G.W., and Trifunac, M.D. (1967). Analysis of Accelerograms – Parkfield Earthquake, *Bull. Seism. Soc. Am.*, **57**, 1193–1220.

Iyengar, R.N. (2001). Probabilistic Methods in Earthquake Engineering, H. Aref, J.W. Philips (eds), Kluwer, Dordrecht, Netherlands.

Kanai, K. (1957). Seismic-empirical Formula for the Seismic Characteristics of the Ground, *Bull. of the Earthquake Research Inst.*, Japan, **35**, 309–325.

Kondorskaya, N.V., Zaharova, A.I., Vandisheva, N.V., and Rautian, T.G. (1990). Analysis of the Instrumental Data of the Earthquake from August 30, 1986 (in Russian).

Kozin, F. (1988). Autoregressive Moving Average Models of Earthquake Records, *Probabilist. Eng. Mech.*, **3**, 58–63.

Levenberg, K. (1944). A Method for the Solution of Certain Non-Linear Problems in Least Squares, *Quart. Appl. Math*, **2**, 164–168.

Liu, S.C. (1970). Evolutionary Power Spectral Density of Strong-motion Earthquakes, *Bull. Seism. Soc. Am.*, **60**, 891–900.

Makris, N. (1997). Rigidity-plasticity-viscosity: can electrorheological dampers protect base-isolated structures from near-source ground motions?, *Earthq. Eng. Struct. Dyn.*, **26**, 571–591.

Marquardt, D. (1963). An Algorithm for Least-Squares Estimation of Nonlinear Parameters. In: *SIAM J. Appl. Math.*, **11**, 431–441.

Mavroeidis, G.P., and Papageorgiou, A.S. (2003). A Mathematical Representation of Near-Fault Ground Motions, *Bull. Seism. Soc. Am.*, **93**, 3, 1099–1131.

McGuire, R.K., and Hanks, T.C. (1980). RMS Accelerations and Spectral Amplitudes of Strong Ground Motion during the San Fernando, California Earthquake, *Bull. Seismol. Soc. Am.*, **70**, 1907–1919.

Radu, C. (1994). Catalogue of Strong Earthquakes Originating in Romania During the Period 994–1900 (Manuscript).

Raileanu V., Bala A, Hauser F., Prodehl C., and Fielitz W. (2005). Crustal properties from S-wave and gravity data along a seismic refraction profile in Romania. *Tectonophysics*, **410**, 251–272.

Shinozuka, M., and Deodatis, G. (1988). Stochastic Process Models for Earthquake Ground Motion, *Probabilist. Eng. Mech.*, **3**, 114–123.

Shinozuka, M. Deodatis, G., Zhang, R., and Papageorgiou, A.P. (1999). Modeling, Synthetics and Engineering Applications of Strong Earthquake Wave Motion, *Soil Dyn. Earthq. Eng.*, **18**, 209–228.

Vanmarcke, E.H., Fenton, G.A., and Heredia-Zavoni, E. (1997). SIMQKE-II: Conditioned Earthquake Ground Motion Simulator, Princeton University, Princeton, N.J., 25 p.

Wen, Y.K., and Gu, P. (2004). Description and Simulation of Nonstationary Processes Based on Hilbert Spectra, *J. Eng. Mech.*, **130**, 942–951.

ESTIMATION OF THE RECURRENCE AND PROBABILITY OF VRANCEA INTERMEDIATE DEPTH EARTHQUAKES

V. GINSARI[*]
Institute of Geology and Seismology, Academy of Sciences of Moldova

Abstract. Aspects of seismic hazard assessments specific to Vrancea intermediate earthquakes are presented in the current paper. Three methods were employed to make estimates: (a) classical Gutenberg-Richter (1956), (b) maximum entropy principle (MEP) modified for estimating the recurrence of strong earthquakes (Berril and Davis, 1980; Dong et al., 1984) and (c) Huo and Hwang's (1994) modification of the recurrence law containing characteristics of a stochastic distribution. An analysis of recurrence relationships was performed for two types of magnitudes, for different time intervals and by assigning alternative values of maximum possible earthquake magnitude. The probability of an earthquake occurring in a specified magnitude range and in a specified time limit was estimated. The dependence of final estimates on the choice of values of M_{min} and M_{max} and on sample size was demonstrated. From an analysis of established intervals of recurrence it follows that the recurrence period of earthquakes with $M = 7.0$ is from 30 to 60 years with a relatively high probability of $R = 0.5$–0.7 (for $T = 50$ years). For magnitude $M_{G-R} = 7.5$ ($M_w = 7.7$) the recurrence interval varies from 100 (most pessimistic estimation) to 380 years (most optimistic estimation) with a probability of $R = 0.1$–0.25 (for $T = 50$ of years).

Keywords: Seismic hazard, Vrancea, recurrence

1. Introduction

A seismic hazard, defined as the probability that various levels of acceleration (velocity or intensity) can be exceeded for a certain period of time, sets a model

---

[*]Institute of Geology and Seismology, Academy of Sciences of Moldova

A. Zaicenco et al. (eds.), *Harmonization of Seismic Hazard in Vrancea Zone,*
© Springer Science + Business Media B.V. 2008

of earthquake occurrence as a function of the attenuation relationship and the numerical values of its parameters. The first and most basic step in this process is estimating recurrence parameters for earthquakes in terms of magnitude.

In spite of the fact that linear relations describing the recurrence of earthquakes, as a rule in the form of the Gutenberg-Richter (G-R) law (1956), are the most general and widely applied in seismic hazard assessments, further developments of this method and a search for alternative or additional methods are under way. Realistic estimates of recurrence intervals using the G-R method imply that a complete and homogeneous catalogue of earthquakes over a time interval spanning sometimes two centuries and including seismic events in a wide magnitude range exists for a region of interest. Thus, the accuracy of the definition of magnitude should be in the range of 0.1 units of the magnitude. However, in practical seismology, catalogues of this quality virtually do not exist. The usual situation is the following: large magnitude earthquakes registered by instruments are scarce, and at the same time, small magnitude events are insufficiently recorded which renders the catalogues incomplete due to the scanty number of registrations.

Detailed investigations of correlation and recurrence graphs carried out during the last decades (Molchan et al., 1997; Pacheco et al., 1992) show that observed data follow the G-R law inside a rather narrow magnitude range, i.e., generally in the middle part of the graph ($4.5 \leq M \leq 6.0$) although for different seismic regions the borders of a linear range vary. It is obvious that extrapolating such correlations in the direction of large magnitudes to define recurrence can be problematic. That is why in a number of publications in the past decade other statistical methods are suggested for more realistic estimates of recurrence intervals.

The b value estimated on a set of earthquakes of observed magnitude according to the magnitude-frequency relation $log\ N\ (m) = a\text{-}bm$, where $N\ (m)$ is the number of earthquakes with a magnitude exceeding m, is strongly correlated with the technique of approximation used. In addition, the parameters a and b depend on (i) magnitude type, (ii) accuracy of definition which at best does not exceed 0.1 for the modern instrumental catalogues compiled in regions equipped with high-quality modern networks and can reach 0.6 by recalculating magnitudes from historical data on the intensity of earthquakes, and (iii) the size of a data set N. The results reported by some researchers (Bender, 1981) and personal experience demonstrate that for $N = 100$, the accuracy of the value for b is close to 10%, and for 25 earthquakes in a data set, the standard deviation of b estimations is close to 25% from the mean of the estimated parameter.

By using the least squares method (MLS) for a linear approximation of both the interval and the cumulative variant of the G-R relation, there is a tendency to underestimate the *b* value as magnitudes above the maximum observed ones are not included in the data set. The formulas for calculating the maximum likelihood (MML) offered earlier (Aki, 1965; Utsu, 1965; and Page, 1968) for discrete or continuous magnitudes give biased estimates if they are applied to the interval data, and the bias increases as the interval size increases.

Other researchers adapt mathematical methods used in other fields on seismological problems. One such method is the maximum entropy principle (MEP). It was first adapted to analyzing seismic hazards by Berrill and Davis (1980). Dong et al. (1984) further developed this approach. One application of MEP for seismic hazard and risk assessment was done by Sun and Pan (1995) to estimate the possible effect of a strong earthquake on Sumatra.

MEP offers an alternative means of estimating strong earthquakes including and supplementing specific results from using traditional methods such as MLS and MML.

In this paper, the results of the application of the above mentioned methods are given for estimating the recurrence of earthquakes of an intermediate depth ($60 \leq h \leq 170$ km) generated in the Vrancea zone.

2. Methods for Estimating the Recurrence of Earthquakes

As mentioned earlier, to assess intervals of earthquake recurrence, a set of methods is applied in which a log-linear G-R relation forms the cumulative and non-cumulative basis. Preference is given to the cumulative form, partially mitigating the disadvantages and gaps in available earthquake catalogues.

2.1. GUTENBERG-RICHTER LAW

Assuming that earthquakes occur randomly, their number decreases exponentially as magnitude increases and is not limited. Gutenberg and Richter (1956) express this relationship in the following equation:

$$\log N\,(m) = a\text{-}bm \tag{1}$$

where $N\,(m)$ is the number of earthquakes per year with a magnitude equal to or greater than m, and a and b are constants. The constant a reflects the seismic activity in the region of interest, and b is the value reflecting a ratio of weak to strong earthquakes. Knowing the parameters of (1), it is possible to calculate the average period of earthquake recurrence (in years) with a magnitude greater or equal to M:

$$T(m \geq M) = 1/n(m \geq M) \tag{2}$$

The rate of recurrence of earthquakes with magnitudes of interest for engineering applications is one of the key points in assessing seismic hazard and risk.

2.2. GUMBEL'S METHOD

Another solution for this problem is Gumbel's theory of extreme values adapted for seismology (Epstein and Lomnitz, 1966; Howell, 1980; Ebel, 1984). This method allows one to compare past and present seismicity on the basis of modern recurrence correlations and to calculate the probability of a strong earthquake occurring within certain limits of magnitude and time.

Assuming that seismicity is a random Poisson process, it was demonstrated that the strongest earthquake magnitude for a unit of time (1 year) follows a cumulative distribution function:

$$G(m) = \exp(-\alpha \exp(-\beta m)) \tag{3}$$

where m is a magnitude $M \geq m$, α and β can also be determined in terms of values of a and b as:

$$\alpha = \exp(a \ln 10) \tag{4}$$

$$\beta = b \ln 10 \tag{5}$$

The most probable or most frequently observed (modal) magnitude $\overline{m}$ during a maximum of 1 year is:

$$\overline{m} = (\ln \alpha)/\beta \tag{6}$$

and a modal maximal magnitude $\overline{m}_T$ for a period of time T would be:

$$\overline{m}_T = \overline{m} + (\ln T)/\beta \tag{7}$$

The seismic hazard $R_T(m)$ – the probability of an earthquake of magnitude m or higher occurring during the period T – can be determined by the following equation:

$$R_T(m) = 1 - \exp(-\alpha T \exp(-\beta m)) \tag{8}$$

This equation is used by the author to estimate seismic hazards.

It is necessary to note that in reality an upper limit of earthquake magnitude exists for any given region. For the Vrancea zone, the maximum possible magnitude varies from 7.5 to 8.0 according to different authors (Ginsari, 2000; Друмя and Степаненко, 1972; Бунэ and Катрих, 1980; Radulian, 1981; Lungu et al., 1999). An assessment of method choice and the validity of these estimates is beyond the scope of this study. The existence of an upper limit for earthquake magnitude has allowed Huo and Hwang (1994); Dong et al. (1984) to offer other mathematical methods adapted for use in seismology.

2.3. HWANG MODIFICATION

Only the key moments of this approach are presented below. In 1994, Huo and Hwang offered this modification of the G-R recurrence relation that satisfies the properties of a probabilistic distribution:

$$n(\geq M) = e^{\alpha - \beta M} \frac{1 - e^{-\beta(M_{max} - M)}}{1 - e^{-\beta(M_{max} - M_{min})}} \tag{9}$$

This equation contains the values of the threshold magnitude in a data set and the value of the maximum possible magnitude for the corresponding source zone. The values α and β are also determined according to the formulas:

$$\alpha = a\, ln10, \qquad\qquad \beta = b\, ln10 \tag{10}$$

The corresponding periods of earthquake recurrences are calculated according to (2).

2.4. MAXIMUM ENTROPY PRINCIPLE METHOD

Dong et al. (1984) have developed in detail a method based on MEP to increase the reliability of estimating earthquake recurrence. The key points of the MEP method are presented below.

Assume that the range of values of magnitude is $[M_{min} - M_{max}]$, where M_{min} is the minimal magnitude in a data set (as a rule, this corresponds to the lowest magnitude level of interest to engineers which can equal 5.0 or 6.0 for Vrancea earthquakes), and M_{max} is the magnitude of the maximum possible earthquake in the region under investigation. For the corresponding data set, the mean magnitude is given as:

$$\overline{M} = \frac{\sum\limits_{alli} n_i M_i}{\sum\limits_{alli} n_i} \tag{11}$$

Uncertainties in the information about magnitude are measured by calculating entropy using the equation:

$$S = -K \int_{M_{min}}^{M_{max}} f_M(m) \cdot \ln(f_M(m)) dm \tag{12}$$

S is entropy, K is a constant and $f_M(m)$ is the probability distribution function for magnitude M. As a result, it is possible to determine $f_M(m)$ as follows:

$$f_M(m) = \frac{\lambda\, e^{-\lambda m}}{e^{-\lambda M_{min}} e^{-\lambda M_{max}}} \qquad (M_{min} \leq m \leq M_{max}) \tag{13}$$

Where parameter λ can be obtained numerically:

$$\frac{1}{\lambda} + \frac{M_{min} e^{-\lambda M_{min}} - M_{max} e^{-\lambda M_{max}}}{e^{-\lambda M_{min}} - e^{-\lambda M_{max}}} = \overline{M}, \tag{14}$$

value of $\overline{M}$ is determined from (11).

After integration of (14) with respect to m and additional calculations, the number of earthquakes with magnitudes exceeding m can be determined as:

$$N(m) = \frac{N(M_{min})(e^{-\lambda m} - e^{-\lambda M_{max}})}{e^{-\lambda M_{min}} - e^{-\lambda M_{max}}} \qquad (M_{min} \leq m \leq M_{max}) \tag{15}$$

Using (15) and (2), it is possible to calculate the corresponding interval of recurrence $T(m)$. Assuming that the distribution of earthquakes over time corresponds to Poisson's process, the probability R of one or more events with $M \geq m$ occurrence can be calculated for t years (Lomnitz, 1974)

$$R = 1 - \exp\left(-t N(M_{min}) \frac{e^{-\lambda m} - e^{-\lambda M_{max}}}{e^{-\lambda M_{min}} - e^{-\lambda M_{max}}} \right) \tag{16}$$

Thus, calculating recurrence parameters can be done using three methods:

- The classical form of the recurrence law by G-R (method of the least squares)
- Huo and Hwang's (1994) modification of the recurrence law
- MEP

After applying the Gumbel formulas, only the probability of an earthquake of a relevant magnitude range occurring is given. In addition, comparisons of current and previous estimates of recurrence are made.

3. Data

The catalogues of earthquakes compiled by C. Radu (1982, 2003) provide values of G-R magnitude. In parallel, the same calculations are done for the catalogue ROMPLUS (Oncescu et al., 1999) using magnitude M_w, obtained from correlation formulas or direct definitions of the seismic moment. Magnitudes M_w are recommended for estimating seismic hazards within the framework of the program GSHAP (1993) as the characteristic that most adequately reflects the size of an earthquake and is not affected by saturation and has plain physical sense. Actually, the G-R magnitudes are identical to magnitudes M_s (determined from surface waves). In general, magnitudes M_s do not pertain to intermediate depth earthquakes (with $h > 60$–70 km); however, such magnitudes were frequently used in previous years to calculate seismic hazards from Vrancea intermediate earthquakes. Therefore, estimates of recurrence parameters are made for both types of magnitudes.

4. Results of Calculations

The numerical values for each of four different data sets are in Tables 1–4.

TABLE 1. Recurrence intervals and the probability of strong Vrancea earthquakes (data set 1)

Data of strong real events	M_{G-R}	Recurrence interval			Probability $R_{T=50}$ (m)	
		By (1)	By (9)	By (15)	By (8)	By (16)
	6.5	18	20	29	0.938	0.813
	6.6	21	24	38		
1990.05.30	6.7	26	30	48	0.858	0.648
	6.8	31	37	61		
	6.9	36	45	78		
1986.08.30	7.0	44	57	101	0.683	0.389
	7.1	52	73	133		
1977.03.04	7.2	62	94	177	0.553	0.245
	7.3	74	125	243		
1940.11.10	7.4	88	172	345	0.432	0.135
	7.5	106	254	520	0.377	0.092

The estimates of recurrence intervals in Table 1 are based on a data set for 1901–2000 that included 96 earthquakes greater than $M_{min} = 5.0$. The value of M_{max} is 7.8 for computing formulas (9) and (15). The increment of the magnitude values in all our calculations ($M = 0.3$) was chosen to allow for possible errors in the definition of magnitude in the first half of the 20th century.

The equation for calculating the approximating line is:

$$lgN (m) = 3.747 - 0.769 \, M \tag{17}$$

The recurrence graphs are plotted in Figure 1.

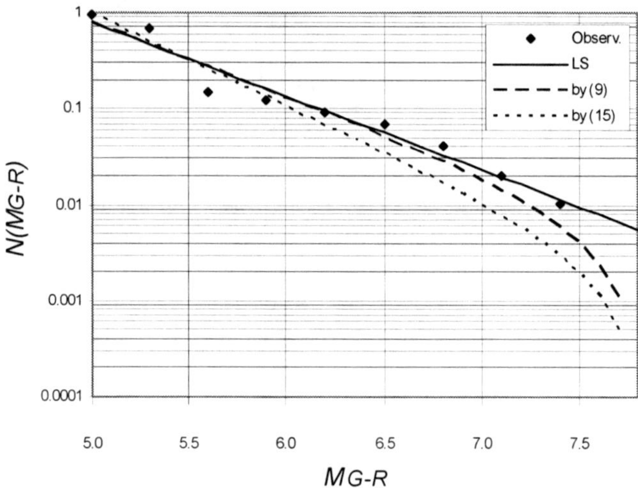

Figure 1: Recurrence Graphs for (1), (9) and (15) for Vrancea Earthquakes with $M \geq 5.0$

The estimates of recurrence intervals in Table 2 were made using a data set for 1901–2000 that included 97 earthquakes with magnitudes greater than $M_w = 5.3$ (which corresponds approximately to $M_{G\text{-}R} = 5.0$). The maximum value of magnitude is $M_w max = 7.8$ (for computing formulas (9) and (15). The choice of the minimum magnitude was dictated by the need to have a data set of a sufficient size to make statistical estimates in a wide magnitude range. The equation for calculating the approximating line for G-R is:

$$lg N (m) = 4.493 - 0.838 \, M_w \tag{18}$$

TABLE 2: Recurrence intervals and the probability of strong Vrancea earthquakes (data set 2)

Data of strong real events	M_w	Recurrence interval			Probability $R_{T=50}$ (m)	
		By (1)	By (9)	By (15)	By (8)	By (16)
	6.8	18	19	16	0.995	0.958
	6.9	19	23	19		
1990.05.30	7.0	23	30	24	0.879	0.872
	7.1	29	38	30		
	7.15	32	44	35		
1986.08.30	7.25	38	58	45	0.729	0.667
	7.3	42	67	52		
1977.03.04	7.4	51	94	72	0.624	0.499
	7.5	62	140	105	0.554	0.377
1940.11.10	7.6	75	233	173	0.486	0.250
	7.7	91	515	379	0.420	0.123

The correlation of M_{GR} and M_w according to GSHAP (1993) is:

$$M_w = 0.92\ M_{G\text{-}R} + 0.81 \tag{19}$$

The recurrence graphs are plotted in Figure 2.

The estimates of the recurrence of Vrancea earthquakes of strong magnitude (G-R types) published by Romanian engineering seismology experts (Lungu et al., 1999) and included in other publications are based on the cumulative recurrence graph for the period 1901–1995 with $M_{Min}= 6.0$ (the minimum level of magnitude of interest to engineering) and includes only 12 earthquakes. The precision of magnitude values is 0.1; in other words, continuous or "exact" magnitude values are used in conjunction with the cumulative data without repeating experimental data on empty intervals.

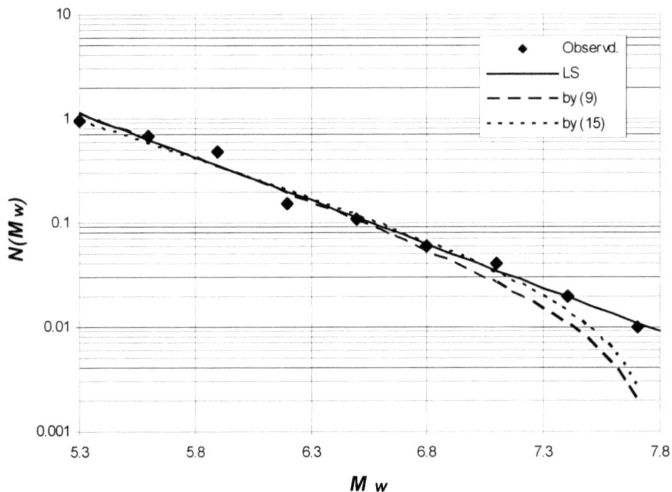

Figure 2: Recurrence graphs for (1), (9) and (15) for Vrancea earthquakes with $M \geq 5.3$

In addition to G-R-based estimates, calculations were made using the Hwang and Huo (1994) method for two values of M_{max}. The results from Lungu et al. (1999) are given in Table 3. The base equation for calculating the approximating line is:

$$\lg N\ (m) = 3.49 - 0.72\ M \tag{20}$$

Recurrence estimates were also done on the same data set but were grouped with a magnitude increment of 0.3 (Figure 3). The results are in Table 4.

TABLE 3: Recurrence intervals of strong Vrancea earthquakes according to Lungu et al. (1999)

Data of strong real events	M_{G-R}	Recurrence intervals		Equation (1)
		Equation (9)		
		M max = 7.8	M max = 8.0	
	7.8	–	457	134
	7.7	704	279	114
	7.6	323	191	96
	7.5	197	139	81
Nov. 10, 1940	7.4	135	105	69
	7.3	98	82	58
Mar. 4, 1977	7.2	75	65	50
	7.1	58	52	42
Aug. 30, 1986	7.0	46	42	36
	6.9	37	35	30
	6.8	30	28	26
May 30,1990	6.7	24	24	22

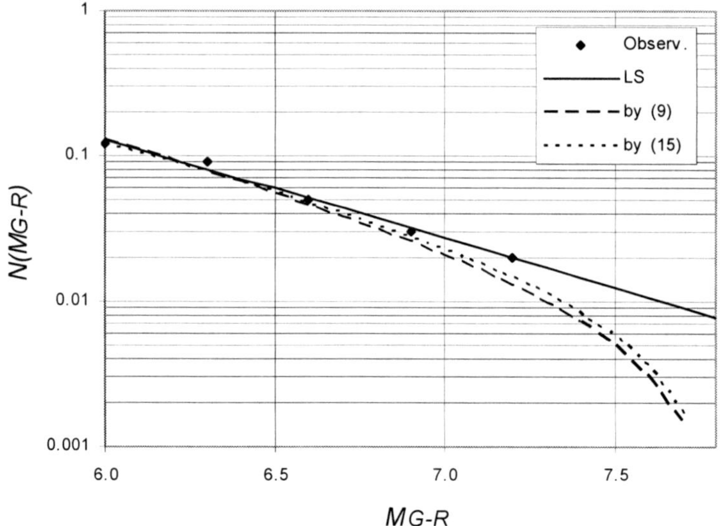

Figure 3: Recurrence graphs for (1), (9) and (15) for Vrancea earthquakes with $M_{G-R} = 6.0$ from 1901 to 2000

In addition to estimates for (1) and (15), estimates for MEP were calculated. The base equation for calculating the approximating line is:

$$\lg N(m) = 3.18 - 0.678\, M \tag{21}$$

TABLE 4: Recurrence intervals and the probability of strong Vrancea earthquakes (data set 3)

Data of strong real events	M_{G-R}	Recurrence interval			Probability $R_{T=50}(m)$	
		By (1)	By (9)	By (15)	By (8)	By (16)
	6.5	17	18	18	0.950	0.937
1986.08.30	7.0	36	49	44	0.740	0.675
	7.1	41	61	55	0.685	0.597
1977.03.04	7.2	50	77	69	0.627	0.514
	7.3	59	102	89	0.570	0.427
1940.11.10	7.4	68	139	120	0.515	0.340
	7.5	80	202	173	0.460	0.250

Estimates of the probability of events of a given magnitude occurring over 50 years were also made.

Considering the poor statistical data on earthquakes with magnitudes starting from $M_{min} = 6.0$ from 1901 to 2000, recurrence estimates were additionally done on a data set for the period 1801–2000 which contains 20 events, including the strongest one observed in the region which occurred in 1802 and had a magnitude of $M = 7.5$. The results are in Table 5 and Figure 4.

TABLE 5: Recurrence intervals and the probability of strong Vrancea earthquakes (data set 4)

Data of strong real events	M_{G-R}	Recurrence intervals			Probability $R_{T=50}(m)$	
		By (1)	By (9)	By (15)	By (8)	By(16)
	6.5	23	25	24	0.880	0.870
1986.08.30	7.0	60	75	67	0.560	0.525
	7.1	72	96	84	0.500	0.445
1977.03.04	7.2	87	125	108	0.430	0.370
	7.3	105	168	143	0.370	0.295
1940.11.10	7.4	127	234	196	0.320	0.225
	7.5	154	347	286	0.270	0.160

The base equation for calculating the approximating line is:

$$\lg N\,(m) = 3.99 - 0.825\,M \tag{22}$$

In 1995, Шумила estimated the recurrence of Vrancea earthquakes using the maximum likelihood method. The data sample included 420 events from a magnitude level of $M > 3.0$ for the period 1800–1990. Earthquakes with $M > 3.0$ were introduced to demonstrate the representative level of magnitude $M = 4.0$. A standardized time unit of 1 year was assigned to each of the magnitude values included in the data set. The approximating line was calculated with the equation:

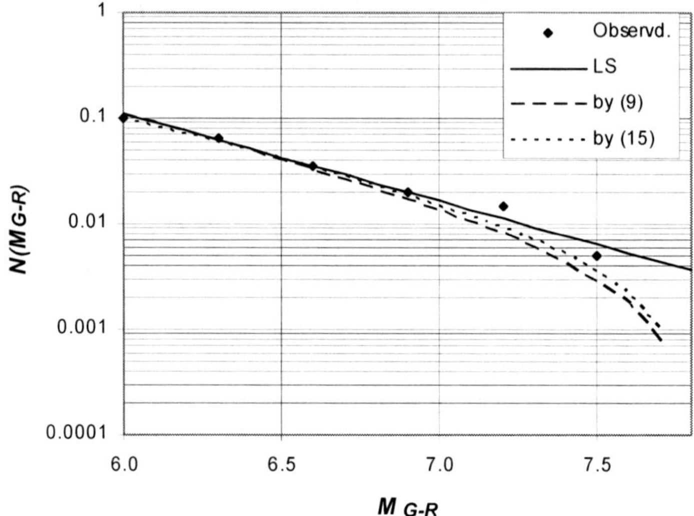

Figure 4: Recurrence graphs for (1), (9) and (15) for Vrancea earthquakes with $M_{G-R} = 6.0$ from 1801 to 2000

$$\ln [N (m > M) /T] = 1.1617 - 1.8629 (M-4.5) \tag{23}$$

where 4.5 is a magnitude level that rates on the sample as a level resulting from uncorrelated estimates of the equation's parameters a and b. According to this equation the recurrence interval for magnitude $M = 7.5$ is 84 years and for $M = 7.0$ is 33 years.

5. Discussion of Results and Conclusions

5.1. RESULTS

Comparing the results in Tables 1 and 2, estimates of Vrancea earthquake recurrence intervals made from virtually the same data set but for different magnitudes and using different methods significantly differ. It is obvious that the results obtained using the classical G-R formula (1), especially for large magnitudes (greater or equal to $M_{GR} = 7.0$ or $M_w = 7.4$), overestimate recurrence and the seismic hazard in general in comparison with what was actually observed.

Estimates using Equation (9) look more acceptable for analyzing recurrence intervals in terms of G-R magnitudes. For the moment magnitude scale, this formula gives excessively large periods of recurrence for the maximum estimated magnitude. Formula (15) yields more credible estimates for magnitudes M_w, than for M_{G-R}. The recurrence intervals calculated from Table 3 according to Hwang and Huo (1994) essentially depend on value assigned to M_{max}. The greater M_{max} is, the smaller the interval of recurrence of large earthquakes. Therefore, the general seismic hazard is higher. It is valid also for calculations done using Hwang's method, and the MEP method.

The estimates in this paper are based, as shown above, on better statistics, than those by Lungu et al. (1999). The chosen values of M_{max} are justified in a number of papers on this subject in recent years (Друмя and Степаненко, 1972; Бунэ and Катрих, 1980; Radulian, 1981; Lungu et al., 1999), where M_{max} varies from 7.5 to 7.8 (in terms of G-R magnitude). Estimates of the probability of the occurrence of earthquakes with certain magnitude in a given time interval (T = 50 years) are in Tables 1, 2, 4 and 5.

The calculations in Tables 4 and 5 demonstrate that the influence of the threshold (or minimum) magnitude on final estimates of recurrence intervals is rather significant. If the value of M_{min} is increased, the recurrence period of an earthquake of magnitude 7.5 is reduced using all three methods. Hence, it can be overestimated in the final seismic hazard assessment.

A comparison of data in Tables 4 and 5 shows that at identical levels of M_{min} and M_{max} but with different samples sizes, the divergence of recurrence estimates for earthquakes of large magnitudes can reach 60–70%.

5.2. CONCLUSIONS

Using three different methods and two catalogues of Vrancea earthquakes with different magnitudes (M_{G-R} and M_w) to assess recurrence intervals of strong earthquakes allows us to get more realistic, complementary estimates that are not mutually exclusive. The advantages and drawbacks of these methods are discussed. The dependence of final estimates on the choice of values of M_{min} and M_{max} and on sample size is shown. From an analysis of established intervals of recurrence it follows that the recurrence period of earthquakes with $M = 7.0$ is from 30 to 60 years with a relatively high probability of $R = 0.5$–0.7 (for $T = 50$ years). For magnitude $M_{G-R} = 7.5$ ($M_w = 7.7$) the recurrence interval varies from 100 (most pessimistic estimation) to 380 years (most optimistic estimation) with a probability of $R = 0.1 - 0.25$ (for $T = 50$ of years).

Acknowledgements

This research was sponsored by NATO's Scientific Affairs Division in the framework of the Science for Peace Programme, project SfP-980468. The author expresses sincere gratitude to E. Isiciko for his constructive assistance in the software design. The author is grateful to Prof. Dieter Mayer-Rosa for valuable input on this paper.

References

M. Aki (1965). Maximum likelihood estimate of b in the formula $\log N = a - bM$ and its confidence limits, Bull. Earthq. R. Inst., Tokyo Univ., 43, 237–239.

B. Bender (1981). Maximum likelihood estimation of b values for magnitude grouped data. Bull. Seism. Soc. Am., 73, 831–851.

J.B. Berril and R.O. Davis (1980). Maximum entropy and the magnitude distribution. Bull. Seism. Soc. Am., 70, 1823–1831.

W.M. Dong, A.B. Bao and H.C. Shah (1984). Use of maximum entropy principle in earthquake recurrence relationships. Bull. Seism. Soc. Am., 74, 725–737.

J.E. Ebel (1984). Statistical aspects of New England Seismicity from 1975 to 1982 and implication to past and future earthquake activity. Bull. Seism. Soc. Am., 74, 1311–1329.

B. Epstein and C. Lomnitz (1966). A model for the occurrence of large earthquakes. Nature, 211, 954–956.

V. Ginsari (2000). Analysis of recurrence law validity by utilization of different fitting techniques (Vrancea zone of intermediate depth seismicity case study). Proceedings of XXVII ESC General Assembly, Lisbon, Portugal, p. 96.

Global Seismic Hazard Assessment Programme. Annali di Geofisica, V. XXXVI, No. 3–4, 1993.

B. Gutenberg and C.F. Richter (1956). Magnitude and energy of earthquakes. Ann. Geophys., 9, 1, 1–15.

B.F. Howell (1980). A comparison of seismic risk estimates in central United States. Earthquake Notes, 51, 13–19.

H.H.M. Hwang and J.R. Huo (1994). Generation of hazard-consistent fragility curves for seismic loss estimation studies. Technical Report NCEER-94-0015, National Center for Earthquake Engineering Research, State University of New York at Buffalo, USA.

C. Lomnitz (1974). Global tectonics and earthquake risk. Elsevier, Amsterdam, p. 131.

D. Lungu, T. Cornea and C. Nedelcu (1999). Hazard assessment and site-dependent response for Vrancea earthquakes. In: Vrancea Earthquakes: Tectonics, Hazard and Risk Mitigation. Kluwer, Dordrecht/Boston/London, pp. 251–267.

G. Molchan, T. Kronrod and G.F. Panza (1997). Multi-scale model for seismic risk. Bull. Seism. Soc. Am., 87, 1220–1229.

M.C. Oncescu, V. Marza and M. Popa (1999). The Romanian earthquake catalogue between 984–1997. In: Vrancea Earthquakes: Tectonics, Hazard and Risk Mitigation. Kluwer, Dordrecht/Boston/London, pp. 43–47.

S.F. Pacheco, C.H. Sholz and L.R. Sykes (1992). Changes in frequency-size relationship from small to large earthquakes. Nature, 355, 71–73.

R. Page (1968). Aftershocks and microaftershocks of the great Alaska earthquake of 1964, Bull. Seism. Soc. Am., 58, 1131–1168.

C. Radu (1982). Catalogul cutremurelor puternice produse pe teritoriul României. Partea I,II. In: Cutremil de pămînt din Romania de la 4 martie 1977. Ed. Academiei Romănia, Bucureşti,78–85.

C. Radu (2003). Catalogul istoric al cutremurelor din Vrancea în perioada 984–1900. In: Construcţii amplasate în zone cu mişcări seismice puternice. Ed. Orizonturi Universitare, Timişoara, p. 25.

M. Radulian (1981). Seismic risk determination for Vrancea intermediate earthquakes. Proceedings of the 2nd International Symposium on the Analysis of Seismicity and Seismic Hazard, Liblice, Czechoslovakia, pp. 545–552.

J. Sun and T.-C. Pan (1995). The probability of very large earthquakes in Sumatra. Bull. Seism. Soc. Am., 85, 1226–1231.

T. Utsu (1965). A method for determining the value of b in a formula $\log n = a - bM$ showing the magnitude-frequency relation for earthquakes (with English summary), Geophys. Bull. Hokkaido Univ. 13, 99–103.

В.И. Бунэ, И.Р. Катрих (1980). Оценка вероятности землетрясений при составлении карты сейсмического районирования. Количественная оценка сейсмических воздействий. Вопросы инженерной сейсмологии, вып, 20, 3–14.

Л.В. Друмя, Н.Я. Степаненко (1972). Карта максимальных возможных землетрясений сейсмического района Вранча. Изв. АН СССР, Физика Земли, 10, 77–78.

Шумила В.И. (1995). Результаты картирования параметров сейсмического режима землетрясений области Вранча. Итоговый отчет по теме: "Исследование сейсмогенеза и сейсмической опасности Карпатского региона". Фонды Института геофизики и геологии АН РМ, с. 47.

SEISMIC PROTECTION AND PROTECTION AGAINST DEMOLITION OF BUCHAREST'S HISTORICAL BUILDINGS

D. LUNGU[*], C. ARION, A. ALDEA
Technical University of Civil Engineering, Bucharest

Abstract. The paper presents (i) historical earthquake damage and lessons learned (ii) how to strengthen tall reinforced concrete fragile buildings in central Bucharest, (iii) international projects for seismic risk reduction in Romania and (iv) the current challenges involved in conserving historical buildings in the center of Bucharest.

Keywords: Earthquakes, Vrancea, damage, protection, heritage buildings

1. Introduction

With more than 2 millions inhabitants and more than 110,000 buildings, Bucharest can be ranked as the megacity with the highest seismic risk in Europe due to (i) soft soil conditions characterized by a long predominant period (1.4 ÷ 1.6 s) of ground vibration during strong Vrancea earthquakes (Aldea, 2002; Arion, 2003) and (ii) the high fragility of tall, reinforced concrete buildings built in Bucharest before WWII and before the 1977 Vrancea earthquake disaster (Lungu, 1999). The city is located on the Romanian Plain between the Danube River and the Carpathian Mountains in the valley of the Colentina and Dambovita rivers which cross the city from NW to SE.

2. Building Damage During Major Historical Earthquakes

2.1. 1802

During the 1802 earthquake (M_{G-R} = 7.4 ± 0.3), in Bucharest many bell towers and towers of churches fell down, and several churches were destroyed (included Cotroceni monastery and St. Spiridon Church). Half the Coltea Tower

---

[*]Technical University of Civil Engineering, Bucharest, 124 Lacul Tei, RO-020396, Romania

A. Zaicenco et al. (eds.), *Harmonization of Seismic Hazard in Vrancea Zone,*
© Springer Science + Business Media B.V. 2008

collapsed, and the remaining part was seriously damaged (Figure 1); most of residences of the wealthy were heavily damaged and some of them collapsed.

Figure 1: Coltea Tower (Photo from Georgescu, 2002)

In Brasov, many chimneys fell and houses and churches were damaged. The earthquake was felt in Transylvania, Sibiu, Sighisoara, Banat, and Timisoara in Romania and in Poland, Bulgaria, Turkey and Russia. In Cernauti some houses were damaged. In Lvov, the walls of an Armenian church cracked and the bells rang by themselves. Light damage occurred even in Moscow.

2.2. 1829

"Wednesday night […] a strong earthquake happened in our capital, […] no house in Bucharest remained without damage; all walls were cracked and in some cases fell down; roofs and chimneys were destroyed." *Curierul Român, Bucuresti*, No.15/27 Nov. 1829.

During the earthquake of 26 November 1829 ($M_S = 6.9$), Bucharest suffered the most. Documents indicated that 115 houses had become unsafe and that 15 of them were heavily damaged and were later demolished. At Campina, a church collapsed. In Sibiu and Iasi, many walls were cracked. The earthquake was felt over a large area including Transylvania and Banat in Romania and in Bulgaria, Poland and Ukraine.

2.3. 1838

The description on page 144 of the effects of the 1838 earthquake ($M_s = 6.9$–7.3) in Bucharest in the book *Voyage dans la Russie Méridionale et la Crimée par la Hongrie, la Valachie et la Moldavie* by M. A. de Démidoff, illustrated by

Raffet and edited by E. Bourdin in Paris in 1841 and 1854 indicates that several houses collapsed and that all houses were damaged to some extent. Also, the document indicates that several deaths occurred in the city. The police report mentioned 8 deaths, 14 injured and 36 collapsed buildings. Many other buildings (especially the larger ones including the Royal Palace) were heavily damaged.

The counselor to the Grand Duke of Saxa, engineer Gustav Schuller who was in Romania at that time, was asked by the Romanian government to make an investigation in the epicentral area. He indicated a maximum intensity of IX in the area of the Vrancea mountains, Focsani and Ramnicu Sarat where many villages were completed destroyed. Schuller concluded that, "All the stone masonry buildings were heavily damaged and some of them, especially the churches and other large buildings, were uninhabitable."

The earthquake was felt over an extended area of Europe including Ukraine, Poland, Bulgaria and Turkey up to (then) Constantinople and in the northeast of Italy.

2.4. 1940

During this 10 November event, in Bucharest the most significant loss was the complete collapse of the Carlton Building, the highest reinforced concrete frame building (47 m, 12 stories) in Romania at that time. By 24 November, 136 people had been found dead in its rubble. Several high-rise reinforced concrete buildings in Bucharest were very severely damaged including the Belvedere, Wilson, Lengyel, Pherekide, Brosteni and Galasescu structures. Other important buildings in Bucharest that suffered important damage were the Justice Palace, the Romanian Atheneum, the Opera, the National Theatre, the CEC Bank and the Postal Palace.

Two zones of maximum seismic intensity were identified: one in Focsani and Panciu and the second one in the area from Campina to Bucharest. In those areas, the seismic intensity was over VIII and was close to IX on the Mercalli-Sieberg scale. In Panciu no building was standing after the earthquake (*Timpul* newspaper on 12 November 1940). In Iasi, all the buildings were damaged, City Hall presented large cracks and the Faculty of Medicine was uninhabitable.

Since the earthquake was a deep event (about 140 kilometers [km]) it was felt over 2,000,000 km^2 i.e. to the east in Odessa, Craiova and Moscow; to the north up to Saint Petersburg; to the west to Tissa river and to the south to Istanbul.

2.5. 1977

The 4 March 1977 (Gutenberg Richter magnitude $M_{G-R} = 7.2$, moment magnitude $M_w = 7.5$) was the most destructive earthquake in the history of Romania (Figures 2–4). This earthquake:

- Killed 1,578 people including 1,424 in Bucharest
- Injured 11,221 people including 7,598 in Bucharest and 3,723 in the rest of the country
- Destroyed or seriously damaged 33,000 housing units in high-rise apartment flats and conventional dwellings (35,000 families, more than 200,000 persons homeless)
- Caused lesser damage to 182,000 other dwellings
- Destroyed 374 kindergartens, nurseries and schools and badly damaged 1,992 others
- Destroyed 6 university buildings and damaged 60 others
- Destroyed 11 hospitals and damaged 2,288 other hospitals and 220 polyclinics
- Damaged almost 400 cultural institutions (theatres, museums, etc.)
- Damaged 763 factories
- Directly affected over 200,000 people (Fattal et al., 1977)

It was felt in an area of 1.3 million km^2 and caused damage over an area of about 80,000 km^2 within which the most frequently occurring intensity did not exceed VII (MM). Much of the damage was caused to old reinforced concrete buildings of 6–12 stories. These structures have a fundamental period on the order of 0.7–1.6 s which places them on the ascending branch of the Bucharest spectrum. Progressive damage during the earthquake should have caused a lengthening of their period and an increase in the lateral forces acting on them. In contrast, rigid structures of large panel or frame construction with shear walls of the same height as well as one to three story masonry dwellings suffered little damage (Ambraseys, 1977).

> About 108 people died and three multistory reinforced concrete apartment buildings (having a natural period of vibration of the order 0.8–0.9 s) collapsed completely in the town of Svistov, northern Bulgaria, approximately 270 km to the south of the epicentre (Tezcan et al., 1978).

Figure 2: Typical collapse of pre-World War II tall reinforced concrete buildings

(i) Wilson building buit in 1930s   *(ii) Lizeanu building, built in 1960s*

Figure 3: Collapse of the building's soft story at ground level

Faculty of Medicine in Bucharest

Effect of bombing during WW II   *1977 earthquake*

Faculty of Chemistry

Figure 4: Bucharest 4 March 1977. Heavy damage to two university buildings built at the end of 19th century (Photo © UTCB D. Lungu)

3. Vulnerability of Buildings in Central Bucharest

3.1. REASONS FOR VULNERABILITY

The explanation for the location in the center of the most fragile or vulnerable tall reinforced concrete buildings in Bucharest as well as of the 29 (out of 32) that collapsed during the 1977 event can be found in the Plan for Urban Development issued in 1935 by the Municipality of Bucharest. The plan recommends locating the tallest buildings in the city center i.e. buildings of six to seven full stories plus two to three setback stories (roof height of the building less than or equal to street width). The collapse of buildings in the city's center is clearly explained by their lack of mass vertical-symmetry, their lack of structure horizontal-regularity, the damage accumulated during the 1940 earthquake, the low strength concrete (mean of compressive strength ≤ 200 daN/cm^2) used in construction, the soft ground floors due to commercial uses (no filled masonry walls) and by their dynamic characteristics (Lungu, 1998, 2000). Currently, 127 tall reinforced concrete buildings built before WWII have been identified by authorities and structural engineers as "seismic risk class 1," i.e. buildings that would likely collapse or be very severely damaged during an earthquake similar to or larger than the 1977 event (Figure 5). The actual number of fragile, high-rise reinforced concrete buildings in the center of Bucharest is probably twice the number of existing buildings identified as seismic risk class 1 because many known to be the most vulnerable have still not yet been identified as such (Lungu, 2005). This is due to the fact that in addition to the high cost of adequately retrofitting fragile buildings, there are architectural and historical constraints, e.g., after the large demolition campaign in 1980, it was suggested that no more demolition be allowed in the city center.

Unfortunately today, the lessons learned in 1977 are still inadequately understood.

- "A systematic evaluation should be made of all buildings in Bucharest erected prior to the adoption of earthquake design requirements, and a hazard abatement plan should be developed." (Fattal et al., 1977)
- "Tentative provisions for consolidation solutions would preferably be developed urgently." (Japan International Cooperation Agency [JICA], June 1977)
- "...Bucharest is sited on deep alluvium.... Much of the damage was due to soil amplification associated with deep layers of silty clay, loess...Such sites would provide sufficient chances for dangerous amplifications in the shaking of such buildings." (Tiedemann, 1992)
- "Bucharest had been microzoned as part of the UNESCO Balkan Project, with microzones denoting three levels of risk. The worst destruction occurred in the lowest-risk microzone." (Berg et al., 1980)
- "A ground motion spectrum should be provided corresponding to each soil condition. A considerable number of strong motion seismographs will be required for this purpose." (Japan International Cooperation Agency, June 1977)

On March 30, 1977 the national strategy for strengthening buildings damaged by the 1977 earthquake was established by the Romanian government in a letter to the Municipality of Bucharest's General Inspector for Construction of Romania as follows: "The retrofitting of buildings must provide (i) for old buildings the same resistance they had before the 1940 earthquake (when they survived!) and (ii) for new buildings the same resistance they had when they

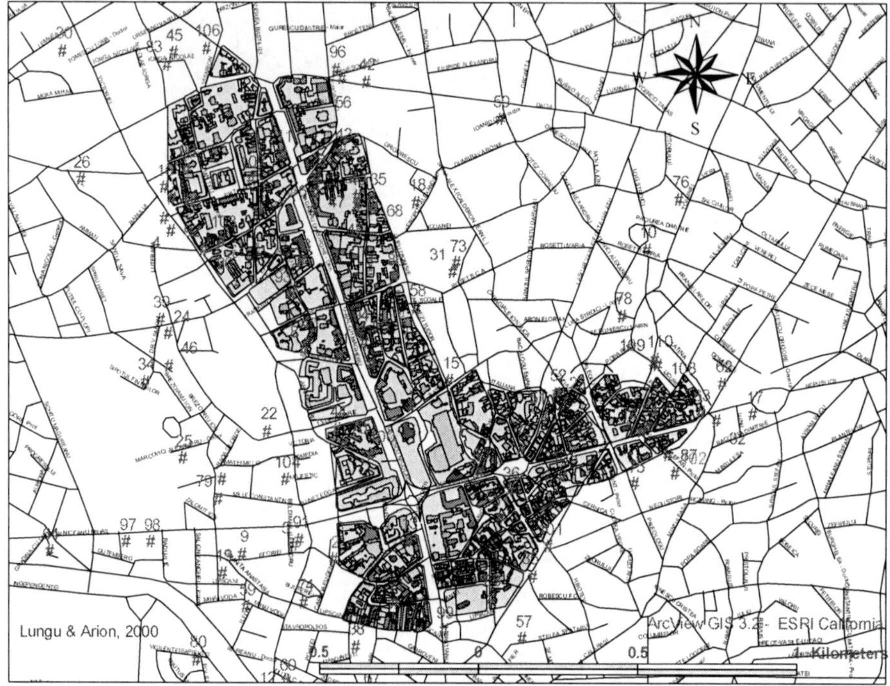

Figure 5: Map of central Bucharest with vulnerable buildings built prior to 1945 and identified as having the highest risk of collapse in case of a strong earthquake ($M_w \geq 7.5$)

were designed." This government order was further explained in a letter to the Technical University of Civil Engineering (UTCB), Bucharest from the General Inspector for Construction of Romania and the General Director of the Central Institute for Research Design and Coordination for Construction dated July 11, 1977 as follows:, "The retrofitting of the buildings damaged by the 1977 earthquake will consist of strictly local repairs of damaged elements. Additional measures for seismic protection are not allowed." This 1977 Romanian government strategy for repairing damaged buildings has proved to be a regrettable mistake.

3.2. SEISMIC RISK REDUCTION STRATEGY AND PROGRAMS

The present day national programs for seismic risk reduction in Romania focus on the following three objectives: (i) strengthening fragile buildings in Bucharest, (ii) upgrading the code for the seismic design of buildings and structures and (iii) seismic instrumentation. Currently in Bucharest, about 20% of the building stock was built before World War II (WWII), less than 40% was built before the 1977 Vrancea earthquake and over 40% was built after the 1977 event. According to existing data, in Bucharest there are more than 400 residential buildings recognized as seismic risk class 1. At the top of the list are the 127 tall reinforced concrete buildings built in central Bucharest prior to WWII: 19 are located on two central boulevards – Calea Victoriei and Magheru and Balcescu (Figures 6 and 7). Most of them are considered to be part of the city's architectural heritage. The risk matrix in Table 1 indicates that there are even class 2 risk buildings of importance.

TABLE 1: Seismic risk matrix indicating seismic risk classes (1, 2 and 3)

Seismic vulnerability/fr agility class	Importance and exposure class			
	I Essential facilities	II Hazardous buildings	III General buildings	IV Minor buildings
1	1	1	1 and 2	3
2	1 and 2	2	3	3
3	3			

Balcescu 25 (Wilson)

Magheru 20

Magheru 27

Calea Victoriei 2–4

Calea Victoriei 101 A-B

Calea Victoriei 124

Figure 6: Architectural heritage buildings in central Bucharest: tall reinforced concrete buildings built before 1940 in city center (examples)

Figure 7: National Romanian Television (TVR) reinforced concrete flexible building built in 1960s and Ministry of Transport, Construction and Tourism (MTCT) building, built in 1930s

The Historical Monuments List of 2004 prepared by the National Institute for Historical Monuments contains 26 cultural heritage structures in Bucharest. Of those, these seismic risk class 1 buildings might be considered priorities for strengthening.

(i) Multistory steel structures: Ministry of Transport, Construction and Tourism.

(ii) Multistory reinforced concrete structures: Romanian Government Building and City Hall of Sector 1, Bucharest.

(iii) Masonry and reinforced concrete buildings: Royal Palace i.e. National Museum of Art of Romania (central building). Only the Palace of Justice and the Telephone Building are now fully retrofitted (Figure 8).

Figure 8: Retrofitting of the Palace of Justice (2006) and Telephone Building (2002) in Bucharest

There are more than 200 orthodox churches in Bucharest and of those almost half are listed as architectural heritage sites (Figure 9). Most of the old churches need structural strengthening and immediate rehabilitation.

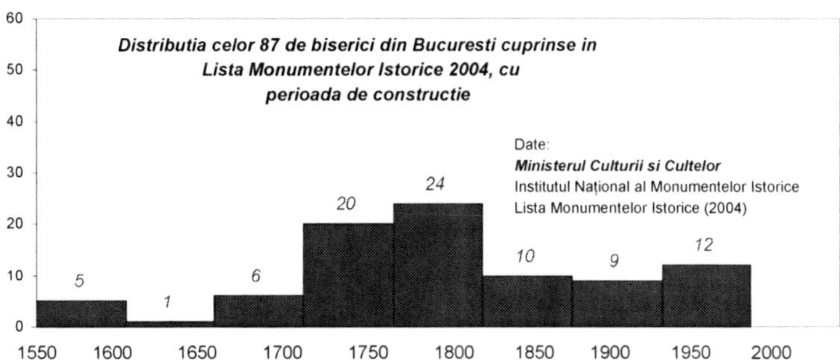

Figure 9: Distribution by age of the churches listed as architectural heritage sites according to the list of Historical Monuments of Bucharest 2004

To seismically upgrade architectural heritage buildings, the central role of the state has to be replaced by new partnership models with the private sector; international donors like UNESCO, the Council of Europe and the World Monuments Fund and civil society at large. A number of historic buildings and cultural assets have deteriorated or have been lost due to the lack of financial resources to prevent further deterioration. In fact, many of Bucharest's architectural heritage buildings located in city center "protected areas" are ready to be demolished to construct new buildings (Figures 10 and 11). Unfortunately, there are plans to demolish even seismic risk class 1 low-rise buildings instead of retrofitting them due to the price of land. Unfortunately, the existing law on protected areas cannot legally prevent the demolition of buildings located therein.

Figure 10: Bucharest heritage building in the city center intentionally prepared for demolition

Figure 11: Bucharest intentional fire and destruction to encourage the collapse of the Asan Mill, industrial heritage, 1853

4. International Projects for Seismic Risk Reduction in Romania

4.1. JAPAN INTERNATIONAL COOPERATION AGENCY PROJECT

The JICA "Technical Cooperation Project on the Reduction of Seismic Risk for Buildings and Structures" started on 1 October 2002. The project was signed in 2002 when 100 years of diplomatic relations between Japan and Romania were celebrated. The scope of the project is to strengthen capacity to deal with and prevent earthquake-related disasters in Romania. The duration of the project was 5.5 years. The implementing agency was the National Center for Seismic Risk Reduction (NCSRR), a public institution of national interest subordinated to the Ministry of Transport, Construction and Tourism.

Figure 12: Reaction frame donated by JICA and NCSRR tests on RC shear wall

Project activities were carried out by NCSRR in partnership with the UTCB and the National Institute for Building Research (INCERC) Bucharest. As part of the project, 29 young Romanian engineers were trained in Japan and 7 Japanese long-term experts and 37 Japanese short-term experts worked in Romania. Equipment for seismic instrumentation, for the dynamic characterization of soil and for structural testing costing 260 million yen (i.e. 2.17 million USD) was donated by JICA to Romania at NCSRR (Figure 12). The total cost of the project was roughly 7 million USD.

4.2. WORLD BANK PROJECT

The World Bank "Hazard Risk Mitigation and Emergency Preparedness" Project for Romania (HREM) has several components: (i) strengthening disaster management capacity; (ii) earthquake risk reduction and (iii) flood, pollution and project management. Component ii has the following subcomponents for a total of 71.2 million USD (i.e. about one third of the total cost of the project): strengthening high priority buildings and lifelines; design and supervision; building code review and studying code enforcement and professional training in cost-effective retrofitting. The project management unit for component ii is located at MTCT.

The World Bank report "Preventable Losses: Saving Lives and Property through Hazard Risk Management: A Strategic Framework for Reducing the Social and Economic Impact of Earthquake, Flood and Landslide Hazards in the Europe and Central Asia Region" published in October 2004 concluded that Romania was one the most seismically active countries in Europe and that Bucharest was one of the ten most vulnerable cities in the world. The following recommendations from the Report concern Romania: (i) upgrade the legal framework for hazard-specific management; (ii) review the existing building code for retrofitting vulnerable buildings; (iii) conduct a comprehensive public awareness campaign for earthquake risk; (iv) invest in hazard mitigation activities in order to reduce the risks caused by earthquakes and (v) develop a financing strategy for catastrophic events.

4.3. RISK-EUROPEAN UNION PROJECT

This project is entitled "An Advanced Approach to Earthquake-Risk Scenarios with Applications for European Cities" had a budget of 2.5 million euros (European Union [EU] 66% and participants 34%). With financial assistance from the European Commission the RISK-EU Project was launched in 2001 and ended in 2006. It developed a general and modular methodology for creating earthquake-risk scenarios that concentrates on the distinctive features of European cities including both modern and historical buildings. In order to identify the weak points in urban systems, it was based on seismic-hazard assessments, a systematic inventory and typology of the elements at risk and an analysis of their relative value and vulnerability. The resulting scenarios provide concrete estimates of direct possible earthquake damage and will lead to action plans for cities. Project participants included Bulgaria, France, Greece, Italy, Macedonia, Romania and Spain. UTCB was responsible for Work Package 1 – distinctive European features, inventory database and typology and for Work Package 7 – seismic risk scenarios (Lungu, 2002). Work Package 1 had two objectives: (i) chart distinctive features of European cities focusing on identity; population characteristics; urban area and elements at risk; impact of past earthquakes on elements at risk; strong motion data in the city and seismic hazard; geological, geophysical and geotechnical information; evolution of earthquake resistant design codes and earthquake risk management efforts and (ii) construct a European inventory database and typology including a classification of building occupancy and the typology matrix. Work Package 1 also contained a comparative study of the seismic characteristics of Barcelona, Bitola, Bucharest, Catania, Nice, Sofia and Thessalonica.

4.4. PROHITECH PROJECT

The 2.4 million euro (EU 88%, participants 12%) "Earthquake Protection of Historical Buildings by Reversible Mixed Technologies" (PROHITECH) project started in 2004 and will finish in 2008. PROHITECH is a research project designed to protect and conserve cultural heritage in the Mediterranean mainly by providing seismic protection for historical buildings and monuments dating from ancient times up to the 1950s. The main objective is to develop sustainable methodologies for the use of reversible mixed technologies. The structures to be protected cover a wide and diversified range of categories including both masonry and reinforced concrete buildings and also some steel constructions that need to be fitted with adequate anti-seismic provisions. Reversible mixed technologies exploit the peculiarities of innovative materials and special devices and allow ease of removal when necessary. Furthermore, combining different materials and techniques optimizes the potential to predict global behavior during seismic events. The ultimate goal of the project is a code for using these technologies to protect existing constructions.

4.5. SFB 461 PROJECT

The research project "Strong Earthquakes: A Challenge for Geosciences and Civil Engineering" (SFB 461) started in 1996 and ended in 2007 (four periods of 3 years each) and involved seven institutes of Collaborative Research Center 461, of the University of Karlsruhe. The Romanian participants were the Romanian Group for Strong Vrancea Earthquakes i.e.: UTCB, the National Institute of Earth Physics, INCERC, the Faculty of Geology and Geophysics of the University of Bucharest and GEOTEC.

The benefits to Romania from the project came from new country-wide free-field instrumentation, the first 100 m borehole with instruments in Bucharest, a test building with seismic rubber bearing isolators, significant bilateral exchanges for Romanian and German scientists, interdisciplinary and multidisciplinary approaches to earthquake engineering and engineering seismology and joint papers in various European and national seismic conferences.

References

Ambraseys, N.N., 1977. Long-period effects in the Romanian earthquake of March 1977, Nature, Vol. 268, July, pp. 324–325.

Aldea, A., 2002. Vrancea source seismic hazard assessment and site effects, Ph.D. thesis, Technical University of Civil Engineering Bucharest, Bucharest, 256p.

Arion, C., 2003. Seismic zonation of Romania considering the soil condition and seismic sources. Ph.D. thesis, UTCB, Bucharest, 181p.

Beles, A.A., 1941. Cutremurul si constructiile, Buletinul Societatii Politehnice din Romania, an LV, nr.10–11, Oct.–Nov., pp. 1045–1211.

Berg, G., Bolt, B., Sozen, M., Rojahn, Ch., 1980. Earthquake in Romania March 4, 1977. An Engineering Report. Edited by National Academy Press, Washington, DC.

Fattal, G., Simiu, E., Culver, C., 1977. Observations on the behavior of buildings in the Romania earthquake of   March 4, 1977, US Department of Commerce, National Bureau of Standards, Special Publication 490, 160p.

Georgescu, E.S., 2002. The partial collapse of Coltzea tower during the Vrancea earthquake of 14/26 October 1802: the historical warning of long-period ground motions site effects in Bucharest. Proceedings of the International Conference "Earthquake Loss Estimation and Risk Reduction" Oct. 24-26, 2002, Bucharest, Romania.

Japan International Cooperation Agency (JICA), 1977. The Romanian earthquake. Survey report by Survey group of experts and specialists dispatched by the government of Japan, June 1977.

Lungu, D., 1999. Seismic hazard and countermeasures in Bucharest-Romania, Bulletin of the International Institute of Seismology and Earthquake Engineering IISEE, Tsukuba, Japan, Vol. 33, pp. 341–373.

Lungu, D., Demetriu, S., Arion, C., 1998. Seismic vulnerability of buildings exposed to Vrancea earthquakes in Romania. In Vrancea Earthquakes. Tectonics, Hazard and Risk Mitigation, Kluwer, Wenzel F., Lungu D. (eds.), Proceedings of First International Workshop on Vrancea Earthquakes, pp. 215–224.

Lungu, D., Arion, C., Baur, M., Aldea, A., 2000. Vulnerability of existing building stock in Bucharest, 6ICSZ Sixth International Conference on Seismic Zonation, Palm Springs, California, Nov. 12–15, pp. 837–846.

Lungu, D., Aldea, A. Arion, C., Cornea, T., Vacareanu, R., 2002. RISK UE, WP 1: European distinctive features, inventory database and typology, earthquake loss estimation and risk reduction interational conference, Bucharest, Romania, Oct. 24–26. Lungu D., Wenzel F., Mouroux, P., Tojo, I. (eds.), Proceedings of International Conference, Vol. 2, pp. 251–271.

Lungu, D., Arion, C., Vacareanu, R., 2005. City of Bucharest seismic profile. EE 21C, International Conference Earthquake Engineering in 21st Century Skppje, Ohrid, Macedonia, Aug. 28–Sept. 1.

Tezcan, S.S., Yerlici, V., Durgunoglu, H.T., 1978. A reconnaissance report for the Romanian earthquake of 4 March 1977, Earthq. Eng. Struct. Dyn. 6, 397–421.
Tiedemann, H., 1992. A Handbook on Risk Assessment. Edited by Swiss Reinsurance Company, CH-8022 Zurich, Switzerland.

SEISMIC MICROZONATION OF CHIŞINĂU: A TOOL FOR REDUCING SEISMIC RISK

V. ALKAZ[*], A. ZAICENCO, E. ISICIKO
Institute of Geology & Seismology, 2028, Academiei 3 str., Kishinev, Moldova

Abstract. The objective of the current study is to investigate the dynamic properties of soft soils in Chişinău City which is exposed to seismic hazards from Vrancea seismic sources. Empirical transfer functions for soft soils were determined using earthquake and ambient noise records. Available geotechnical surveys (measuring shear wave velocities v_s) allowed employing 1-D and 2-D numerical methods for considering soft soil effects on the parameters of ground shaking. Observations of structural damage during the 1977 and 1986 seismic events allowed the development of relationships between the parameters of ground shaking and MSK intensities. A study of the amplification capacity of sites provided the basis for locating zones with varying seismic intensities and for developing a new seismic microzonation map of Chişinău. The iso-intensity map using the MSK scale shows that variations in the expected intensity values ranged between VII and VIII for the city's area. A methodology of seismic microzonation based on a complex study of soil dynamic properties was developed.

Keywords: Seismic hazard, building damages, site effects, seismic microzonation

1. Introduction

In studying seismic risks, an assessment of the influence of local soil conditions on seismic effect is of primary importance. This effect on earthquake motion has been described in numerous studies (e.g., Borcherdt, 1970; Ambraseys,

---
[*]Institute of Geology & Seismology, 2028, Academiei 3 str., Kishinev, Moldova, e-mail: alcazv@yahoo.com

1977; Hartzell, 1979; Aki, 1988, Borcherdt et al., 1989, Hough et al., 1990; Field et al., 1992; Borcherdt, 1994; Duval et al., 2001; Fitzko et al., 2007). Regardless of the results achieved, the problem of soil condition effects nowadays is deemed as extremely important but is yet only partially solved. Both empirical and theoretical methods have been widely used to evaluate the peculiarities of seismic ground motion in different regions. Assessing their influence on earthquake ground motion is usually carried out through a site effect analysis. Site response studies on Chişinău's urban area commenced in the 1970 with instrument observations at sites featuring different geological and geotechnical properties (Roman and Alkaz, 1996). Strong Carpathian (Vrancea) earthquakes in the 20th century occurred on 4 March 1977, 30 August 1986 and 30 May 1999 and provided abundant macroseismic damage data. Macroseismic data were used to validate the results of the site effect analysis obtained from the strong earthquake records. Currently, numerical modeling methods are extensively employed. The objective of this paper is to highlight the results of multidisciplinary studies on the influence of soil conditions on the ground response during Vrancea earthquakes.

2. Data and Methodology

2.1. DATA

The following data and methods were employed in this study.

- Geological and geotechnical data for 1,210 sites.
- Records of Carpathian earthquakes (Gutenber-Richter magnitude $2.8 \leq M_{GR} \leq 7.2$ and focal depth $100 < H \leq 150$ km) at 14 sites. Strong earthquakes ($M_{GR} > 6.0$) were recorded by accelerometers operating in the triggering mode, and weak events were recorded by six permanently operational portable seismic stations.
- Microtremor measurements at 85 sites. Data were recorded for 180 seconds (s) three times at each site. The measurements were made during the night using C5C seismometers when the contaminating effects of traffic and industrial noise were minimal.
- Measurements of shear wave velocity at 118 points using the down-hole method.
- 1-D (Ratnikova, 1984) and 2-D (Zahradnik, 1982) numerical modeling of the amplification capacity of soils.
- Macroseismic (damage) data during three strong Vrancea earthquakes: 1,845 inspected buildings after the 1977 event, 2,496 after the 1986 quake and 660 after the 1990 event.

2.2. INSTRUMENT DATA PROCESSING

The first method used by the authors is based on a comparative analysis of seismic records obtained at stations with different soil conditions: one of them is the reference station located on hard rock while the others are situated on soft soils (the reference site method in Borcherdt, 1970). The distances between the stations were much smaller compared with hypocentral distances. Thus the combined effect of source and seismic wave propagation was practically the same. A comparison of seismic records could be done in either the time or the frequency domain. Considering the large amount of data, a simplified characteristic of accelerograms was used, namely the average amplitude of the oscillations $\bar{A}$.

The seismic intensity increment ΔI at a particular location was determined using the formula (Guide for Seismic Microzoning, 1988):

$$\Delta I = 3.3 \cdot \lg(\bar{A}_i / \bar{A}_0) \tag{1}$$

where $\bar{A}_i$, $\bar{A}_0$ are average amplitudes of oscillations for soft-soil sites and hard-rock sites during the earthquake.

The second method, which has become widely used, could be named "the single station method." It evaluates site amplification as the ratio between horizontal and vertical spectra for the same station. It was used for the first time by Langston (1979) to evaluate the velocity profile from P-waves, and after that it was used by Nakamura (1989) to evaluate ambient noise excitation. The ratio of response spectra for horizontal acceleration and response spectra for vertical acceleration at a seismic station can be used to study site dynamic amplification (Nakamura ratio). The ratios between the horizontal and vertical response spectra of the motions recorded at the ISS-2 seismic station in Chişinău and a seismic station in Cahul (about 150 km south of Chişinău) are shown in Figure 1.

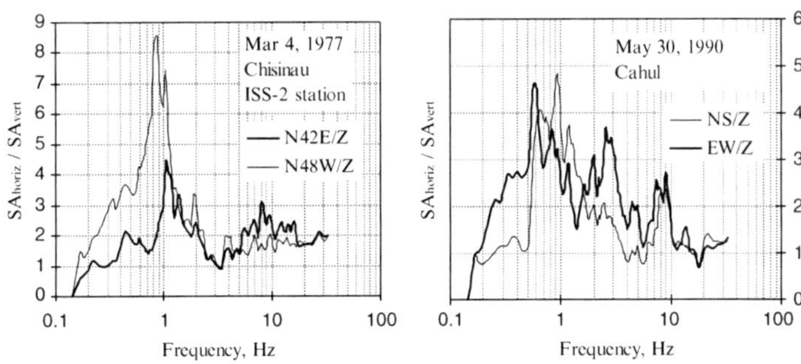

Figure 1: Spectral ratios for seismic stations in Chişinău and Cahul

2.3. MACROSEISMIC DATA PROCESSING

Interpreting macroseismic data implies their conversion into seismic effects in terms of intensity values for a particular area. A methodology combining macroseismic and geotechnical data has been used (Alkaz et al., 1996). As a seismic effect indicator, the degree of building damage has been used. Building data were divided into types A, B and C according to the Modified MSK-84 scale (Djuric et al., 1988). Compared with the MSK-64 scale, the MSK-84 scale allows an extended classification of structure types for seismic-resistant buildings up to intensities 7, 8 and 9 (types C7, C8 and C9). Macroseismic data processing was performed in three consecutive steps – correction, estimation and analysis.

The correction step provides a better estimate of the degree of damage when the data sample includes buildings that suffered k number of earthquakes or when the structures studied were considerably damaged prior to the last earthquake. To remove this pre-existing information, the difference between the mean values of damage $\Delta\mu = (\mu_{k+1} - \mu_k)$ was introduced in the data to make $k = 1$.

The estimation step computes average seismic intensity values (Ershov and Shebalin, 1984) for selected areas with equal target parameter properties (water table, densification capacity, soil category according to building code classification, number of strong earthquakes suffered by the structure, etc.):

$$\bar{d} = \sum_i d_i n_i / \sum_i n_i \tag{2}$$

where n_i is the number of buildings with damage degree $d_i = 0 \div 5$. Based on the average degree of damage, Ershov and Shebalin have proposed a set of curves for all types of buildings that provides estimates of seismic intensity values.

The analysis step is a statistical curve-fitting analysis using a second-order polynomial. The influence of target parameters from the estimation step on the degree of damage is explored. The degree of influence of different factors (geological, geotechnical, etc.) on seismic effect is computed in this final step.

2.4. NUMERICAL METHOD FOR WAVE PROPAGATION IN AN ELASTIC MEDIUM

An approach for wave field computation using the finite difference method has been used (Zahradnik, 1982). In this method, the model is divided into 2-D blocks or layers with different geophysical parameters. The physical parameters of each block are considered constant. The wave field of seismic response is computed under the excitation of the plane wave given by the SH impulse. As the spectrum of the impulse wave has almost constant amplitude on a relatively broad range of frequencies, the ratio of the amplitude spectrum of seismic response on a free surface of the model to the impulse spectrum at bedrock represents the spectral characteristic (or transfer function) of the 2-D model at the specified point.

3. Geological Conditions of the Area Studied

Chişinău is the capital of the Republic of Moldova and has a population of more than 700,000. It is situated at ≈200 km from the Vrancea epicentral zone. During the last century, the city has experienced several strong earthquakes (Table 1). One of the first attempts to produce a seismic microzonation map of Chişinău dates back to 1941 (Figure 2).

TABLE 1: Recorded PGA and observed intensities from strong Vrancea earthquakes in Chişinău

Earthquakes	Magnitude, M_{GR}	Hypocentral distance (km)	Max PGA (cm/s^2)	Max MSK intensity
November 10, 1940	7.4	210	–	8
March 4, 1977	7.2	200	99.16	7
August 30, 1986	7.0	230	232.5	8
May 30, 1990	6.9	210	204.3	6

Detailed geological and geotechnical data were collected to determine soft soil variations at 1,210 sites.

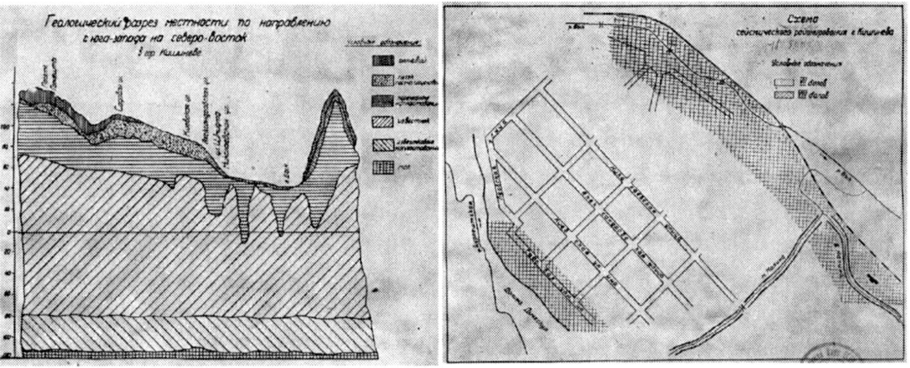

Figure 2: Map of seismic microzonation of Chişinău proposed in 1941 (Tshoher and Tischenko, 1941)

From a geological viewpoint, the study area is resting on sedimentary rocks. The surface deposits are represented by alternating loam, sand loam, sands, clays and silts. These sandy clay loose deposits overlay rocks of Sarmathian limestone (seismic bedrock of Chişinău urban area).

A special distinctive feature of the soil structure of this area is the stratification and substantial variations of the thickness of sandy clay deposits within a small area: from 0 m in the flood area of the city to 200 m on the watersheds. The depth of the water table exhibits variations in different locations in the city. On watersheds and the edges of slopes, the depth is 10–15 m or more while at low-altitude sites it is 0–5 m. The values of shear wave velocity, V_s, in covering loose deposits are 130–500 m/s and on bedrock (limestone) 700–1,300 m/s.

The study area is characterized by sharp variations in elevation: the steep slopes (>6°) of the valleys of the Bik river and ravines occupy a significant area of the city. A typical geotechnical cross-section is provided in Figure 3.

4. Results and Discussion

4.1. SITE EFFECT ASSESSMENT

Based on geological and geotechnical information, lithologically homogenous units for instrument measurements were selected. For each point, the empirical site response was evaluated using different techniques (sediment vs. bedrock ratio, Nakamura's method). Also, a set of analytical amplification functions for horizontal ground motion was computed based on log data and 1-D modeling (Ratnikova, 1984). Validating the results was done with the help of 2-D models (Zahradnik, 1982).

At the majority of the points measured, a satisfactory fit between the empirical and analytical predominant periods and amplification levels was established. The site response spectra exhibited peaks between 0.5 and 8.0 Hz, and the amplification factors ranged from 3 to 7. Examples of soil properties used in calculations are given in Tables 2 and 3; examples of calculated (1-D model) and empirical (Nakamura's method) transfer functions are provided in Figure 4. Examples of 2-D structure and transfer functions calculated for specified points where the calculations were made are given in Figures 5 and 6. As a comparison, transfer functions calculated for the same points with 1-D modeling are also provided.

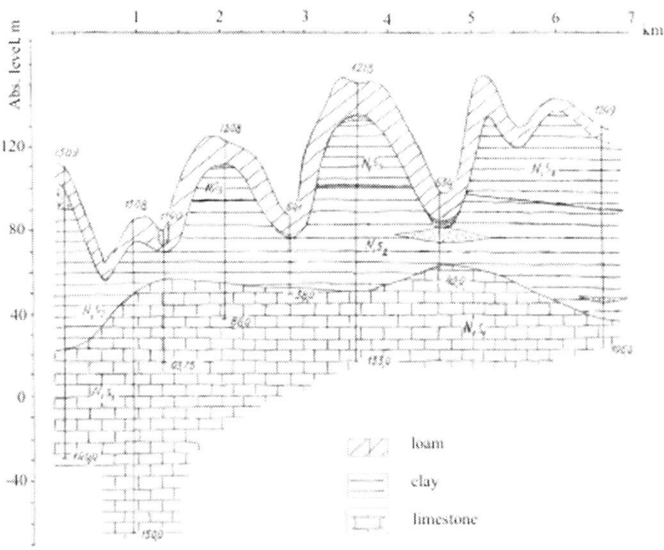

Figure 3: Chişinău: east-west geotechnical cross-section

The results of processing earthquake records demonstrate that the increment of seismic intensity at the sites with respect to hard rock ranges from 1.5 to 2.5 on the MSK-64 scale.

The ratios between the horizontal and vertical response spectra of the ground motions recorded at seismic stations in Chişinău have also shown amplification for low and high frequencies, irrespective of earthquake magnitude.

TABLE 2: Soil properties: Chişinău Site 1

Vp (km/s)	Vs (km/s)	ρ (g/cm³)	H (km)	Δp	Δs	Lithology
0.29	0.18	1.69	0.004	0.7	0.55	Loamy ground
0.48	0.18	1.72	0.004	0.55	0.5	Friable clay
2.00	0.24	2	0.004	0.2	0.3	Water-saturated clay
0.84	0.40	1.95	0.006	0.3	0.3	Neogene clay
0.84	0.52	2	0.021	0.3	0.25	Sands, clay
1.86	0.53	2.1	0.02	0.2	0.2	Clay, loam
1.10	0.44	1.97	0.011	0.25	0.25	Clay, sand
1.76	0.63	2.25	0.017	0.15	0.15	Clay, sand, silts
2.80	1.37	2.5				Lime-stone

TABLE 3: Soil properties: Chişinău site 2

Vp (km/s)	Vs (km/s)	ρ (g/cm³)	H (km)	Δp	Δs	Lithology
0.22	0.125	1.6	0.002	0.8	0.7	Made ground
0.38	0.20	1.7	0.002	0.65	0.4	Loam
1.5	0.26	1.9	0.008	0.6	1.2	Clay
1.6	0.48	2.1	0.008	0.25	0.25	Aleuritic clay
2.5	1.1	2.4				Limestone

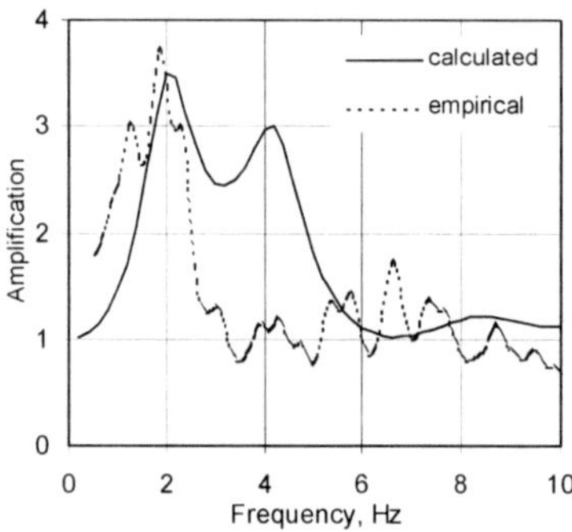

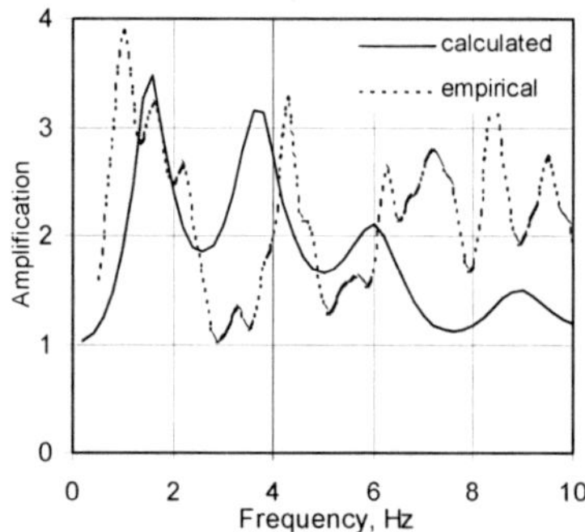

Figure 4: Examples of empirical and calculated amplification functions for different sites in the area studied: 1 and 2 = sites described in Tables 2 and 3

The presence of geotechnical boreholes in the vicinity of the seismic stations in Chişinău allows an evaluation of the influence of the local soil conditions on the spectral content of seismic motions.

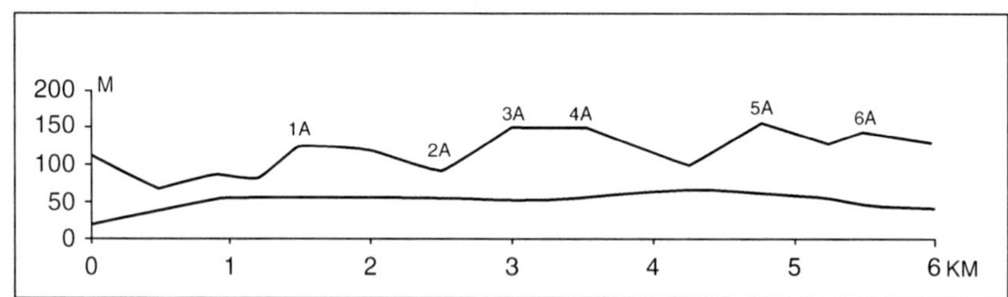

Figure 5: 2-D geotechnical cross-section model (Figure 3) used in the assessment of the amplification capacity of soils (Zahradnik, 1982). 1A–6A – the location of characteristic (representative) points where the calculation of amplification functions was made

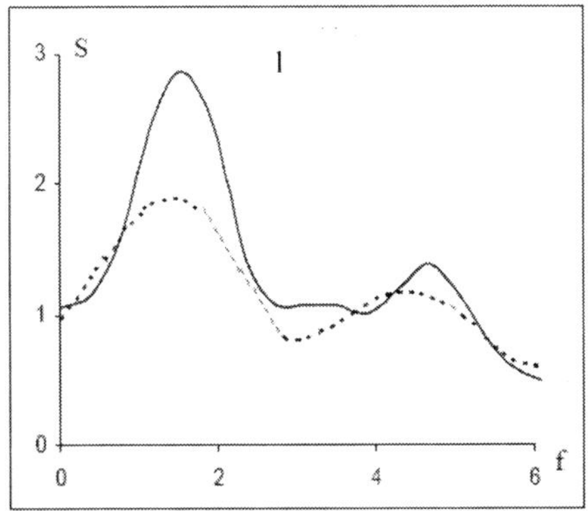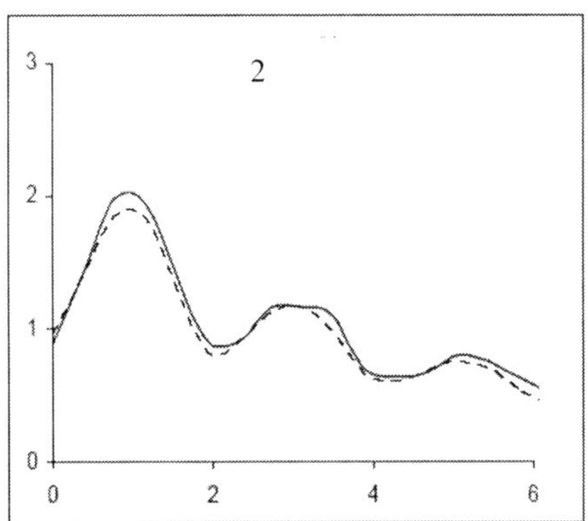

Figure 6: Examples of amplification functions calculated for different sites in the study area: 1 and 2 = sites 1A and 4A in Figure 5 (dotted line = 1-D modeling; thick line = 2-D modeling)

Based on knowledge of geometry and the elastic properties of soils, the averaged five-layer geotechnical model of the study area was created (Table 4) and average amplification factors for different sites were determined. It is noted that layers 2, 3 and 4 have variable parameters determined by site location.

TABLE 4: Geotechnical model of the study area

	Type	Thickness (m)	Density (g/cm^3)	Velocity (m/s)
1	Soil	0.5	1.5	120
2	Quaternary clay-sandy conglomerate	0–20	1.52–1.78	140–230
3	Quaternary clay-sandy water-saturated conglomerate	0–22	1.71–2.03	200–350
4	Neogene clay	0–200	1.86–2.17	260–450
5	Neogene limestone (bedrock)		2.45	1,300

The map of the average amplification factor for the central part of Chişinău is in Figure 7. A combination of different unfavorable soil conditions usually results in high values of structural response and greater damage during strong Vrancea earthquakes. For example, the August 30, 1986 Vrancea earthquake with magnitude 7.0 and epicentral distance of ~230 km to the southwest of Chişinău resulted in devastating damage to numerous buildings including seismic-resistant structures in some areas of the city (the fault rupture propagated almost directly towards Chişinău). Figure 7 shows part of the map of damage distribution in the city that was prepared on based on macroseismic investigations of 3,068 buildings. The distribution of buildings investigated immediately after the earthquake by degree of damage and number of stories is presented in Figure 8. The probability density functions for damage degree are provided in Figure 9.

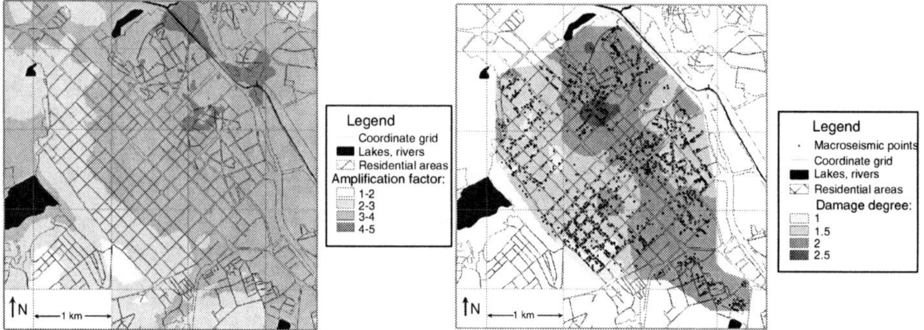

Figure 7: Average amplification factor for the central part of Chişinău and map of damage distribution (Chişinău center, 30.08.1986 earthquake)

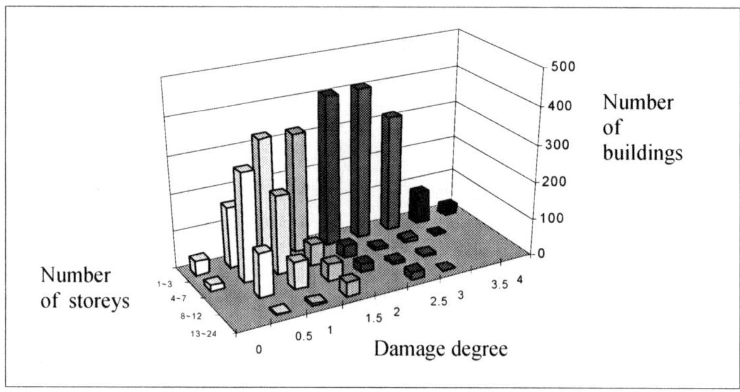

Figure 8: Distribution of buildings studied after the 30.08.1986 Vrancea earthquake by number of stories and degree of damage

According to previous studies (Alkaz, 1994; Alkaz and Boguslavski, 1990; Alkaz et al., 1987) several factors were considered to understand damage intensity variations more completely. The amplification of low frequencies by soft soils was one of the most important reasons for the significant damage observed in modern high-rise buildings. Consequently, the "low frequency zones" in the city were the most "hazardous" during the August 30, 1986 earthquake. The main factors controlling the damage to low-rise structures were the number of earthquakes suffered (accumulated damage) and the depth of water table.

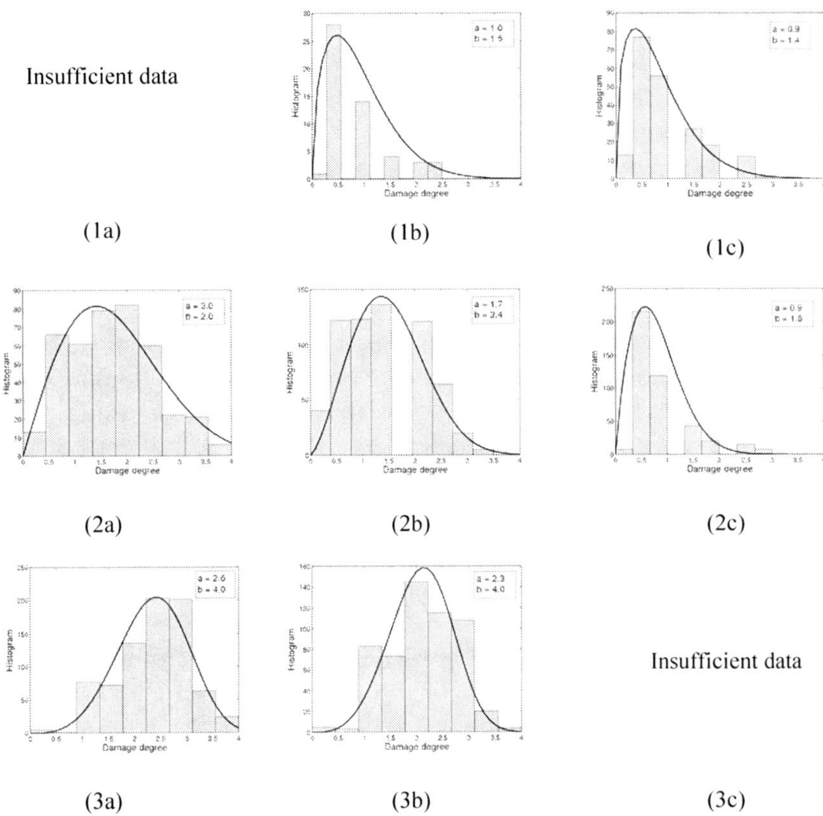

Figure 9: Statistics of damage data after 30.08.1986 Vrancea earthquake. Weibull distribution parameters (a, b) and histograms for structural damage: (**1a**) type A without previous damage, (**1b**) type B without previous damage, (**1c**) type C without previous damage, (**2a**) type A suffered 04.03.1977 earthquake, (**2b**) type B suffered 04.03.1977 earthquake, (**2c**) type C suffered 04.03.1977 earthquake, (**3a**) type A suffered 04.03.1977 and 10.11.1940 earthquake, (**3b**) type B suffered 04.03.1977 and 10.11.1940 earthquake, (**3c**) type C suffered 04.03.1977 and 10.11.1940 earthquake

The map of amplification and the map of the degree of structural damage for 30 August 1986 earthquake were compared to validate the results obtained. In spite of the high dispersion of values of degree of damage, a correlation coefficient of $\rho = 0.6$ with soil amplification values was obtained.

5. Seismic Microzonation of Chişinău

The quality of seismic microzonation is basically determined by the degree to which the methodology applied adequately considers the peculiarities of the region. As mentioned in other papers (Alkaz et al., 1996, 2005), the application of microzonation methods based on empirical ratios obtained for shallow earthquakes proves ineffective for the Vrancea region. For intermediate-depth earthquakes, the methodology should include (i) seismic properties of soils for the upper part of the cross-section as well as for larger depths (down to layers with $V_s \geq 1,400$ m/s), (ii) a distribution of shear wave velocity with depth, (iii) the quantitative contribution of different soil parameters to seismic effect, (iv) expected response spectra for each intensity zone on the map of microzones. It might be difficult to combine all these requirements in a single seismic microzonation methodology; therefore, we used a set of methods that successfully revised and complemented each other (Table 5).

This combined methodology was applied to seismically microzone the urban area of Chişinău. In addition to drawing the current microzonation map, a set of supplementary maps was developed including a map of the thickness of soft soil deposits, geotechnical and geomorphological maps, a map of soil amplification capacity and a map of the dominant periods of soil (Figure 10).

TABLE 5: Methodology for seismic microzonation

Method	Output
1. Geological and geotechnical studies	Map of geotechnical zonation of territory

2. Instrumental seismological studies:
  • Earthquakes, explosions
  • Microtremors
  • Seismic logging:
    shallow ($\leq$50 m), deep (50–200 m)
	Increment of seismic intensity ΔI
	Predominant periods of soil vibration
	Velocity Vp, Vs distribution

3. Macroseismic studies	Quantitative contribution ΔI of different factors on seismic effect
4. Theoretical methods	(a) Transfer functions of soils (b) Increment of seismic intensity ΔI (c) Influence of surface, internal topography (d) Recorded accelerograms (response spectra)

According to this map, the surface soil layer is thin in the central part of the city and thick at the watersheds. The dominant periods of soil range from 0.1 to 1.6 s and gradually increase from the center to the north and south of the city.

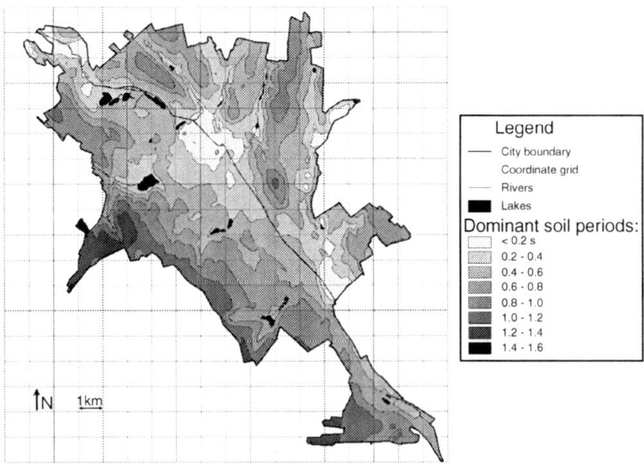

Figure 10: Map of dominant periods of soil in the territory of Chişinău

To control the values of seismic intensity increment ΔI obtained by various methods and to accept the final version of the seismic microzonation map, eight reference points were selected. These points were assigned to locations where results from different methods were available and provided good geographical coverage of the city. The results of the comparative analysis showed good agreement with the seismic records and numerical simulation method. The current building code SNiP-II-7-81, however, provided results inconsistent with the observations and simulations which is a natural outcome of the approach adapted by the code which was to compute the seismic intensity increment only on the basis of the mechanical properties of the soils in the upper 10 m. This approach is not valid for intermediate-depth earthquakes and for the soft soils in Chişinău when free-field ground motion is seriously influenced by the soil deposits overlaying the bedrock with a thickness of dozens and hundreds of meters. Considering the fact that seismic intensity increments ΔI computed from numerical models were close to the observed values, it was decided to use these models to create a seismic microzonation map of Chişinău. This map with seismic intensities is provided in Figure 11.

It should be emphasized also that authors support the idea of representing seismic risk in terms of damped response spectra. Nevertheless, the map is designed using MSK intensity values that are still in use in the current building code (SNiP-II-7-81).

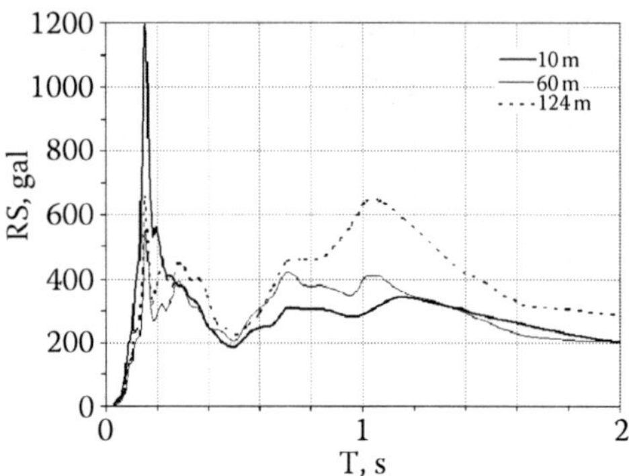

Figure 11: Acceleration response spectra for different thicknesses of soft soil in Chişinău (earthquake return period T = 475 years)

The iso-intensity map of microzonation (MSK scale) shows variations in the intensity values between VII and VIII. According to building code SNiP II-7-81 and the manual *Soil Condition and Seismic Hazard* published in 1988, the microzonation map is helpful in land use planning and aseismic design, in seismic risk assessment and in developing mitigation measures.

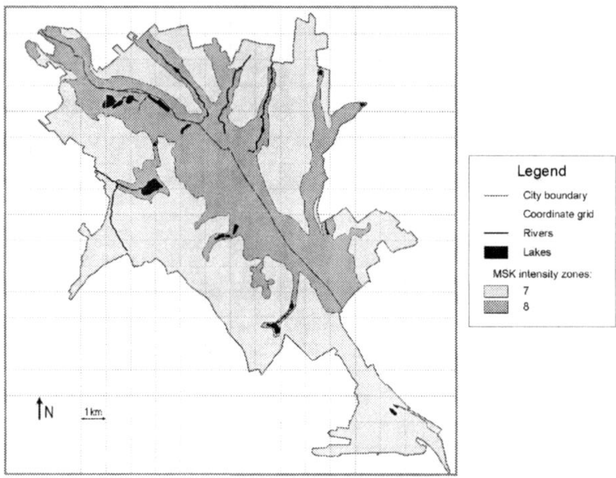

Figure 12: Map of seismic microzonation for the territory of Chişinău

6. Conclusions

The combined use of instrumental, macroseismic and numerical methods revising and cross-checking each other has revealed significant site effects on the urban area of Chişinău. Soil amplification factors correlate well with the structural damage observed after three strong Vrancea earthquakes.

The iso-period map shows zones of short predominant periods in the central part of the city; the period increases gradually to the northern and southern parts of the city.

This assessment of amplification capacity and of predominant periods of soils provided a new seismic microzonation map for Chişinău (Figure 12). The iso-intensity map of microzonation using the MSK scale shows that variations in the expected intensity distribution ranged between VII and VIII. The microzonation map is designed for land use planning and seismic-resistant design, seismic risk assessment and for developing mitigation measures.

Acknowledgements

This research was sponsored by NATO's Scientific Affairs Division in the framework of the Science for Peace Programme, project SfP-980468. Support from INTAS project 05-104-7584 is gratefully acknowledged.

The authors are greatly indebted to Prof. Dieter Mayer-Rosa for his valuable suggestions.

References

Aki, K. (1988). Local site effects on strong motion. Proc. Earthq. Eng. Soil Dyn., II, 103–155.

Alkaz, V. (1994). Effects of soil conditions upon damage to structures in Kishinev, Moldova. August 30, 1986 Earthquake (Abstract), 24 General Assembly of ESK, Athens, p. 126.

Alkaz, V. and Boguslavski, F. (1990). The effects of water level and category of grounds onto the intensity response of Carpathian earthquakes at the territory of the MSSR. Proceedings of the IX European Conference on Earthquake Engineering, Moscow, 4-A, 134–139.

Alkaz, V., Boguslavski, F. and Boldirev, O. (1987). An experience of seismic microzoning in the conditions of multilayered soils. J. Seismol. Res., 10, 54–73 (in Russian).

Alkaz, V., Isiciko, E. and Pavlov, P. (1996). Methodology of macroseismic data interpretation concerning microzonation, (Abstract). First Congress of the Balkan Geophysical Society, Athens, p. 528.

Alkaz, V., Drumea, A., Isiciko, E., Ginsari, V. and Bogdevici, O. (2005). Methodology of microzonation and it application for Kishinev City territory. Chisinev, Elena, 115 p. (in Romanian).

Ambraseys, N.N. (1977). Long-period effects in the Romanian earthquake of March 1977. Nature, 268, 324–325.

Borcherdt, R.D. (1970). Effects of local geology on ground motion near San Francisco Bay. Bull. Seismol. Soc. Am. 60, 29–61.

Borcherdt, R.D. (1994). Estimates of site dependent response spectra for design (methodology and justification). Earthq. Spectra, 10(4), 617–655.

Borcherdt, R.D., Glassmoyer, G., Der Kiureghian, A. and Cranswick, E. (1989). Results and data from seismologic and geologic studies following earthquakes of December 7, 1988 near Spitak, Armenia, S.S.R., U.S. Geol. Surv. Open-File Rept. 89-163A.

Djuric, V.I., Sevastianov, V.V. and Potapov, V.A. (1988). Soil Condition and Seismic Hazard Estimation, Nauka, Moscow (in Russian).

Duval, A.M. Bard, P.V., Lebrun, B., Lacave-Lachet, C., Riepl, J. and Hatzfeld, D. (2001). H/V technique for site response analysis. Synthesis of data from various surveys. Bull. Geofis. Theor. Appl., 42(3–4), 267–280.

Ershov, I.A. and Shebalin, N.V. (1984). The problem of seismic scale from seismological point of view. In: V. Steinberg (ed.), Assessment of Seismic Actions, Nauka, Moscow, pp. 78–96 (in Russian).

Field, E.H., Jacob, K.H. and Hough, S.H. (1992). Earthquake site response estimation. A weak-motion case study. Bull. Seismol. Soc. Am., 82, 2283–2307.

Fitzko, F., Costa, G., Delise, A. and Suhadolc, P. (2007). Site effects analyses in the Old City Center of Trieste (NE Italy) using accelerometric data. J. Earthq. Eng., 11(1), 33–48.

Guide for Seismic Microzoning (1988). Moscow , Stroiizdat (in Russian).

Hartzell, S. (1979). Analysis of the Bucharest strong ground motion record for the March 4, 1977 Romanian earthquake. Bull. Seismol. Soc. Am. 79, 513–530.

Hough, S.E., Borcherdt, R.D., Friberg, R.A., Busby, R., Field, E. and Jacob, K.H. (1990). The role of sediment-induced amplification in the collapse of the Nimitz freeway during the October 17, 1989 Loma Prieta earthquake. Nature, 344, 853–855.

Langston, C.A. (1979). Structure under Mount Rainier, Washington, inferred from teleseismic body waves. J. Geophys. Res., 84, 4749–4762.

Nakamura, Y. (1989). A method for dynamic characteristics estimations of subsurface using microtremor on the ground surface. QR Rail. Tech. Res. Inst., 30(1), 25–33.

Ratnikova, L.I. (1984). Computation of oscillations at the free surface and internal points of horizontally-layered damped soil. In: Moscow, Nauka (ed.), Seismic Microzoning, 235 p. (in Russian).

Roman, A. and Alkaz, V. (1996). Peculiarities of ground seismic reaction in Kishinev, Moldova. In: V. Schenk (ed.), Earthquake Hazard and Risk, Kluwer, Dordrecht, pp. 281–287.

SNiP II-7-81 (1982). Building Norms and Rules, Stroiizdat, Moscow (in Russian).

Tshoher, V. and Tischenko, V. (1941). Report "The Carpathian Earthquakes of October, 22 and November, 10, 1940". Archives of the Institute of Geology and Seismology, Academy of Sciences, Republic of Moldova (in Russian).

Zahradnik, J. (1982). Seismic response analysis of two-dimensional absorbing structures. Stud. Geophys. Geod., 26, 24–41.

RELEVANCE OF H/V SPECTRAL RATIO TECHNIQUE FOR BUCHAREST CITY

B. GRECU[1*], M. RADULIAN[1], N. MANDRESCU[1], G. PANZA[2]
[1]National Institute for Earth Physics, Romania
[2]Department of Earth Sciences – University of Trieste and the
Abdus Salam International Centre for Theoretical Physics – ESP
SAND Group, Italy

Abstract. We investigate the validity and relevance of the H/V spectral ratio technique in the particular case of Bucharest City which was strongly affected by the 4 March 1977 ($M_w = 7.4$) damaging earthquake. Our study shows that the coincidence of the predominant frequency response of the entire succession of the Quaternary unconsolidated deposits above the Frăteşti layers (f approx. 0.7 Hz) underlying the Bucharest urban area with the main frequency component in the Vrancea seismic source spectrum for the largest shocks ($M_w > 7$) is critical in controlling the strong ground motion properties. When applied to noise data, the H/V spectral technique reproduces the predominant frequency for the sedimentary cover beneath the city and the relatively uniform distribution of this structure over the city's area. An analysis of data on earthquakes of different sizes shows that the resonance response of the sedimentary cover becomes effective only for major Vrancea shocks and is practically negligible for events below magnitude 7. Consequently, we recommend avoiding any extrapolations from small and moderate earthquakes to large shocks when making the seismic microzonation map for strong Vrancea earthquakes in the Bucharest area.

Keywords: H/V spectral ratio, Vrancea earthquakes, seismic microzonation of Bucharest

*National Institute for Earth Physics, Romania

A. Zaicenco et al. (eds.), *Harmonization of Seismic Hazard in Vrancea Zone,*   133
© Springer Science + Business Media B.V. 2008

1. Introduction

The level of ground motion caused by earthquakes is mainly controlled by three factors: source, path and site response. In many cases, site effects have been recognized as being responsible for the damage caused during earthquakes (e.g., Northridge, USA 1994; Kobe, Japan 1995; Athens, Greece 1999; Izmit, Turkey 1999) and special attention has been paid to modeling these effects. Thus in the last three decades, various methods for characterizing site effects have been developed and applied in different regions of the world in order to understand soil behavior during an earthquake. Due to its low cost and its simplicity, the most popular technique is the H/V ratio computed for ambient seismic noise data. The spectral ratio of the horizontal-to-vertical component of ambient seismic noise in most cases shows a peak whose frequency is related to the fundamental frequency of the site under investigation (Nakamura, 1989).

Debates on the reliability of this method have generated many experimental and numerical studies. One significant issue was a comparison of the H/V spectral ratios from noise data with those from earthquake data. This comparison allowed some researchers (Chavez-Garcia et al., 1990; Yamanaka et al., 1993; Field and Jacob, 1995; Bindi et al., 2000; LeBrun et al., 2001) to conclude that ambient noise H/V spectral ratios provide reliable estimates of the fundamental frequencies of soil deposits. Regarding amplification, the question is still open; several authors found a good correlation between noise and earthquakes H/V spectral ratios (Lermo and Chavez-Garcia, 1994; Zaslavsky et al., 2000) while others showed unsatisfactory results (Zare et al., 1999; Satoh et al., 2001).

Bucharest is one of the most vulnerable cities in the world to seismic action, although the main threat comes from earthquakes located rather far away (the Vrancea intermediate-depth earthquakes located at about 140–160 kilometers [km] epicentral distance to the north). Events above magnitude 7 are dangerous or extremely dangerous for the city. Statistics based on historical records show that on average, three large earthquakes occur each century in Vrancea. For example, during the last century four events with $M_w \geq 7.0$ were recorded; two of them caused heavy damage and casualties in Bucharest (in 1940, $M_w = 7.7$ and in 1977, $M_w = 7.4$). The largest Vrancea shocks have a strong radiation in the period range of 1–2 s which matches the resonance domain for 10–20 storey buildings. This explains why the extreme damage reported in Bucharest was mainly to tall buildings (32 tall buildings collapsed during the 1977 shock).

At present, the dispute among seismologists and civil engineers about which of the two key factors – source or surface geology – prevails in controlling ground motion intensity in the Bucharest area is still open. A ground motion computation based on physical laws rather than on empirical techniques like

H/V spectral ratios or array techniques would be preferable, but they may not be able to reproduce the variability observed, and some authors consider that the details of ground motion are practically unpredictable and that they can be handled only in a probabilistic way (e.g., Sokolov et al., 2004). In contrast, a neo-deterministic approach can well explain ground motion variability with relatively simple models for source and local site structure (e.g., Moldoveanu and Panza, 2001; Cioflan et al., 2004; Panza et al., 2002; Field et al., 2000) provided adequate computer codes for wave propagation modelling are used (e.g. Panza et al., 2001).

The purpose of the present paper is to check the validity and relevance of the H/V spectral ratio technique in the particular case of the Bucharest urban area. This is an essential and imperative matter since considerable efforts have been and are currently being made to produce reliable maps of seismic microzonation in Bucharest using a large data set of noise and moderate-to-low seismicity records (e.g., Ritter et al., 2005). As we shall demonstrate, in the case of Bucharest City, the H/V spectral ratio analysis is useless if not accompanied by a careful interpretation of what we know about strong ground motion (observed and modelled) characteristics.

2. Local Structure

Bucharest City is situated on a plain (Vlasia Plain, part of a bigger morphological unit, Romanian Plain) slightly dipping toward the southeast, following the direction of the valleys of the Dambovita and Colentina rivers (Figure 1). The lithological complex of Lower and Upper Quaternary deposits (Figure 2) is characterized by a succession of six main units (Mândrescu et al., 2004).

The Frătești complex consists of three layers of sand and gravel (A, B and C in Figure 2), separated by two intercalated layers of clay. The layers have a similar structure with coarse sands and gravel at the bottom and medium-fine sands transforming gradually to clays in the upper part. The entire complex is gently dipping from south to north and becomes thicker along the same direction.

The marl complex is represented by a succession of marl and clay, sometimes sandy marl with an intercalation of fine sands. The rocks composing the marl complex correspond from a granulometric point of view to some lacustrine formations in facies of shallow depth in which the determinant material is represented by the pelitic fraction.

The upper part of the marl complex is continuously covered by a bank of sands of brown-black color with rust-colored intercalations (Mostiștea sands).

The intermediate clay deposits are developed between the Mostiştea sands and the Colentina sands and gravel. These clays have a variable thickness of between 5 and 20 meters (m).

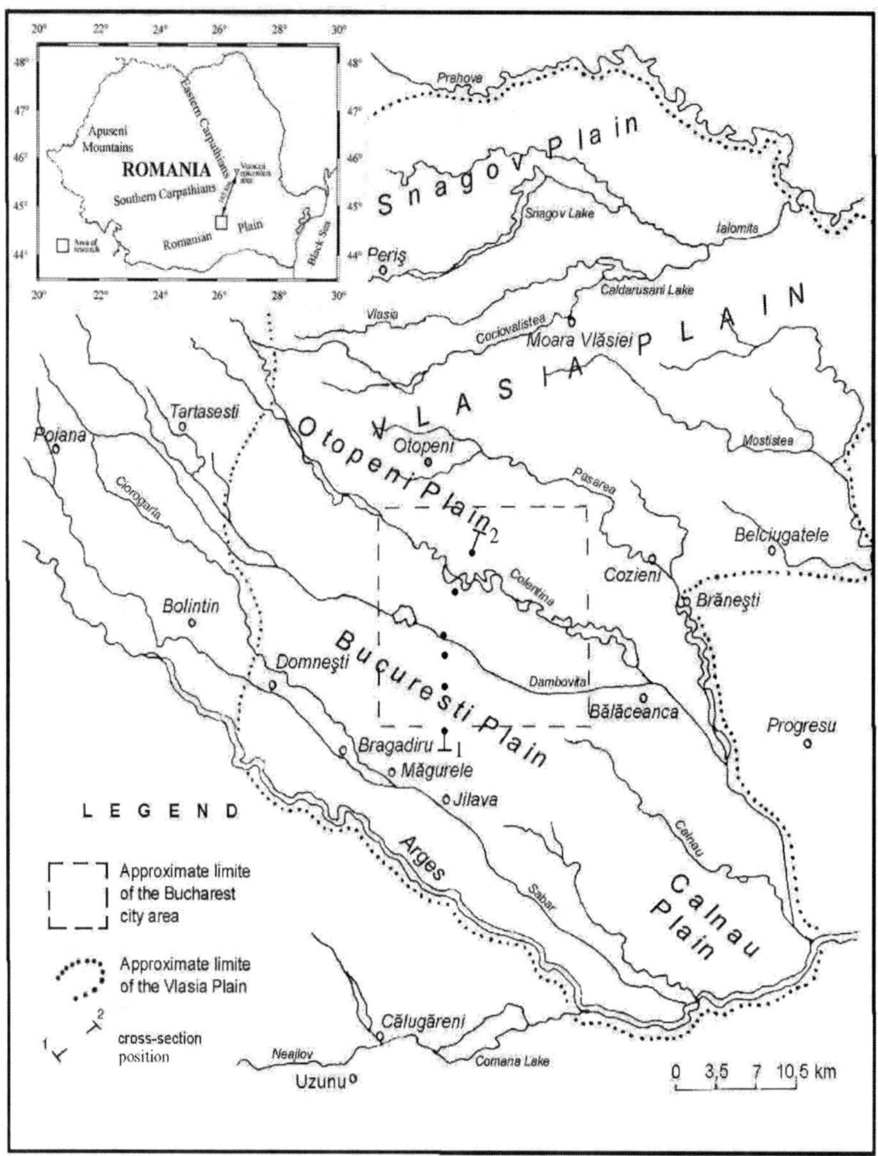

Figure 1: Location of Bucharest urban area (marked by the rectangle) and Vrancea epicentral zone

The Colentina gravels and sands show a transition from gravels located at the basement of this complex to sands at the upper part.

This unit is loess-like deposits with a lithology characterized by a large variety of granulation of the component elements from clays and silt to fine sands and sometimes even coarse sands.

The following average shear wave velocities were adopted: loess-like deposits – 0.215 km/s; Colentina gravels and sands – 0.315 km/s; intermediate clays – 0.325 km/s; Mostiştea sands – 0.360 km/s; marl complex – 0.420 km/s and Frăteşti layers – 0.650 km/s (Mândrescu et al., 2007). The predominant periods of oscillation, T, of the subsurface layers over the territory of Bucharest computed using a simple relation $T = 4h/\beta$ where β is the S-wave velocity and h the layer thickness range between 1.0 and 1.9 s, and increase from south to north as a consequence of the constant increase of the thickness of the Quaternary cohesionless deposits (Mândrescu, 1978; Mândrescu et al., 2004).

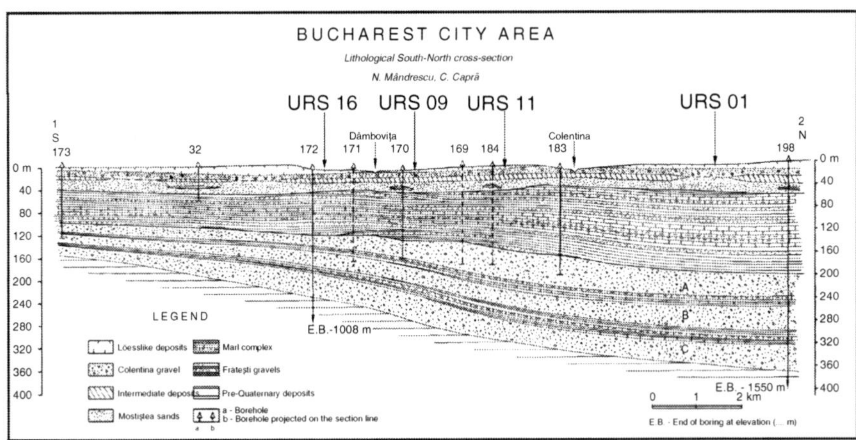

Figure 2: Lithological south-north vertical cross section across Bucharest urban area. The spectral ratios for the four Urban Seismology Project stations in the figure are given as examples in Figure 3

3. Application of the H/V Spectral Ratio Technique

An experiment for recording seismic noise in the Bucharest area deployed 16 recording stations for 2 months in 1997 (Bonjer et al., 1999). The sites were selected to be adjacent to boreholes with drilling reaching the bedrock, i.e., to sites with precisely known thicknesses of the sedimentary cover. Three-component 2-Hz Mark Products velocity sensors were installed at each site and were operated for at least 30 min. Recently, as part of the URS (Urban Seismology) project carried out by the Collaborative Research Center, 461

Strong Earthquakes of Karlsruhe University and the National Institute for Earth Physics of Bucharest, 32 broadband stations operated between October 2003 and August 2004 (Ritter et al., 2005).

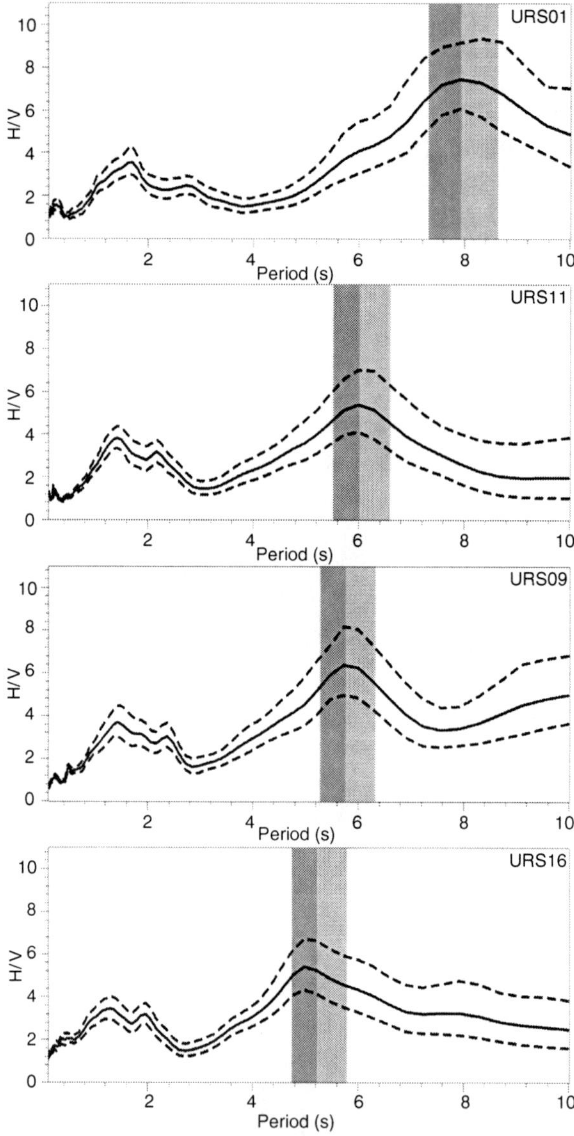

Figure 3: H/V spectral ratios for the ambient noise recorded in different sites in Bucharest area during the URS experiment. The site locations are given in Figure 2

The application of Nakamura's technique (Nakamura, 1989) to estimate H/V spectral ratios from the data of the 1997 experiment shows a relatively stable predominant peak between 1 and 2 s (except at one site) with an average of 1.4 s. It was interpreted as broad and stable soil resonance. In the simple horizontal layer resonance interpretation ($T = 4h/\beta$), a resonance peak around 1.4 s corresponds to a ~120 m thick layer of unconsolidated sediments with a shear wave velocity $Vs \sim 0.35$ km/s. The spatial variation of the resonance period at the measurement sites (in the range of 1.15–1.60 s) has apparently no correlation with the variation of the thickness of the sedimentary cover indicated by geological data (shift toward lower frequencies from south to north).

We applied the same technique to the data recorded by the URS experiment. Examples of H/V spectral ratios for several sites in Bucharest are given in Figure 3. Our results reproduce the resonance in the period range from 1 to 2 s as previously obtained (see diagram for site 16 in Figure 2 of Bonjer et al., 1999). Extending the frequency band allowed us to identify a second resonance at large periods (>5 s) that was even better defined than the peak at short periods. The two peaks are visible in all cases independent of the particular location of the site. These results do not contradict the general uniform subsoil structure beneath the city described in the local structure section: the peak between 1 and 2 s corresponds to the Frăteşti layer resonance, while the peak around the 6 s period corresponds to the resonance of the entire sedimentary cover above the cretaceous limestone basement. In H/V spectral ratios, the peak at long periods is more pronounced than the one at 1–2 s since the contrast of impedance is significantly higher at the cretaceous limestone basement than at the transition between young and old sediments. In addition, since the sites selected in Figure 3 are aligned roughly in a north-south direction, we can notice a gradual decrease of the predominant frequency from south to north in correlation with the slight dipping of the subsurface layering in the same direction (see Figure 2).

In parallel, we applied the H/V spectral ratio method to earthquakes to test if the results obtained using ambient noise would be confirmed. Certainly in this case, the source information and the geometry of the problem differ significantly from the noise case. The earthquakes are located at intermediate depths (91–154 km) in a confined volume located at about 140 km epicentral distance from Bucharest City (Table 1). The fault plane solutions show reverse faulting processes that are characteristic for Vrancea intermediate-depth seismicity. The solutions for the largest shocks (1977, 1986 and 1990) are close to each other while some variations are noticed for the smaller events. In all cases, the radiation patterns indicate the rather similar behavior of the radial and transversal components.

Let us first look at the results of the H/V analysis applied to ambient noise and to different sizes of Vrancea earthquakes recorded at the INCERC station, the only site that has recorded all the strong Vrancea earthquakes since the 1977 event. The horizontal component H of the ratio is the spectrum of the resultant of the two horizontal components or the spectrum of the maximum horizontal component. The representation of H/V for ambient noise (Figure 4) is similar with that given in Figure 3. Two distinct peaks are visible, one around 6 s and the other around 1.4 s. Stationarity tests (using different day and night time windows and different time intervals during 1 day) show that the period band and amplitude of the peaks are stable.

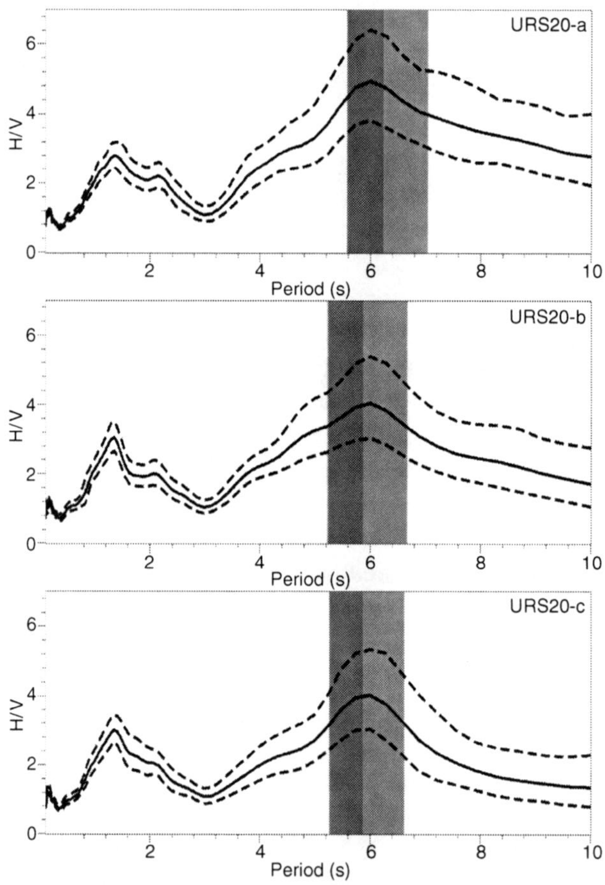

Figure 4: H/V spectral ratios for the ambient noise recorded at INCERC station and computed for (a) day period, (b) night period, (c) one day. Solid line: average; broken lines: ± standard deviation

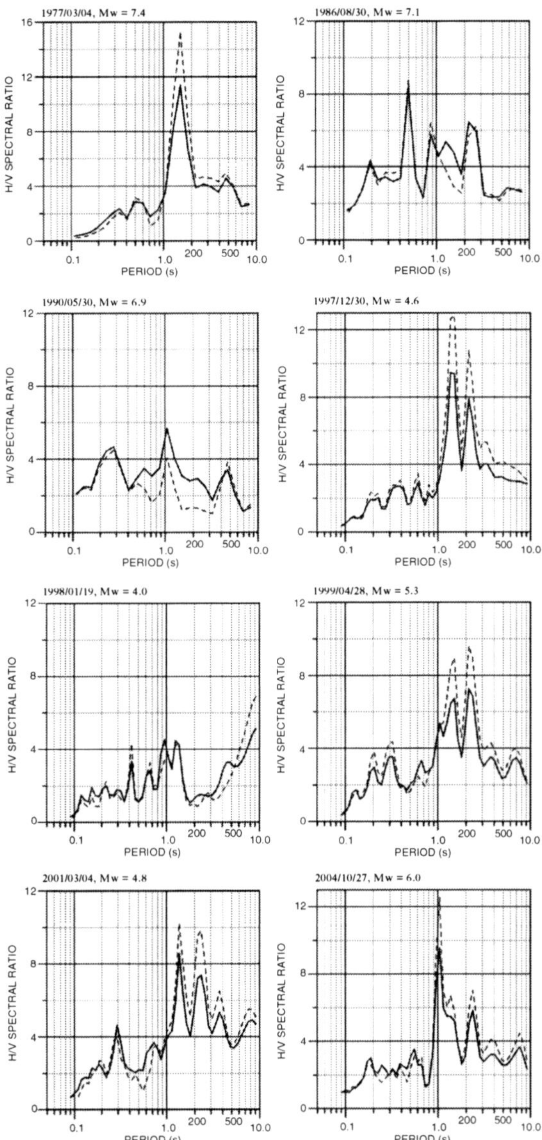

Figure 5: H/V spectral ratios for earthquakes (see Table 1) recorded at INCERC station. Solid line takes H as the resultant of the horizontal components; broken line takes H as the maximum among the horizontal components

To identify possible source effects, we selected the different sizes of Vrancea earthquakes listed in Table 1 as examples. The associated H/V spectral ratios estimated using the recordings at the INCERC station are plotted in Figure 5. The shape of the H/V ratios changes significantly from one event to the other. The prominent and remarkably confined peak around 1.5 s observed for the largest earthquake (March 1977) coincides with the smallest peak shown in Figure 4. For the two events with a magnitude around 7 (August 1986 and 30 May 1990), the dominating peaks at 2.3 and 1 s respectively do not match at all the prediction based on noise analysis. As shown above, focal mechanism characteristics seem not to influence the spectral ratios significantly. A better agreement with noise-based predictions is obtained for the smallest earthquakes (January 1998 and March 2001), even though some differences in amplitude level are observed. The peak around 6 s is easily observed in the ambient noise case (Figures 3 and 4) and for the smallest, $M_w = 4.0$, earthquake (Figure 5), but it is not seen in the case of the stronger signals. The noise measurements were performed with an STS2 sensor that performs satisfactorily down to 0.083 Hz while the accelerometers are unstable and very poor at low frequencies (SESAME project report, 2001, www.obs.ujf-grenoble.fr). Therefore, any earthquake analysis above 5 s is doubtful due to instrumental noise (decrease in the amplitude of the seismic signal recorded by accelerometer sensors toward the electronic noise level of the acquisition system with increasing period).

The physical meaning of the H/V spectral ratios is still doubtful. For the relatively large-scale shallow structure beneath the Bucharest urban area, a plausible interpretation of the H/V technique is directly related to the ellipticity curve of Rayleigh waves, and its efficiency relies on its ability to identify the fundamental frequency of the soft soils since the vertical component of Rayleigh wave motion systematically vanishes around the fundamental S-wave resonance frequency (e.g. Panza, 1985). Therefore, the H/V ratios for ambient noise (Figure 4) and for earthquakes (Figure 5) are mapping different phenomena and comparing them is not straightforward and should be done with care.

For earthquakes below and above magnitude 7, the absolute Fourier spectra (Figure 6) have completely different shapes: the amplitude in the period range from 1 to 2 s is practically negligible in the case of small and moderate earthquakes, and it becomes progressively more important as the magnitude increases as can be seen in Figure 6. The resonant amplification by the sedimentary cover is visible only for the largest earthquakes ($M_w > 7$), i.e., when the source radiates also at periods close to the fundamental resonance range of the sedimentary cover, and it clearly depends on magnitude (~1.6 s for an M_w 7.4 event, ~1.3 s for an M_w 7.1 event, ~1.2 s for an M_w 6.9 event and ~1.0 s for an M_w 6.0 event). This result supports the idea that the source size (M_W) controls the ground motion characteristics in the Bucharest urban area, the sedimentary layer response being a secondary effect that becomes relevant only if properly excited by the seismic source. The dependence of the frequency content of the seismic ground motion on earthquake magnitude is revealed also by a neo-deterministic microzoning analysis (Cioflan et al., 2004).

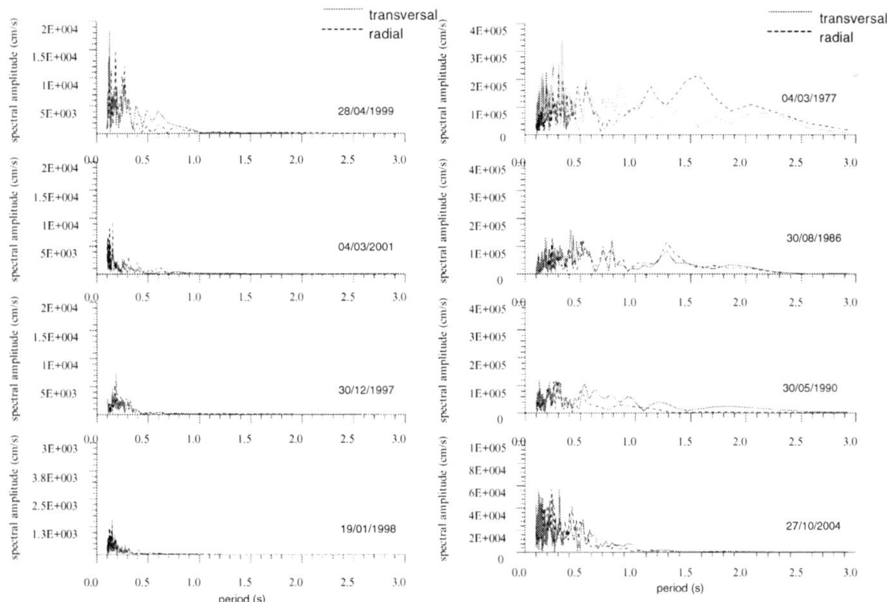

Figure 6: Fourier amplitude spectra of S-wave radial and transversal components for studying earthquakes (see Table 1) recorded at INCERC station. For the events of 27 October 2004 and 19 January 1998 the vertical scale is expanded by a factor of 4.

Dependence on the magnitude of the ground motion spectral shape and H/V ratios can in principle be due not only to source effects but also to variations in the incidence angles and non-linear behavior of soils during strong shaking. Taking into consideration focus-site geometry, the fluctuations of the incidence angle should not be important. To evaluate the strain level in Bucharest due to Vrancea earthquakes we use the ratio between the recorded peak ground velocity (PGV) and Vs velocity close to the surface according to Newmark and Rosenblueth (1971) and Aki and Richards (2002). If we consider the PGV values recorded during the last significant Vrancea events (1977 M_w = 7.4, 1986 M_w = 7.1, 1990 M_w = 6.9 and 2004 M_w = 5.9) and the average shear wave velocity in the uppermost 100 m (~350 km/s), we obtain a maximum strain value of 0.002 for the 1977 event and smaller values for the other events (0.0005 for the 1986 and 1990 events and 0.0001 for the 2004 event). As mentioned by Paskaleva et al. (2004), ground failure initiates at strains ≈0.001 and reaches completion for strains equal to and larger than ≈0.0056. These estimates suggest possible non-linear effects for the largest earthquakes but are not too important.

Consequently, we consider that the main factor explaining the change in spectral shapes is the coincidence of the source's predominant frequency with the resonant frequency response of the sedimentary layer beneath the city.

To test if source radiation control on the shape of the H/V ratios becomes less important for small earthquakes, in Figure 7 we compare the spectral ratios computed for noise windows recorded just before the P-wave arrivals with the spectral ratios computed from the records of the smallest earthquakes (19 January 1998 M_w = 4.0 and 30 December 1997 M_w = 4.6) of our selection.

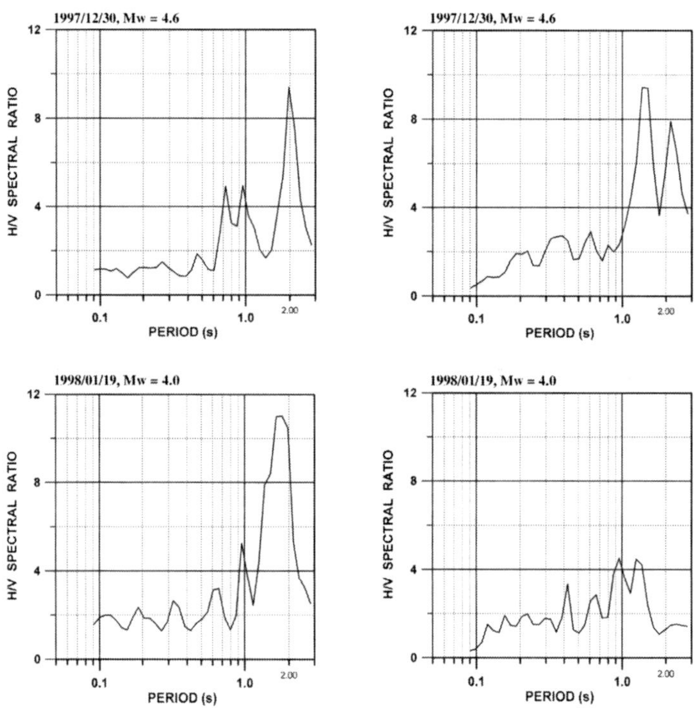

Figure 7: H/V spectral ratios computed at INCERC station for the earthquakes of 30 December 1997 and 19 January1998 for noise (*left*), and S-wave signal (*right*). The windows for S-wave trains and noise prior to P-wave arrival time are of 10 s each

Most of the peaks in the H/V ratios can be associated with noise contribution (since for these earthquakes the seismic radiation is negligible above 1 s), although some source influence is still noticeable as secondary peaks around 0.4 s for the 1998 event and around 0.3 s for the 1997 event, which agrees with the source scaling laws for Vrancea subcrustal sources (e.g., Gusev et al., 2002). Despite the fluctuations and differences in amplification, in all cases the spectral ratios using noise and earthquake windows suggest the presence of a peak within the 1–2 s range close to the average 1.4 s resonance period determined from microseisms (see Figure 4).

The major result of our analysis – progressive source radiation control on the shape of the H/V ratios as earthquake magnitude increases – is in agreement with the numerical simulation of strong ground motion in Bucharest for Vrancea earthquakes made by the neo-deterministic approach (Moldoveanu and Panza, 1999, 2001; Moldoveanu et al., 2000). Given the fact that the damage pattern cannot be controlled by the building stock alone, observed damage distribution shows that the characteristics of observed motion change significantly when source parameters change, while the geometry and structure of the medium between the focus and the site are practically the same. Therefore, site amplification effects cannot be considered as separate from source effects.

4. Conclusions

The impact of Vrancea earthquakes on Bucharest City represents an atypical case due to the unusually high damage they cause despite the city's relatively large distance from the hypocenter. Our main goal is first to identify what the determining factor is that controls ground motion intensity in the Bucharest area – the shallow sedimentary cover versus the seismic source – and second to investigate the stability and reliability of the H/V technique and to test its meaning and its actual relevance for seismic microzoning purposes.

The test for local geological control on the seismic ground motion level and spectral content for different sizes of Vrancea events recorded at the same site illustrates the increasing effect of the source with increasing magnitude prevailing upon the local site effect as magnitude exceeds 7. This effect cannot be explained by the focal mechanism or focus-site geometry variations as the analysis of our dataset shows.

The fundamental phenomenon responsible for the amplification of seismic ground motion over soft sediments is the trapping of seismic waves in a given frequency band due to the impedance contrast between sediments and the underlying bedrock with the possible formation of local surface waves (for 2-D or 3-D structures like sedimentary basins). Many studies show that the amplification patterns (the fundamental resonance period) revealed by the H/V technique can be well correlated with surface geology if the geology is simple and has a strong impedance contrast to the bedrock, but the absolute level of amplification cannot be determined in a straightforward way (e.g., Lachet and Bard, 1994). In the case of the Bucharest urban area, the impedance contrast between the sedimentary layers and the bedrock is rather small; therefore, we should always consider coupling source and site factors. The earthquake magnitude that controls the frequency content of the source spectrum is, in turn, controlling the excitation of the sedimentary layers. This explains why the H/V ratio technique cannot be used to reliably define site effects using earthquake data.

As our results suggest, the long-period seismic wave amplification in the Bucharest area is the joint effect of the deep soil deposits and seismic source radiation. Only the largest Vrancea shocks ($M_w > 7$) are able to effectively excite the resonance period of the sedimentary shallow layer (in the range 1–2 s) that are responsible for the collapse of tall buildings in the city. Due to this reason, extrapolations from small-to-moderate earthquake analyses are practically of no use for modeling strong motion characteristics for microzoning purposes. For the same reason, we can explain why the damage in Bucharest dramatically increases when the earthquake size is above a critical value ($M_w \sim$ 7) and relatively minor effects are reported for magnitudes below 7.

The H/V spectral ratio technique was originally used for microtremors, not for earthquake data; its physical meaning is not yet clear for earthquake data. In its initial form (Nakamura, 1989), it was based on the too simple assumption that the vertical component of motion generated by microtremors is free from near-surface influence (e.g. Lokmer et al., 2002; Panza et al., 2002, 2004; Romanelli et al., 2003; Kouteva et al., 2004; Ding et al., 2004; Hamzehloo et al., 2006). Although the technique has been recently applied by some researchers to earthquake data for site effect estimation (if earthquake recordings are used, the H/V technique is usually called the receiver function technique, see e.g. Lermo and Chávez-Garcia, 1993), as our results show, using the H/V technique for site effect estimation in the case of Vrancea earthquakes is questionable. If the H/V ratios for Bucharest mainly reflect the ellipticity of Rayleigh waves and if surface wave generation is clearly dependent on the incident waves at the site, we may conclude that the application of the H/V technique is not adequate for intermediate-depth events.

Therefore, we recommend that, in the particular case of strong Vrancea earthquakes impact in the Bucharest area, to avoid any extrapolations from small and moderate earthquakes to large shocks when making the seismic microzonation map.

TABLE 1: List of the earthquakes with recordings available selected at the INCERC station. S/N ratio is computed for 1 Hz when noise window before P arrival is available (not possible for SMAC-B instruments)

No.	Date	Time	Lat. (°N)	Lon. (°E)	Depth (km)	Epicentral distance (km)	M_w	S/N ratio	PGA EW comp.	PGA NS comp.
1	1977/03/04	19:21	45.77	26.76	94	155	7.4	–	163.1	194.9
2	1986/08/30	21:28	45.52	26.49	131	123	7.1	–	88.7	95.3
3	1990/05/30	10:40	45.83	26.89	91	165	6.9	–	76.6	98.7
4	30/12/1997	04:39	45.54	26.32	139	123	4.6	43	2.6	2.6
5	19/01/1998	00:53	45.64	26.67	105	139	4.0	7	0.3	0.5
6	28/04/1999	08:47	45.49	26.27	151	117	5.3	510	8.3	16.6
7	04/03/2001	15:38	45.51	26.24	154	119	4.8	110	3.2	3.0
8	27/10/2004	20:34	45.84	26.63	105	160	6.0	1,535	23.7	22.0

Acknowledgements

Our study benefited from data recorded by the K2 network installed as part of the joint German-Romanian research programme CRC461 (Bonjer and Rizescu, 2000) and the strong motion data provided by the National Institute for Building Research of Bucharest. The present work was supported as part of the PN II-TD Programme project contract no. 281/1.10.2007.

References

Aki, K. and Richards, P.G. (2002). *Quantitative Seismology*, 2nd edition. University Science Books, Sausalito, CA.

Bindi, D., Parolai, S., Spallarossa, D. and Catteneo, M. (2000). Site effects by H/V ratio: Comparison of two different procedures. J. Earthq. Eng., 4, 97–113.

Bonjer, K.-P. and Rizescu, M. (2000), Data Release 1996–1999 of the Vrancea K2 Seismic Network. Six CD's with evt-files and KMI v1-,v2-, v3-files. Karlsruhe-Bucharest, July 15, 2000.

Bonjer, K-P., Oncescu, M.-C., Driad, L. and Rizescu, M. (1999). A note on empirical site response in Bucharest, Romania, *Vrancea Earthquakes: Tectonics, Hazard, and Risk Mitigation* (edited by F. Wenzel and D. Lungu), Kluwer, pp. 149–162.

Chavez-Garcia, F.J., Pedoti, G., Hatzfeld, D. and Bard, P.-Y. (1990). An experimental study near Thessaloniki (Northern Greece). Bull. Seismol. Soc. Am., 80, 784–800.

Cioflan, C.O., Apostol, B.F., Moldoveanu, C.L., Panza, G.F. and Mărmureanu, G. (2004). Deterministic approach for the seismic microzonation of Bucharest. Pure Appl. Geophys., 161, 1149–1164.

Ding, Z., Chen, Y.T. and Panza, G.F. (2004). Estimation of site effects in Beijing City. PAGEOPH, 161, 1107–1123.

Field, E. and Jacob, K. (1995). A comparison of various site-response estimation techniques, including three that are not reference-site dependent. Bull. Seismol. Soc. Am., 85, 1127–1143.

Field, E.H., the SCEC Phase III Working Group (2000). Accounting for site effects in probabilistic seismic hazard analyses of Southern California: Overview of the SCEC Phase III report. Bull. Seismol. Soc. Am., 90, S1–S31.

Gusev, A., Radulian, M., Rizescu, M. and Panza, G.F. (2002). Source scaling for the intermediate-depth Vrancea earthquakes. Geophys. Int. J., 151, 879–889.

Hamzehloo, H., Vaccari, F. and Panza, G. F. (2006). Toward a reliable seismic microzonation in Tehran, Iran. Eng. Geol., 93, 1–16.

Kouteva, M., Panza, G.F., Romanelli, F. and Paskaleva, I. (2004). Modelling of the ground motion at Russe site (NE Bulgaria) due to the Vrancea earthquakes. J. Earthq. Eng., 8(2), 209–229.

Lachet, C. and Bard, P.-Y. (1994). Numerical and theoretical investigations on the possibilities and limitations of Nakamura's technique. J. Phys. Earth, 42, 377–397.

LeBrun, B., Hatzfield, D. and Bard, P.-Y. (2001). Site effect study in urban area: Experimental results in Grenoble (France). Pure Appl. Geophys., 158, 2543–2557.

Lermo, J. and Chavez-Garcia, F. (1993). Site effect evaluation using spectral ratios with only one station. Bull. Seismol. Soc. Am., 83, 1574–1594.

Lermo, J. and Chavez-Garcia, F.J. (1994). Are microtremor useful in site response evaluation? Bull. Seismol. Soc. Am., 84, 1350–1364.

Lokmer, I., Herak, M., Panza, G.F. and Vaccari, F. (2002). Amplification of strong ground motion in the city of Zagreb, Croatia, estimated by computation of synthetic seismograms. Soil Dyn. Earthq. Eng., 22, 105–113.

Mândrescu, N. (1978). The Vrancea earthquake of March 4, 1977 and the seismic microzonation of Bucharest. Proc. 2nd Inter. Conf. Microzonation, San Francisco, CA, 1, 399–411.

Mândrescu, N., Radulian, M. and Mărmureanu, G. (2004). Microzonation of Bucharest: Geology of the deep cohesionless deposits and predominant period of motion. Rev. Roum. Geophysique, 48, 120–130.

Mândrescu, N., Radulian, M. and Mărmureanu, G. (2007). Geological, geophysical and seismological criteria for local response evaluation in Bucharest urban area. Soil Dyn. Earthq. Eng., 27, 367–393.

Moldoveanu, C.L. and Panza, G.F. (1999). Modelling for microzonation purposes of the seismic ground motion in Bucharest, due to the Vrancea earthquake of May 30, 1990, *Vrancea Earthquakes: Tectonics, Hazard, and Risk Mitigation* (edited by F. Wenzel and D. Lungu), Kluwer, pp. 85–97.

Moldoveanu, C.L. and Panza, G.F. (2001). Vrancea source influence on local seismic response in Bucharest. Pure Appl. Geophys., 158, 2407–2429.

Moldoveanu, C.L., Marmureanu, G., Panza, G.F. and Vaccari, F. (2000). Estimation of site effects in Bucharest caused by the May 30–31, 1990, Vrancea seismic events. Pure Appl. Geophys., 157, 249–267.

Nakamura, Y. (1989). A method for dynamic characteristics of subsurface using microtremor on the ground surface, Quarter. Rep. Rail. Tech. Res. Inst., 30(1), 25–33.

Newmark, N. and Rosenblueth, E. (1971). *Fundamentals of Earthquake Engineering*, Prentice Hall, Englewood Cliffs, NJ.

Panza, G.F. (1985). Synthetic seismograms: The Rayleigh waves modal summation. J. Geophys., 58, 125–145.

Panza, G.F., Romanelli, F. and Vaccari, F. (2001). Seismic wave propagation in laterally heterogeneous anelastic media: Theory and applications to seismic zonation. Adv. Geophys., 43, 1–95.

Panza, G.F., Alvarez, L., Aoudia, A., Ayadi, A., Benhallou, H., Benouar, D., Bus, Z., Chen, Y., Cioflan, C., Ding, Z., El-Sayed, A., Garcia, J., Garofalo, B., Gorshkov, A., Gribovszki, K., Harbi, A., Hatzidimitriou, P., Herak, M., Kouteva, M., Kuznetzov, I., Lokmer, I., Maouche, S., Marmureanu, G., Matova, M., Natale, M., Nunziata, C., Parvez, I.A., Paskaleva, I., Pico, R., Radulian, M., Soloviev, A., Suhadolc, P., Szeidovitz, G., Triantafyllidis, P. and Vaccari, F. (2002). Realistic modeling of seismic input for megacities and large urban areas (the UNESCO/IUGS/IGCP project 414). Episodes, 25(3), 160–184.

Panza, G.F., Romanelli, F., Vaccari, F., Decanini, L. and Mollaioli, F. (2004). Seismic ground motion modelling and damage earthquake scenarios: A possible bridge between seismologists and seismic engineers, *Earthquake: Hazard, Risk, and Strong Ground Motion* (edited by Y.T. Chen, G.F. Panza and Z.L. Wu), Seismological Press, Beijing, pp. 323–349.

Paskaleva, I., Panza, G.F., Vaccari, F. and Ivanov, P. (2004). Deterministic modelling for microzonation of Sofia – An expected earthquake scenario. Acta Geol. Geophys. Hung., 39(2–3), 275–295.

Ritter, J.R.R., Balan, S.F., Bonjer, K.-P., Diehl, T., Forbinger, T., Marmureanu, G., Wenzel, F. and Wirth, W. (2005). Broadband urban seismology in the Bucharest metropolitan area. Seismol. Res. Lett., 76(5), 574–580.

Romanelli, F., Vaccari, F. and Panza, G.F. (2003). Realistic modelling of the seismic input: Site effects and parametric studies. JSEE, 5(3), 27–39.

Satoh, T., Kawase, H. and Matsushima, S. (2001). Differences between site characteristics obtained from Microtremors, S-waves, P-waves, and codas. Bull. Seismol. Soc. Am., 91, 313–334.

Sokolov, V., Bonjer K.-P. and Wenzel, F. (2004). Accounting for site effect in probabilistic assessment of seismic hazard for Romania and Bucharest: A case of deep seismicity in Vrancea zone. Soil Dyn. Earthq. Eng., 24, 929–947.

Yamanaka, H., Dravinski, M. and Kagami, H. (1993). Continuous measurements of microtremor on sediments and basement in Los Angeles, California. Bull. Seismol. Soc. Am., 63, 1227–1253.

Zare, M., Bard, P.-Y. and Ghafory-Ashtiany, M. (1999). Site characterizations for the Iranian strong motion network. Soil Dyn. Earthq. Eng., 18, 101–123.

Zaslavsky, Y., Shapira, A. and Arzi, A. (2000). Amplification effects from earthquakes and ambient noise in the Dead Sea rift (Israel). Soil Dyn. Earthq. Eng., 20, 187–207.

A CONTRIBUTION TO THE ASSESSMENT OF THE SEISMIC VULNERABILITY OF LARGE STRUCTURES IN SOFIA CITY

I. PASKALEVA[1][*], G. KOLEVA[1], F. VACCARI[2], E. ZUCCOLO[2], G. PANZA[3]

[1]*Central Laboratory for Seismic Mechanics and Earthquake Engineering, Bulgarian Academy of Sciences, 3 Acad. G. Bonchev str., 1113 Sofia, Bulgaria, Tel: +359-2-9793341, Fax: +359-2-9712407*

[2]*DST – University of Trieste, E. Weiss 4, 34127 Trieste, Italy, e-mail: vaccari@.units.it, Tel: +39-040-5582119, Fax: +39-040-5582111, Trieste, Italy*

[3]*DST – University of Trieste, E. Weiss 4, 34127 Trieste, Italy, e-mail: panza@units.it, Tel: +39-040-5582117, Fax: +39-040-22407334, ICTP-SAND group, Trieste, Italy*

Abstract. An advanced modeling technique that allows us to compute realistic synthetic seismograms was used to create a database of realistic synthetic accelerograms in a set of selected sites in the Sofia urban area. The accelerograms can be used to assess the local site response in terms of the response spectra ratio (RSR). The results of this study, i.e. time histories, response spectra and other ground motion parameters, can be used for various earthquake engineering analyses. Finally, with the help of 3-D finite element modeling, a building's structural performance is assessed.

Keywords: Sofia, seismic hazard, synthetic seismograms, damage estimates

1. Introduction

Seismic hazard and seismic risk are high in Sofia City and the surrounding area (Bonchev et al., 1982; Bulgarian Code, 1987). Tall buildings, big dams and

---

[*]Central Laboratory Seismic Mechanics and Earthquake Engineering (CLSMEE), Bulgarian Academy of Sciences (BAS), Acad. G. Bonchev block 3, 1113 Sofia, Bulgaria, e-mail: paskalev@geophys.bas.bg

A. Zaicenco et al. (eds.), *Harmonization of Seismic Hazard in Vrancea Zone,*   151
© Springer Science + Business Media B.V. 2008

other structures are subject to considerable social and environmental risks from strong seismic events that could generate casualties and huge economic losses. Seismic loading is one of the approaches to seismic risk reduction in engineering and building practices. The retrofitting of structures can be properly designed by means of seismic loading computations (Foutch, 2000) based on models derived from accurate geological and geo-technical site studies.

2. Site Dependent Seismic Motion Computation

According to the Bulgarian seismic code of 1987 (Bulgarian Code, 1987), Sofia is in a seismic category characterized by intensity IX (MSK) which corresponds to a horizontal acceleration of 0.27g for anchoring the elastic response spectrum. Detailed ground-motion modeling was done for possible earthquakes chosen according to the seismotectonic regime of the area. Broadband synthetic seismograms were computed in a laterally heterogeneous anelastic medium using the hybrid approach (Fäh et al., 1993; Panza et al., 2001). The modal part (Panza, 1985; Florsch et al., 1991) was used to model wave propagation from the source to the beginning of the local profile of interest. It is based on the computation of eigenvalues (phase velocity) and eigenfunctions associated with Rayleigh (P-SV motion) and Love (SH motion) waves for a laterally homogeneous medium (layered anelastic halfspace). The phase velocity-frequency space was efficiently explored (Panza et al., 2001) also at high frequencies where nearby modes get extremely close to one another. Anelasticity was taken into account for each layer considering a frequency-independent Q value that expressed the attenuation term in space and time. In the hybrid approach, the seismograms obtained at bedrock with the modal summation were the input for the propagation of the wavefield along the selected local profiles where the finite difference technique allowed us to deal with complicated lateral heterogeneities and therefore to adequately estimate site effects.

Using the specific knowledge about geology and geotechnical properties described in cartographic material available for the Sofia area, three profiles (local 2-D sections) were considered to compute several ground shaking scenarios with magnitude equal to 7 and with different source positions. One bedrock model was taken into account for propagating the wavefield from the source to the beginning of the 2-D sections, (Paskaleva et al., 2004a). This model was used for computing the reference signal (seismograms at bedrock) to be compared with the results of the detailed modeling for defining the site effects described by the ratios of the response spectra (2-D/bedrock). Three-component synthetic seismograms computed in the domains of displacement, velocity and acceleration were processed to extract parameters significant from the engineering point of view.

3. Background

Based on the earthquake history of Sofia and on available seismic hazard assessments provided in the literature, a computation was made with respect to a local seismic source that can strike just beneath the city. The earthquake epicenter corresponds to a real seismic event that struck Sofia in 1858. A generalized scheme of the profiles adopted for the numerical experiments is shown in Figure 1. The seismic wave propagation path consists of the path traveled between the source and the target site ("bedrock structure") and the target local cross sections. The data used to build up the local structural models, up to 1,000 m below the surface were obtained from a large set of boreholes and geological cross sections (Ivanov, 1997; Ivanov et al., 1998; Kamenov and Kojumdjieva, 1983; Petkov and Iliev, 1970). The deep part of the local model describing the structure below 1 km coincides with the bedrock velocity model available in the literature and is assumed to be the same for the whole Sofia Kettle (Stanishkova and Slejko, 1991).

For such a large event at relatively small distances – 10 km – the simple scaling by Gusev (1983, 2003) as reported by Aki (1987) is not capable of modeling the influence of the rupture process (Douglas, 2003). Therefore, the algorithm for simulating the source radiation from a fault of finite dimensions named PULSYN25 (PULse-based wide band SYNthesis) was applied (Gusev and Pavlov, 2006). The source was modeled as a rectangular fault plane with a grid of point subsources; the arrival of the rupture front at a sub source switches its slip. Using random parameters like the spatial distribution of slip and rupture velocity, the program PULSYN25 generates a source (phase and amplitude) spectrum that is close in amplitude to Gusev's (1983) empirical curves and reproduces directivity effects. To compute the synthetic seismogram at a specific site, the PULSYN25 spectra were used to scale the seismograms generated by the hybrid method. This approach produces broadband signals that reproduce directivity effects and the rupture process of the source.

Realistic synthetic seismograms were computed for all the sites of interest along the profiles Sofia 2 (2-2) and Sofia 3 (3-3) shown in Figure 1. By "realistic" we mean that the modeling simultaneously takes into account the geotechnical properties of the site, the position and the geometry of the seismic source and the mechanical properties of the propagation medium following basic laws of physics and avoiding standard convolutive approaches that, as described in Paskaleva et al. (2004b, 2007), have quite questionable validity. Acceleration, velocity and displacement time histories were obtained for transverse (TRA), radial (RAD) and vertical (VER) components considering three different azimuths (0°, 90°, 180°) with respect to rupture propagation.

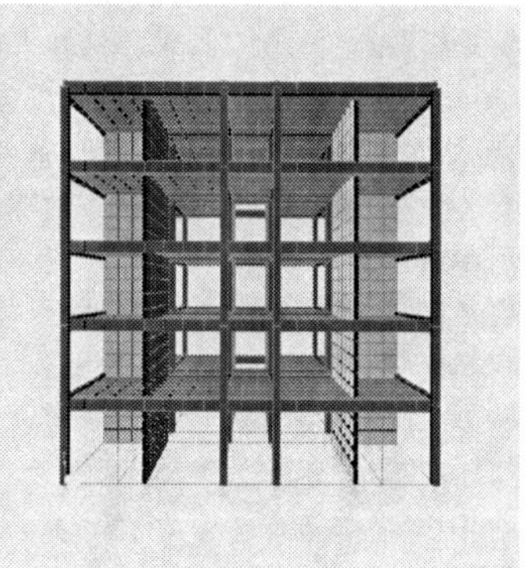

Figure 1: Sofia area with the three profiles superimposed ("Sofia 1", "Sofia 2", "Sofia 3"), (Koleva et al., 2008)

Figure 2: Toggle view of the 3-D building model

This work is focused on the response of a 3-D building (Figure 2) placed on profiles Sofia 2 (Figure 3) and Sofia 3 (Figure 4) (2-2, 3-3 see also Figure 1).

Building calculations were carried out for two characteristic sites (at epicentral distances 13 km [Rec. 30] and 15 km [Rec. 50]). For building response calculations, the acceleration time histories scaled according to the Bulgarian code of 1987 were used.

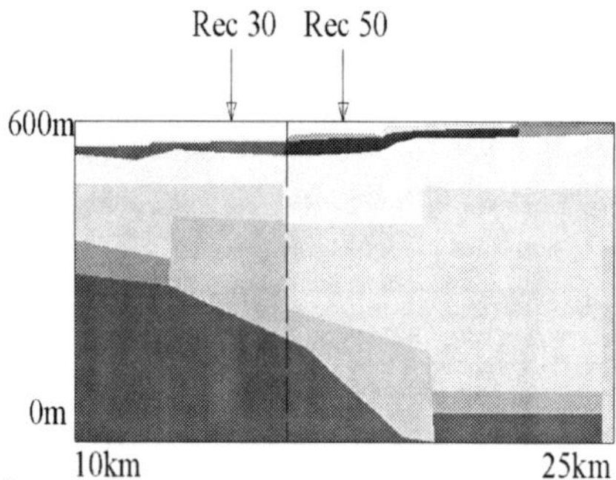

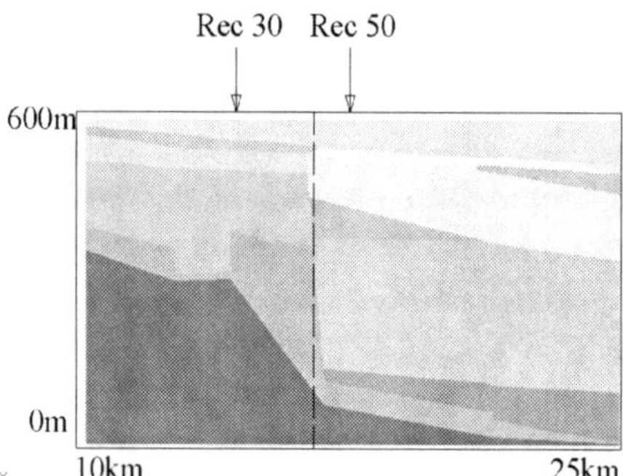

Figure 3: Laterally varying 2-D model Sofia 2 that crosses the town in NW-SE direction (profile 2-2 see Figure 1)

Figure 4: Laterally varying 2-D model Sofia 2 that crosses the town in NE-SW direction (profile 3-3 see Figure 1)

4. Regional Structural Model

The input data necessary for simulating ground motion using the hybrid approach consist of the regional bedrock model, the laterally heterogeneous local model and the earthquake source model. Following the investigations carried out on the recent geodynamic features of the Sofia complex, a geological and geophysical 3-D model of the Earth's crust for this region was derived (Shanov et al., 1998). The analysis of the seismotectonic setting and of the structure of the basement of the Sofia depression was used to specify the reference structural model. The P-wave velocities of the 1-D regional structural model were taken from Shanov et al. (1998), and to be conservative, for the S-wave velocities we assumed $V_P = 2V_S$. This leads to slightly lower S-wave velocities than the ones valid for Poissonian solids; therefore higher ground motion is expected at the free surface.

No specific information exists about the quality factor for P-waves (Qp) for the region, so the values adopted were taken from Dziewonski and Anderson (1981), and the widely applied rule Qp = 2.2*Qs, Stein and Wysession (2003) was used to derive the quality factor for S-waves (Qs).

5. Ground Shaking Scenario for the Profiles

The maximum macroseismic intensity at Sofia – I = IX (MSK) observed in 1858 (Bonchev et al., 1982) – can be expected to occur within a period of 150 years (Christoskov et al., 1989), i.e. it could correspond to a strong scenario earthquake. Recently, seismic hazard maps of the Circum-Pannonian Region (Gorshkov et al., 2000), Panza and Vaccari (2000) show that Sofia is in a node with the potential for an earthquake with M > 6.5 and that it could suffer macroseismic intensity up to X. The intensity expected in the Sofia region is associated with earthquakes occurring in the upper 17–20 km of the lithosphere. Macroseismic intensity I of about IX can be expected in Sofia (Glavcheva, 1990) if an earthquake with magnitude M = 7 located at 10–15 km from the center of the city in the direction west-south (Alexiev and Georgiev, 1997; Christoskov et al., 1989; Slavov et al., 2004; Matova, 2001; Solakov et al., 2001) occurs. In modeling, the source parameters were chosen to approximate the seismic event that hit Sofia in 1858. The parameters of the source mechanism adopted were strike angle 340°, dip angle 78° and rake angle (with respect to strike) 285° (Paskaleva et al., 2007).

6. Building Model

The building considered is a five-story reinforced concrete building (H = 15 m) with fundamental period T = 0.40 s. The construction is regular both in plan and in elevation (L = 22 m; B = 15 m). Modeling and calculations were performed with the help of SAP2000 (Computers and Structures, 2000). Concrete B20 with material properties Young's modulus $E = 2.75 \times 10^7$ kN/m^2 and Poisson's ratio 0.2 was chosen.

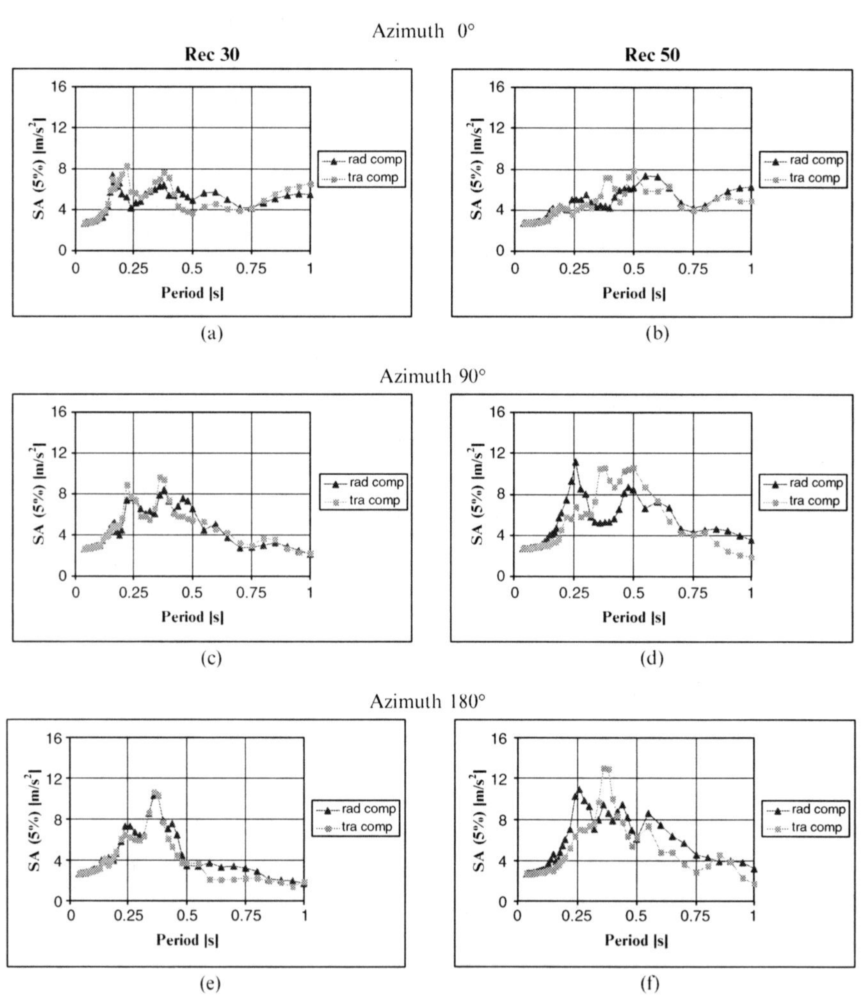

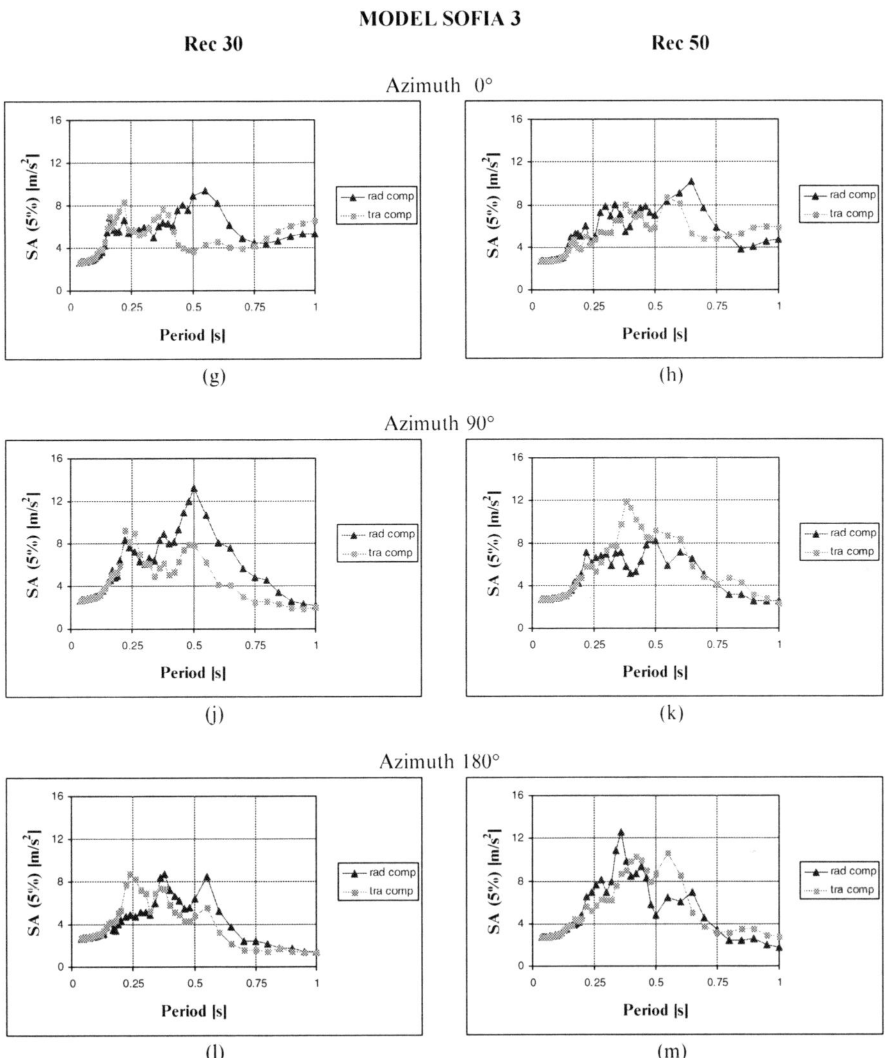

Figure 5: Input spectral acceleration response (PULSYN approach) at the base level at Rec 30 (**a**), (**c**), (**e**), (**g**), (**j**), (**l**) and Rec 50 (**b**), (**d**), (**f**), (**h**), (**k**), (**m**)

The analysis case was defined as a linear modal time history. The load is applied as acceleration time histories in both horizontal directions (RAD Figure 4 and TRA Figure 8). The time histories are first normalized to the maximum of the accelerograms and then normalized to the Bulgarian code of 1987 (Bulgarian Code, 1987) with maximum horizontal acceleration for the area of

Sofia City equal to 2.7 m/s^2. The respective acceleration response spectra of the input accelerograms are given in Figures 5 and 6 The output extracted from the building response represents the influence of the geological site conditions along Model 2 on one building representative of Sofia building stock. The output response spectrum curves with a damping value of 5% (spectral acceleration versus period), for each of two sites are shown in Figure 6a, c, e, g, j, l for Rec. 30 and Figure 6b, d, f, h, k, m for Rec. 50. It can be seen that the peak value is approximately equal to 0.4 s which matches the fundamental period of the construction.

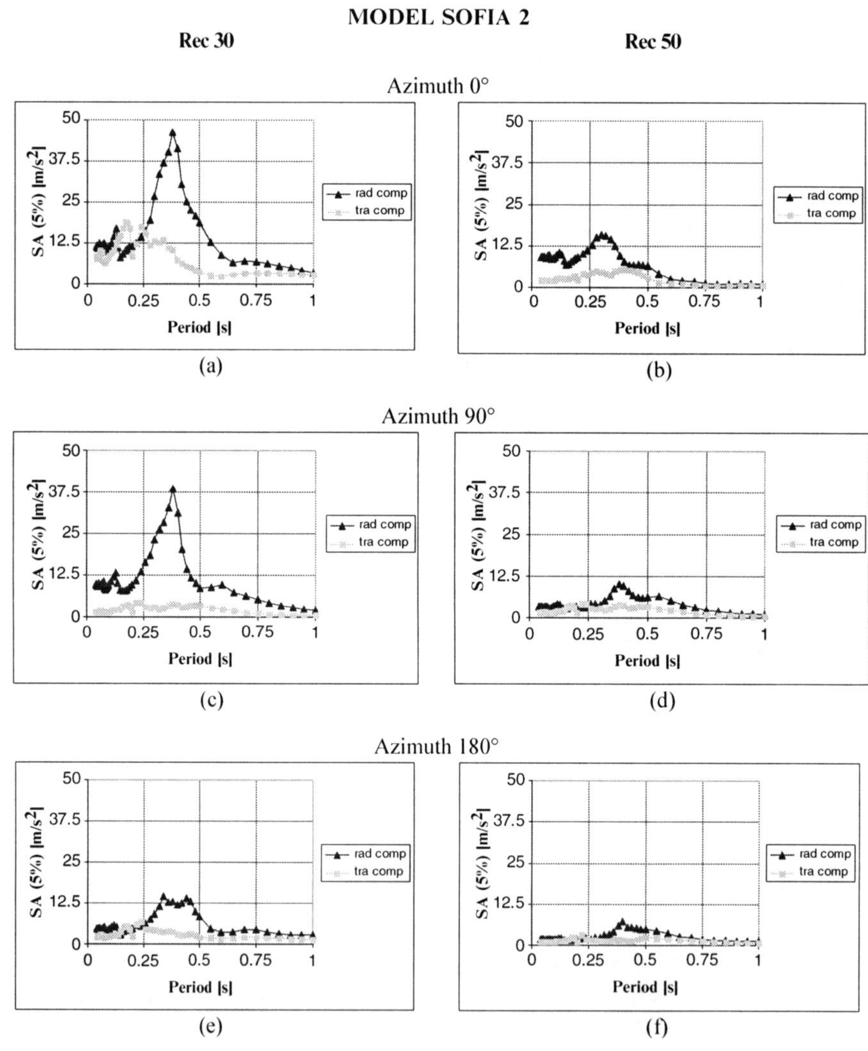

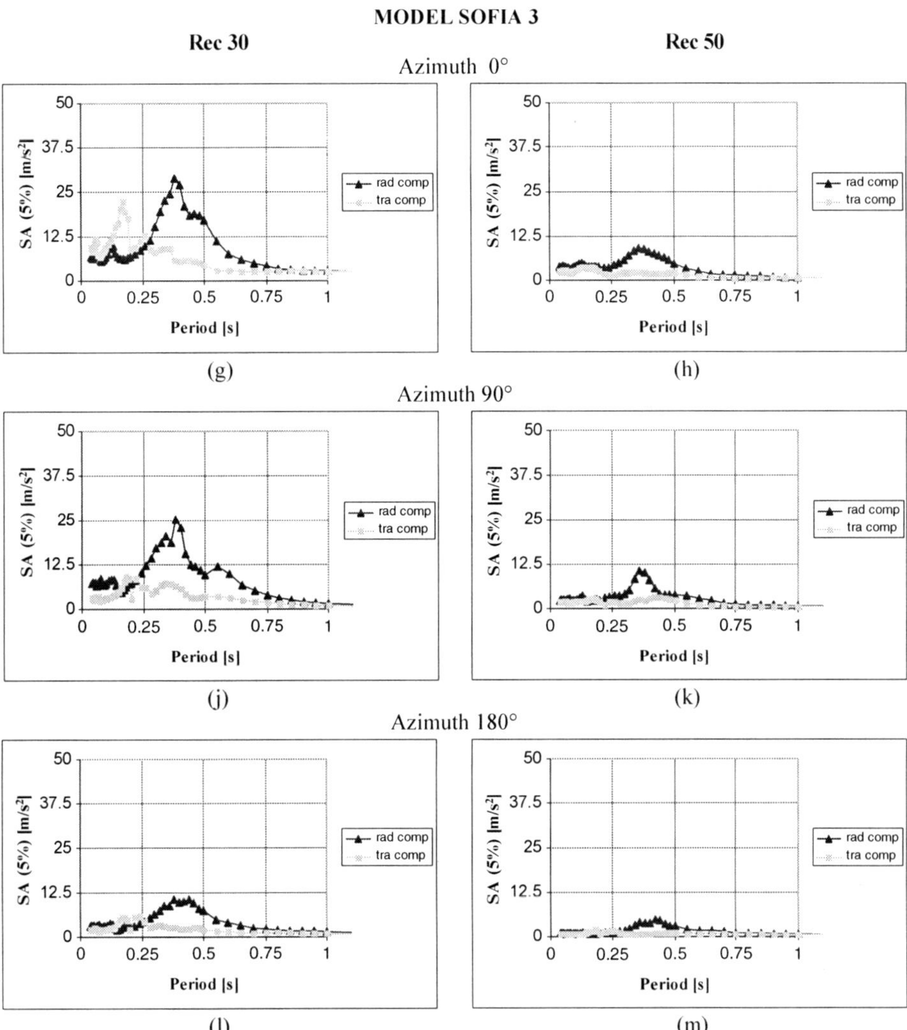

Figure 6: Output roof spectral acceleration response from 3-D model at Rec 30 (**a**), (**c**), (**e**), (**g**), (**j**), (**l**) and Rec 50 (**b**), (**d**), (**f**), (**h**), (**k**), (**m**)

Figure 7 is a map of the central part of Sofia. The distribution of the maximum dynamic factor (DF = SA(T)roof/SA(T)base) over the area of about 25 km^2 is shown in Figure 8 for azimuth 0° (a), 90° (b) and 180° (c). On the vertical axis Z, the maximum dynamic factor (DF) is limited to 7 for the three azimuths. Looking at Figure 8a, b, c we can see that the maximum dynamic factor (DF) is detected for azimuth 0° while the minimum dynamic factor is reached for azimuth 180°. The dynamic factor of 7 is expected value for linear behavior of structures.

7. Conclusions

Synthetic accelerations generated in a laterally heterogeneous anelastic model representative of the local site conditions in Sofia were used as seismic input for a typical building in the city.

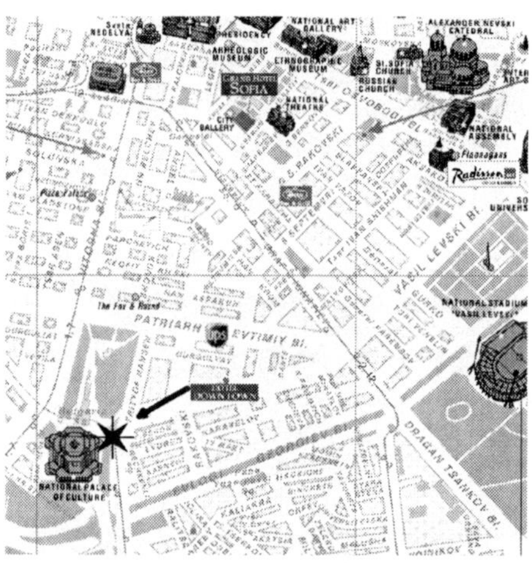

Figure 7: Map of central part of the city covered by the models Sofia 2 and Sofia 3

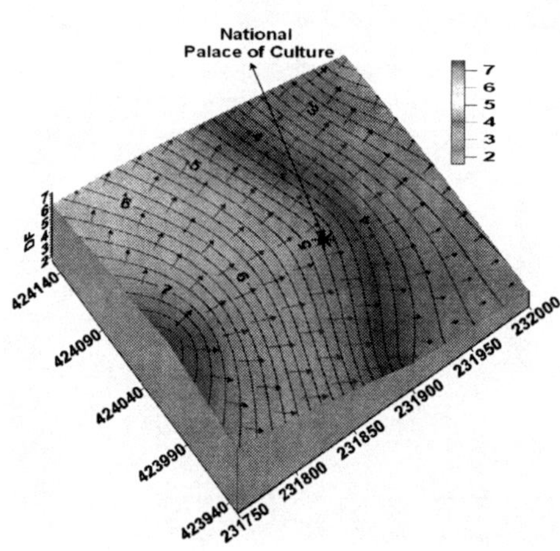

Figure 8a: Dynamic factor (DF) distribution for azimuth 0°

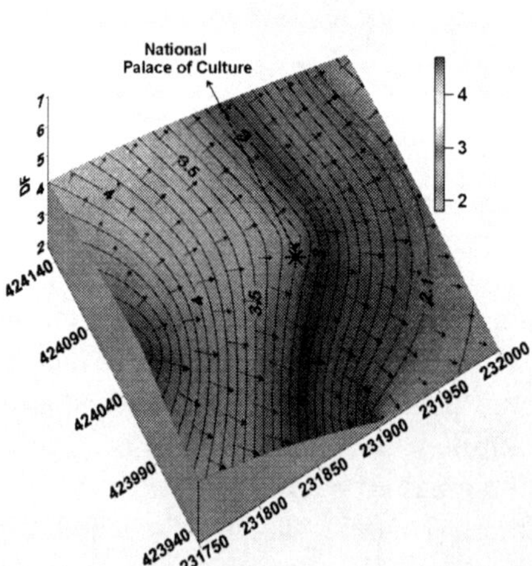

Figure 8b: Dynamic factor (DF) distribution for azimuth 90°

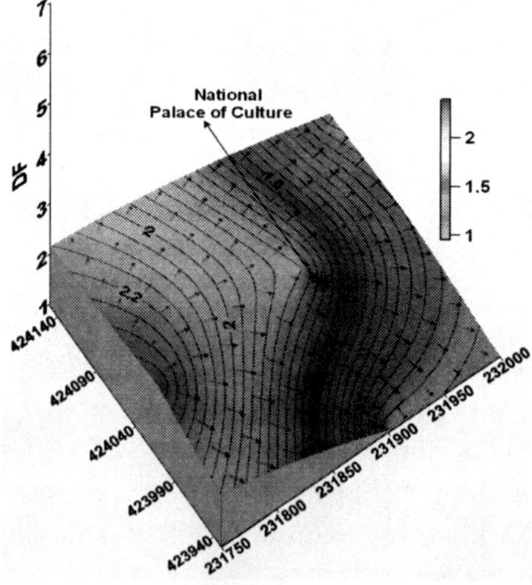

Figure 8c: Dynamic factor (DF) distribution for azimuth 180°

The database constructed from generated time histories and seismic response spectra can assist local authorities in minimizing future losses as it helps properly design earthquake-resistant buildings. For building use, it is necessary to map seismic hazard zones to regulate the seismic safety of new constructions. Local governments must require site-specific seismic hazard evaluations to validate the hazard level to allow appropriate recommendations for mitigation.

Acknowledgements

This research was performed in the framework of the bilateral cooperation between DST-UNITS, Trieste, Italy and CLSMEE-BAS, Sofia, Bulgaria. Financial support from CEI-SAND fellowship, ICTP, Trieste and NATO project SfP-980468 is gratefully acknowledged.

References

Aki, K., Strong motion seismology. In: Strong ground motion seismology, NATO ASI Series, Series C: Mathematical and physical sciences (M. Erdik and M. Toksoz, eds.), D. Reidel, Dordrecht, Vol. 204, pp. 3–39, 1987.

Alexiev, G., Georgiev, Tz., Morphotectonics and seismic activity of the Sofia depression, J. Prob. Geogr., 1–2, 60–69, 1997.

Bonchev, E., Bune, V., Christoskov, L., Karagyuleva, J., Kostadinov, V., Reisner, G., Rizikova, S., Shebalin, N., Sholpo, V., Sokerova, D., A method for compilation of seismic zoning prognostic maps for the territory of Bulgaria, Geol. Balkanica, 12, 3–48, 1982.

Bulgarian Code for Design of Structures in Seismic Regions, Sofia, Bulgaria, 1987.

Christoskov, L., Georgieva, Tzv., Deneva, D., Babachkova, B., Proceedings of the 4th International Symposium on the Analysis of Seismicity and Seismic Risk, Bechyne castle, CSSR, pp. 448–454, 1989.

Computers and Structures, SAP Manual Integrated Finite Element analysis and design of structures, Inc, Berkeley, CA, 2000.

Douglas, J., Earthquake ground motion estimation using strong-motion records: A review of equations for the estimation of peak ground acceleration and response spectral ordinates, Earth-Sci. Rev., 61(1–2), 43–104, 2003.

Dziewonski, A., Anderson, D., Preliminary reference earth model, Phys. Earth Planet. Int., 25, 297–356, 1981.

Fäh, D., Suhadolc, P., Panza, G.F., Variability of seismic ground motion in complex media: The Friuli area (Italy). In: Geophysical exploration in areas of complex geology, II (R. Cassinis, K. Heibig and G.F. Panza, eds.), J. Appl. Geophys., 3, 131–148, 1993.

Florsch, N., Fäh, D., Suhadolc, P., Panza, G.F., Complete synthetic seismograms for high-frequency multimode SH-waves, PAGEOPH, 136, 529–560, 1991.

Foutch, D.A., State-of-the-Art Report on Performance Prediction and Evaluation of Moment-Resisting Steel Frame Structures, FEMA 355, Federal Emergency Management Agency, Washington, DC, 2000.

Glavcheva, R. Parametrization of the isoseismals from Bulgarian earthquakes. Bulg. Geoph. Jurn., 16(4), 38–45, 1990. (in Bulgarian).

Gorshkov, A., Kuznetzov, I., Panza, G.F., Soloviev, A., Identification of future earthquake sources in the Carpatho-Balkan orogenic belt using morphostructural criteria. Seismic hazard of the circum-pannonian region, PAGEOPH Topical Vols., Birkhauser Verlag, 79–85, 2000.

Gusev, A.A., Descriptive statistical model of earthquake source radiation and its application to an estimation of short period strong motion, Geophys. J. R. Astron. Soc., 74, 787–800, 1983.

Gusev, A.A., A program PULSYN02 for wide-band simulation of source radiation from a finite earthquake source/fault, Abdus Salam ICTP, Trieste, Italy, 2003.

Gusev, A.A., Pavlov, V., Wideband simulation of earthquake ground motion by a spectrum-matching, multiple-pulse technique. First European Conference on Earthquake Engineering and Seismology, Geneva, Switzerland, 3–8 September, Paper Number: 408, 2006.

Ivanov, Pl., Proceedings of the International IAEG Conference, Athens, Balkema, Rotterdam, pp. 1265–1270, 1997.

Ivanov, P., Frangov, G., Yaneva, M., Distribution, composition and properties of quaternary deposits in Sofia kettle. XVI Congress Carpathian-Balkan Geological Association, August 30–September 2, Vienna, Austria, Abstracts, p. 328, 1998.

Kamenov, B., Kojumdjieva, N., Stratigraphy of the Neogene in Sofia Basin. Paleontology, stratigraphy and lithology, 18, 69–85, 1983.

Koleva, G., Vaccari, F., Paskaleva, Iv., Zuccolo, E., Panza, G., An approach of microzonation of Sofia city, Acta. Geod. Geoph. Hung. 2008 (in print).

Matova, M., Recent manifestations of seismotectonic activity in Sofia region and their land subsidence potential. Proceedings of the Final Conference of UNESCO-BAS Project on land subsidence, June 27–30, 2001, Sofia, pp. 93–98, 2001.

Panza, G.F., Synthetic seismograms: The Rayleigh waves model summation. J. Geophys., 58, 125–145, 1985.

Panza, G.F., Vaccari, F., Introduction in seismic hazard of the Circum-pannonian region (G.F. Panza, M. Radulian and C. Trifu, eds.), PAGEOPH Topical Volumes, Birkhauser Verlag, 5–10, 2000.

Panza, G.F., Romanelli, F., Vaccari, F., Seismic wave propagation in laterally heterogeneous media: Theory and applications to seismic zonation, Adv Geophys., 43, 1–95, 2001.

Paskaleva, I., Panza, G.F., Vaccari, F., Ivanov, P., Deterministic modeling for microzonation of Sofia – An expected earthquake scenario, Acta. Geod. Geoph. Hung., 39(2–3), 275–295, 2004a.

Paskaleva, I., Matova, M., Frangov, G., Expert assessment of the displacement provoked by seismic events: Case study for the Sofia metropolitan area (Panza, Paskaleva, Nunziata, eds.), PAGEOPH Topical Volume 161, N5/6, Birkhauser Verlag, 1265–1283 (19) 2004b.

Paskaleva, I., Dimova, S., Panza, G.F., Vaccari, F., An earthquake scenario for the microzonation of Sofia and the vulnerability of structures designed by use of the Eurocodes, J. Soil Dyn. Earthq. Eng., 27(11), 1028–1041, 2007.

Petkov, P., Iliev, I., Proceedings of the 3rd European Symposium on EE, Sofia, pp. 79–86, 1970.

Shanov, S., Tzankov, Tz., Nikolov, G., Bojkova, A., Kurtev, K., Character of the Recent Geodynamics of Sofia Complex Graben, Rev. Bulg. Geol. Soc., 59, part I, 3–12, 1998.

Slavov, Sl., Paskaleva, I., Kouteva, M., Vaccari, F., Panza, G.F., Deterministic earthquake scenarios for the city of Sofia, PAGEOPH, Birhauser Verlag, Basel, 161, 1221–1239, 2004.

Solakov, D., Simeonova, S., Christoskov, L., Seismic hazard assessment for the Sofia area, Annali di Geofisica,44, 541–556, 2001.

Stanishkova, I., Slejko, D., Seismic Hazard of Bulgarian Cities, Atti Del 10 Convegno Annuale del Gruppo Nazionale di geofisica della solisda, Roma, 1991.

Stein, S., Wysession, M., An introduction to seismology, earthquakes, and earth structure, Blackwell, Oxford, 2003.

HYBRID MS-BIEM FOR SEISMIC SITE-RESPONSE PHENOMENA: A CASE STUDY OF SOFIA

P. DINEVA[1], I. PASKALEVA[2*], C. LA MURA[3], G. PANZA[4]

[1]*Institute of Mechanics, Bulgarian Academy of Sciences (BAS), Acad. G. Bonchev block4, 1113 Sofia, Bulgaria*
[2]*Central Laboratory Seismic Mechanics and Earthquake Engineering (CLSMEE), Bulgarian Academy of Sciences (BAS), Acad. G. Bonchev block 3, 1113 Sofia, Bulgaria*
[3]*DST-University of Trieste, via E. Weiss 4, 34127 Trieste, Italy*
[4]*The Abdus Salam ICTP, Trieste, Italy*

Abstract. The study presents and solves the 2-D elastodynamic model for seismic in-plane wave propagation in laterally inhomogeneous geological profiles imbedded in a vertically inhomogeneous half-space in which an earthquake source is buried. To this end, an efficient hybrid modal summation-boundary integral equation method (MSM-BIEM) is developed and applied. The MSM is used as a tool for simulating wave propagation from the source position to the multilayered laterally inhomogeneous geological profile where the BIEM is applied. The proposed model and the hybrid tool are used to investigate the phenomena of site effects. In fact, such a methodology has the potential to investigate the combined effects of different physical phenomena like surface topography, lateral inhomogeneity and the existence of water saturation in soils on the estimation of site effects. The model and hybrid computational tool developed are applied to contribute to the seismic risk analysis of the Bulgarian capital Sofia.

Keywords: Lateral inhomogeneity, saturated soils, viscoelastic isomorphism, hybrid technique, site effects

---

*Central Laboratory Seismic Mechanics and Earthquake Engineering (CLSMEE), Bulgarian Academy of Sciences (BAS), Acad. G. Bonchev block 3, 1113 Sofia, Bulgaria, e-mail: paskalev@geophys.bas.bg

A. Zaicenco et al. (eds.), *Harmonization of Seismic Hazard in Vrancea Zone,*
© Springer Science + Business Media B.V. 2008

1. Introduction

A major factor influencing the extent of damage on structures is the phenomenon of local site effects on earthquake ground motion. The term site effect is used here to represent the differences in earthquake ground motion between two nearby sites. In most situations, local amplification can be inferred reasonably by one-dimensional (1-D) models. However, lateral heterogeneity may give rise to very complex wave patterns like, for instance, focusing/ defocusing and locally generated surface waves.

A way to shed some light on the understanding of the site-response phenomena is to develop high-performance methods for simulating seismic wave propagation in complex sites. In these so-called hybrid computational schemes which are currently very popular, the numerical methods are applied only in the laterally heterogeneous part of the medium that is a small part of the total model, whereas the laterally homogeneous part, the bedrock, is treated by analytical methods as ray methods, modal summation methods (MSM) and their modifications. To the authors' knowledge, in the literature the hybrid computational schemes used are mainly based on the finite difference method and MSM, e.g. see Fah et al. (1993a, b) and Panza et al. (2001). Regarding deterministic modeling for the seismic microzonation of Sofia (see Figure 1), three main references can be cited: Paskaleva (2002), Paskaleva et al. (2003) and Slavov et al. (2004). A hybrid approach based on the boundary integral equation method (BIEM) that permits dealing with the properties of wave paths and laterally inhomogeneous local multilayered profiles is discussed by Dineva et al. (2003) where the accuracy and validation study of the proposed method is discussed by solving different reference 1-D and 2-D wave propagation problems. In spite of the well-known advantages of the BIEM, there is a lack of hybrid methods that are based on it.

The aim of this study is (i) to evaluate the phenomena of site effects due to impedance contrasts, surface topography and lateral inhomogeneity of layers, the body wave conversion into surface waves and the existence of saturated soil layers by the hybrid modal summation-BIEM (MS-BIEM) developed and (ii) to apply the proposed hybrid tool to possible seismic scenarios thus contributing to the seismic risk assessment of the city of Sofia.

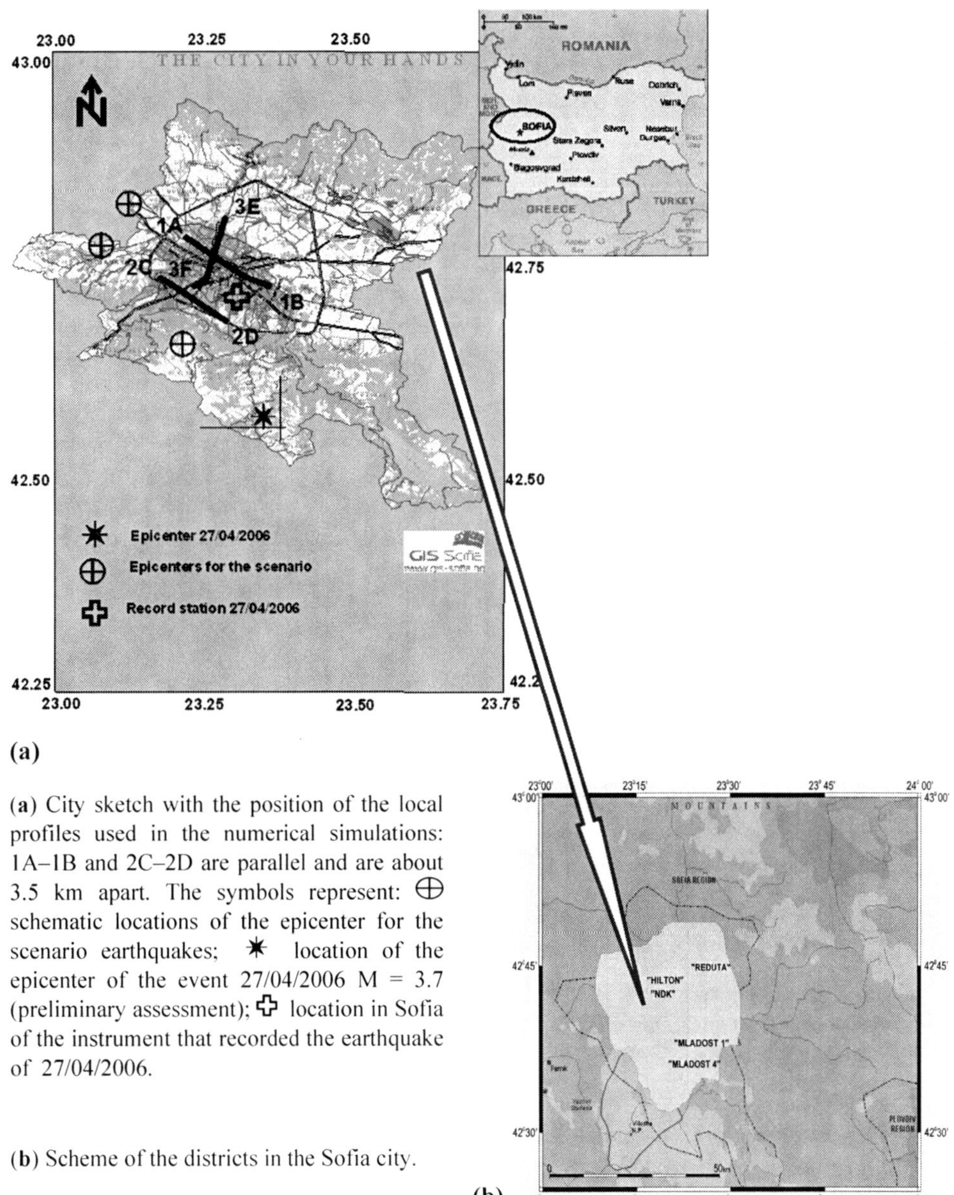

(a)

(a) City sketch with the position of the local profiles used in the numerical simulations: 1A–1B and 2C–2D are parallel and are about 3.5 km apart. The symbols represent: $\oplus$ schematic locations of the epicenter for the scenario earthquakes; $*$ location of the epicenter of the event 27/04/2006 M = 3.7 (preliminary assessment); ✚ location in Sofia of the instrument that recorded the earthquake of 27/04/2006.

(b) Scheme of the districts in the Sofia city.

(b)

Figure 1: (**a**) City sketch with the position of the local profiles used in the numerical simulations: 1A–1B and 2C–2D are parallel and are about 3.5 km apart. The symbols represent: $\oplus$ schematic locations of the epicentre for the scenario earthquakes; $*$ location of the epicentre of the event 27/4/2006 M = 3.7 (preliminary assessment); ✚ location in Sofia of the instrument that recorded the earthquake of 27/4/2006; (**b**) Scheme of the districts in the Sofia city

2. Problem Formulation

Consider a 2-D wave propagation problem in a finite saturated geological region:

$$\Omega = \bigcup_{i=1}^{N} \Omega_i$$

with N non-parallel layers, Ω_i and free surface topography imbedded in a vertically inhomogeneous half-space, Ω_0, where a seismic source is buried as shown in Figure 2. The common boundary between the finite region Ω and the half-space Ω_0 is denoted by Λ .

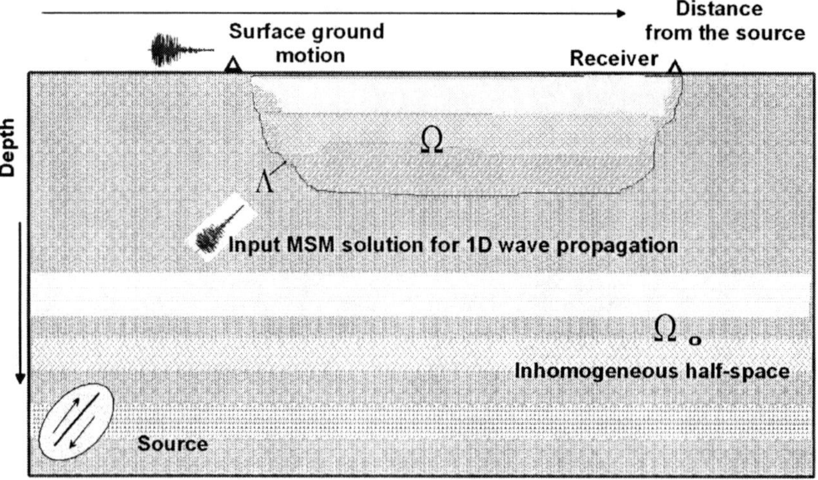

Figure 2: Schematic representation of the hybrid technique. The MSM is applied considering the bedrock model to compute the input signals for the laterally varying part where the signals are obtained by BIEM at a set of sites

We consider the vertical variation of the mechanical properties within Ω_0 modeled with a series of M homogeneous flat layers Γ_i, parallel to the free surface overlaying the homogeneous half-space i.e.

$$\Omega_0 = \bigcup_{k=1}^{M} \Gamma_k$$

The local region considered occupies part of the first layer Γ_1 of the vertically inhomogeneous half-space. All soil layers in the domain Ω_0 are isotropic, anelastic and homogeneous, while the non-parallel layers in the finite geological area Ω are water saturated and their poroelastic properties are described by the visco-elastic isomorphism of the Bardet (1992) model.

The aim is to obtain synthetic seismograms at some points (sites) along the free surface of the local region Ω.

The problem can be split into two sub-problems: (a) the first (external) sub-problem concerns wave propagation in a vertically inhomogeneous domain from the seismic source to the boundary Λ between the finite region Ω and the inhomogeneous half-space Ω_0, (see Figure 2); (b) the second (internal) sub-problem deals with seismic wave propagation in the finite soil stratum with non-parallel layering and surface or subsurface relief.

A plane strain state is assumed in the (x, y) plane, i.e. two displacement components are nonzero and independent of the coordinate z : $u_x = u_x(x,y,t)$, $u_y = u_y(x,y,t)$, $u_z = 0$. The governing equations are Lame-Navier partial differential equations:

$$\left(\alpha_l^2 - \beta_l^2\right)u_{j,ji}\left(x,y,t\right) + \beta_{(l)}^2 u_{i,jj}\left(x,y,t\right) = \ddot{u}_i\left(x,y,t\right) \quad \text{in } Q_B = \Omega_B \times (0,T) \quad (1)$$

where

$$\Omega_B = \Omega \bigcup \Omega_0$$

and commas indicate spatial derivatives; α_l, β_l are complex-valued body wave velocities of: (*) each anelastic l-th layer in Ω_0 (see Panza et al., 2001), (**) each l-th soil saturated layer in Ω, denoted as C_P^{*l}, C_S^{*l} (see Bardet, 1992); T is the duration of the seismic load; u_i is the displacement; $\ddot{u}_i$ is the acceleration.

In order to exclude the time variable and to solve the boundary-value problem in the frequency domain, the Fourier transform is applied to the time variable. The governing partial differential equation becomes an elliptic one.

Based on Simon's et al. (1984) idea that poroelastic and viscoelastic materials have similar dynamic behavior, Bardet (1992) introduced a poroelastic-viscoelastic isomorphism by equating the wave numbers in Biot's (1956) poroelastic model with the viscoelastic ones. It results in equivalence between the poroelastic material and the linear Kelvin-Voigt viscoelastic material with velocities and attenuations of longitudinal and shear waves related to the poroelastic parameters of Biot's model as follows:

$$C_P = \sqrt{\frac{P + 2Q + R}{\tilde{\rho}}} \qquad\qquad C_S = \sqrt{\frac{\mu}{\tilde{\rho}}} \qquad\qquad (2)$$

$$\xi_P = \frac{\tilde{\rho}}{b}\left(\frac{Q+R}{P+2Q+R}\,\frac{n\rho_f}{\tilde{\rho}}\right)^2 \qquad\qquad \xi_S = \frac{\tilde{\rho}}{b}\left(\frac{n\rho_f}{\tilde{\rho}}\right)^2 \tag{3}$$

Here: ξ_P and ξ_S are the corresponding attenuation coefficients that represent the small hysteretic damping ratios for P- and SV-wave velocities due to soil saturation, and P, Q and R are Biot's elastic constants for saturated soils:

$$P = \frac{3(1-\upsilon)}{1+\upsilon}K_{dry} + \frac{Q^2}{R}; \quad \tilde{\rho} = (1-n)\rho_g + n\rho_f; \quad K_{dry} = \frac{2}{3}\frac{\mu(1+\upsilon)}{1-2\upsilon} \tag{4}$$

$$Q = \frac{n\left(1-n-\dfrac{K_{dry}}{K_g}\right)}{\left(1-n-\dfrac{K_{dry}}{K_g}+n\dfrac{K_g}{K_f}\right)}K_g \qquad R = \frac{n^2}{\left(1-n-\dfrac{K_{dry}}{K_g}+n\dfrac{K_g}{K_f}\right)}K_g \tag{5}$$

where: υ is Poisson's ratio, n is porosity, K_{dry}, K_g, K_f, are the bulk modulus of dry solid-skeleton, solid grain and fluid, ρ_g and ρ_f are the solid grain and fluid densities,

$$\mu = \mu_{dry} = \mu_{sat} = \frac{3(1-2\upsilon)}{2(1+\upsilon)}K_{dry}$$

is the shear modulus,

$$b = n^2\frac{g\rho_f}{\hat{k}}$$

is the dissipative coefficient, g is the gravity acceleration, k is the hydraulic conductivity that varies for most of the soils in the interval $10^{-10}:10^{-2}[m/s]$.

Morochnik and Bardet (1996) obtained the approximated expressions (2)–(3) when the frequencies satisfy the condition $(\omega\tilde{\rho}/b) << 1$. Due to the fact that the hydraulic conductivity k has small values for most soils (for sand $k = 10^{-6}:10^{-4}$), this condition is fulfilled for the frequencies encountered in earthquake engineering problems. The proposed equivalent viscous material has the possibility to account for a stiffener of the soil as far as the pore pressure induced by the seismic load resists compression and stiffens the rock in its dilatational deformation.

The applicability of Bardet's (1992) viscoelastic isomorphism to the solution of seismic wave propagation problems in a fluid saturated half-space is discussed in Dineva et al. (2006) by comparing the results for free-field motion in a homogeneous half-space subject to plane longitudinal P-waves when soil properties are described by both the Bardet model and the Biot model.

The limitations of the Bardet model are: (a) it is valid only for the frequency diapason $\omega < 0.1(b/\tilde{\rho})$; (b) it cannot account for the second (slow) longitudinal wave; (c) it is not possible to account for the boundary conditions for the pore fluid pressure since one-phase material is considered.

The boundary-value-problem consisting of the governing wave equation and the boundary conditions in the frequency domain need to be solved for a finite set of discrete values of the frequency ω. Once this is done for a sufficient number of values of ω, a numerical inversion of the relevant variables must be performed to obtain the time-dependent solution.

The boundary conditions are the following: (a) the traction at the free surface is zero; (b) the displacement and traction between layers are continuous; (c) compatibility and dynamic equilibrium conditions for displacement and traction at the boundary Λ between Ω_0 and Ω hold; (d) the Sommerfeld radiation condition is at infinite.

3. Hybrid MS-BIEM for Problem Solution

The total wave field can be written as $u_i = u_i^{fr} + u_i^{sc}$ and $p_i = p_i^{fr} + p_i^{sc}$ where u_i^{fr}, p_i^{fr} denote the free-field displacements and tractions that solve the 1-D wave propagation problem in a vertically inhomogeneous, anelastic half-space. Inside the local geological profile Ω we have $u_i^{sc} = u_i$ and $p_i^{sc} = p_i$, while outside the local soil stratum in the half-space Ω_0, the scattered wave field is $u_i^{sc} = u_i - u_i^{fr}$ and $p_i^{sc} = p_i - p_i^{fr}$.

The equivalent BIEM formulation of the above posed with a partial differential equation boundary-value problem is presented by the following system of boundary integral equations:

$$C_{ij}u_j(x,y,\omega) =$$

$$\int_{\Gamma_{\Omega_m}} U_{ij}^*(x,y,x_0,y_0,\omega)p_j(x_0,y_0,\omega)d\Gamma - \int_{\Gamma_{\Omega_m}} P_{ij}^*(x,y,x_0,y_0,\omega)u_j(x_0,y_0,\omega)d\Gamma$$

$$(6)$$

and

$$C_{ij}\left(u_j\left(x,y,\omega\right)-u_j^{fr}\left(x,y,\omega\right)\right)=$$

$$\int_\Lambda U_{ij}^*(x,y,x_0,y_0,\omega)\left(p_j(x_0,y_0,\omega)-p_j^{fr}(x_0,y_0,\omega)\right)d\Gamma - \tag{7}$$

$$\int_\Lambda P_{ij}^*(x,y,x_0,y_0,\omega)\left(u_j(x_0,y_0,\omega)-u_j^{fr}(x_0,y_0,\omega)\right)d\Gamma$$

here: $m = 1,2,3...N$; C_{ij} are jump terms that depend on the local geometry at the collocation point (x,y), while nodal points (x_0, y_0) are position vectors for the field points on boundary Γ_{Ω_m} of layer Ω_m; u_j and p_j are the unknown total displacements and total tractions on the boundaries and Γ_{Ω_m}; U_{ij}^*, P_{ij}^* are the displacement and traction frequency-dependent fundamental solutions of 2-D elastodynamics given in many papers and books, e.g. see Dominguez (1993).

In order to solve the system of BIEs (6)–(7), the wave free-field motion u_i^{fr}, p_i^{fr} has to be known. Here the solution of the 1-D wave propagation problem for the vertically inhomogeneous half-space with a buried earthquake source represents the free-field motion, and it is determined analytically by means of the MSM (for a review, see Panza et al., 2001).

The proposed hybrid MSM-BIEM combines the advantages of both methods: (a) the inhomogeneity of the wave path from the source position to the finite local region of interest can be approximated by a stack of flat homogeneous layers; MSM allows the treatment of many layers and also the calculation of the wavefield in the inhomogeneous half-space from a seismic event, both for small (a few kilometers) and large (a few hundred kilometers) epicentral distances; (b) the wavefield computed by MSM contains body waves and surface waves, thus it can be used as a realistic incident wavefield that is the input in the BIEM computations; (c) the BIEM permits the modeling of complex laterally varying media with surface and subsurface topography in the local part of the propagation path; (d) the advantages of BIEM in comparison with other numerical methods are (1) the reduction in the dimension of the problem, (2) the semi-analytical character of the method, (3) the solution at each internal point in the domain is expressed in terms of boundary values without recourse to domain discretization, (4) the fundamental solution used in the construction of the boundary integral equations obeys the radiation condition and (5) the possibility to obtain directly, with no other intermediate source of error, the dynamic regime – displacements and tractions.

The hybrid MSM-BIEM developed in combination with the viscoelastic isomorphism used in Biot's dynamic poroelasticity has the additional advantage of giving approximate solutions for complex boundary-value problems concerning seismic wave propagation in laterally inhomogeneous fluid-saturated media.

4. The Earthquake Scenario for the City of Sofia

The city of Sofia is the main administrative center of Bulgaria with the densest population in all of the country (Figure 1). Macroseismic intensities up to X (MSK-76) can be expected in the city. A strong earthquake in Sofia's seismic area could produce disastrous damage over a large area, followed by numerous serious consequences for the whole country on communications and lifelines.

Three types of local geological conditions can be recognized according to data from the geological map. To a certain extent they correspond to the three groups defined in the Bulgarian code (1987). The space distribution of the average shear wave velocity $V_{S,30}$, which is shown in Figure 3, is computed according to the following expression:

$$V_{S,30} = \frac{30}{\displaystyle\sum_{i=1,N} \frac{h_i}{V_i}}$$

where h_i and V_i denote the thickness and shear-wave velocity (at low strain levels) of the i-th formation or layer in a total of N existing in the topmost 30 m. As a regional bedrock model we used the structural model reported in Paskaleva et al. (2003).

In Figure 3, each site (from borehole data) is classified as A, B, C or D according to the value of $V_{S,30}$, and the sub-soils classes are A – rock with Vs > 750 m/s; B – deposits with Vs = 400–750 m/s; C – deposits with Vs = 360–400 m/s; D – deposits with Vs < 360 m/s and E – superficial (depth less than 5 m) sand deposits in Sofia. The geotechnical parameters for the area "E" in the central part of Sofia are derived from in-situ measurements and laboratory tests (Paskaleva, 2002, Slavov et al., 2004).

Two local geological profiles (Figure 4 called "canyon" and Figure 5 called "flat") with layers Ω_i , $i=,1,2,3,$ have been considered in the computations where the coordinates of the corner points (in meters) indicating the geometrical boundaries of the layers are $T_1(-100,0)$, $T_2(-90,0)$, $T_3(-60,0)$, $T_4(-30,0)$, $T_5(30,0)$, $T_6(60,0)$, $T_7(90,0)$, $T_8(100,0)$, $C(-110,270)$, $B(-90,180)$, $A(-60,90)$, $C_1(110,270)$, $B_1(90,180)$, $A_1(60,90)$. The layers Ω_2 and Ω_3 are water saturated. The layer (Ω_1) is not saturated and it has the following mechanical properties: ρ = 2,650 kg/m^3, V_s = 2,700 m/s and V_P = 5,200 m/s. Points S1, S2, S3 and S4 are the sites on the free surface where solutions in the frequency and time domain are computed and shown in the figures.

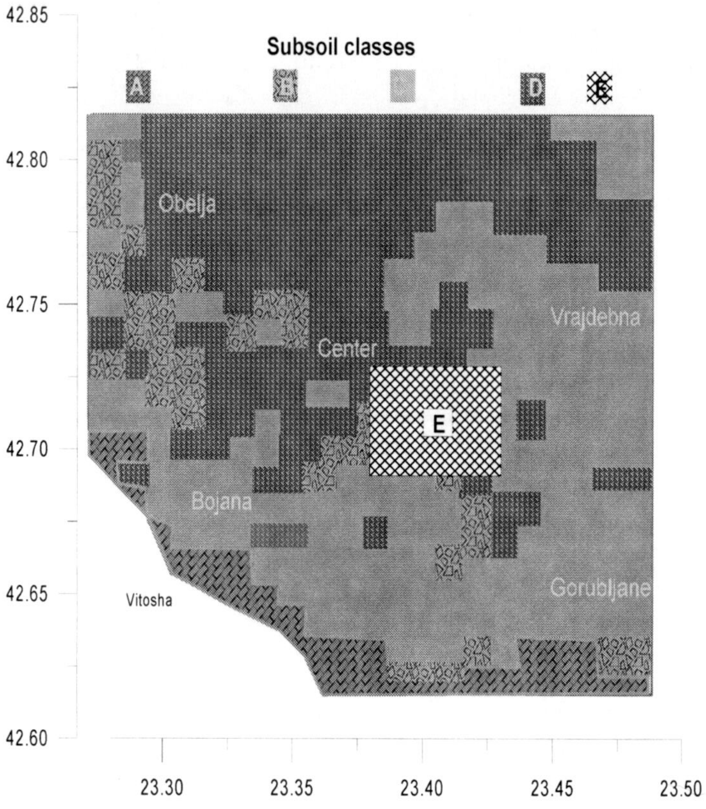

Figure 3: A 30 m deep model of the geological subsoil classes for the central part of Sofia valley: A – rock Vs > 750 m/s; B – deposits Vs = 400–750 m/s, C – deposits Vs = 400–360 m/s; D – deposits Vs < 360 m/s; E – superficial (–uppermost 5 m) sand deposits distribution in Sofia city

The local finite soil region, i.e. the stack of the three layers Ω_i, $i = 1,2,3$, (see Figures 4 and 5) has different mechanical properties if it is located in the five different residential districts of Sofia considered (Figure 1b): Reduta, Hilton, NDK, Mladost 1 and Mladost 4. These districts are situated in the central part of Sofia crossed by the profiles 1A–1B, 2C–2D, 3E–3F given in Figure 1a (Paskaleva, 2002). The following values for fluid (water) density $\rho_f = 1,000\ kg/m^3$ and bulk modulus $K_f = 2,000\ MPa$ are assumed in the calculations.

The geological profiles in Figures 4 and 5 are imbedded into a vertically inhomogeneous half-space (bedrock model of the Sofia region) with an earthquake source (double couple) buried in it. The earthquake scenarios derived from Slavov et al. (2004) used in the numerical examples have the following source properties: (a) M = 3.7, strike angle 21°, dip angle 44°, rake angle 309°, source depth 12 km and epicentre distance 20 km; (b) M = 6.5, same source parameters as (a).

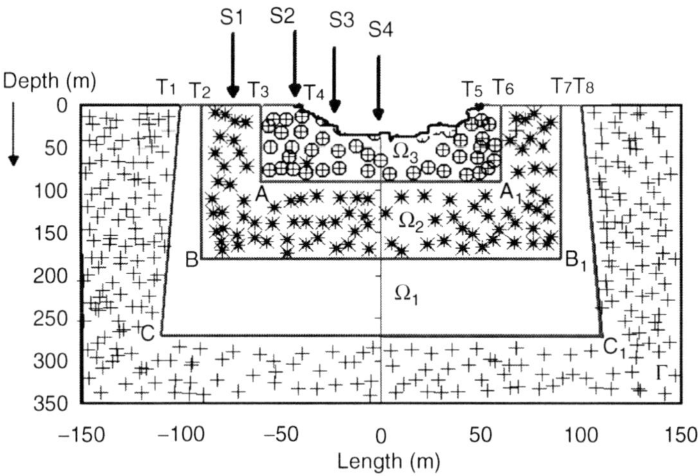

Figure 4: Geological local profile with a relief called variant "canyon" at the free-surface

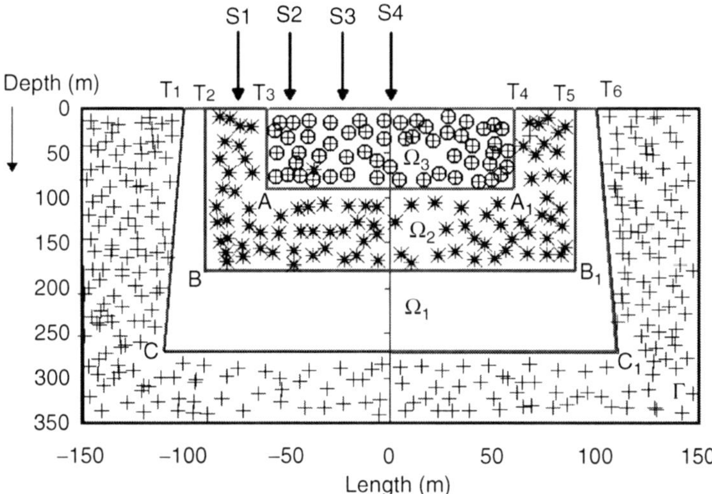

Figure 5: Geological local profile with no relief on the free surface called variant "flat"

5. Simulation Study

Frequency- and time-dependent solutions of the BVP defined in Section 2 are obtained with the hybrid MSM-BIEM code developed. The effect of water saturation in soil as described by the Bardet model can be seen in Figures 6 and 7. The results shown in these figures concern the geological profile given in Figure 4 imbedded in an inhomogeneous half-space with elastic properties representative of the bedrock structural model of the Sofia region and the earthquake scenario (a). The poroelastic material properties are K_g = 3,600 MPa, ρ_g = 2,650 kg/m^3, $k = 10^{-6}$ m/s, $v = 0.2$.

The second water-saturated layer (Ω_2) has porosity $n = 0.36$ and dry bulk modulus K_{dry} = 200 MPa. The third water-saturated layer (Ω_3) has porosity $n =$ 0.3 and dry bulk modulus K_{dry} = 6,167 MPa (results shown in Figures 6a, b and 7a, b). The normalized amplitude of the displacement components at site S2 (−30, 0) versus frequency are shown in Figure 6a, b for dry and saturated soils. Normalization is made with respect to the amplitude of the corresponding displacement component, synthesized for the bedrock reference regional model. Examples of the normalized time series for displacement and velocity at the same site S2 (−30,0) obtained for dry and saturated soil are shown in Figure 7a, b, where normalization is made with respect to the maximum displacement and velocity of the corresponding signals obtained for the bedrock reference regional model.

The effects of soil saturation can be seen in Figure 6a, b: the main peak of the frequency-dependent dynamic response of saturated soil is shifted to lower frequencies with respect to dry soil. The value of the main peak of the dynamic response usually decreases in saturated soil with respect to dry soil. This effect is frequency dependent. Figure 7 also shows very clearly that due to soil saturation, the duration of the seismic signal for saturated soils is shorter than that for dry ones.

The parametric study of the site effects in two different water saturated geological profiles (Figure 4: canyon and Figure 5: flat) representing different zones in the central part of Sofia is made in terms of the acceleration response spectral ratio (RSR), i.e. the response spectra computed from the signals synthesized along the laterally varying section and normalized by the response spectra computed from the topologically corresponding signals synthesized for the bedrock reference model.

The RSRs are obtained for the sites at the free surface: S1 (−75,0), S2 (−30,0) and S4 (0,30). The single degree-of-freedom damped oscillator used for the computation of the response spectra is with 5% damping.

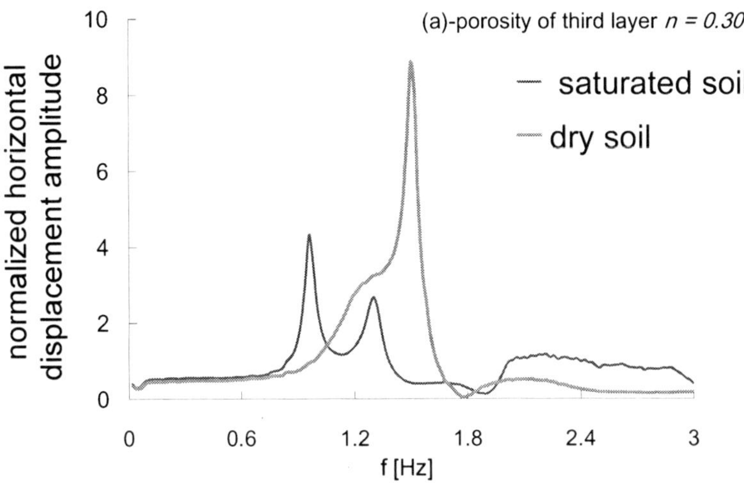

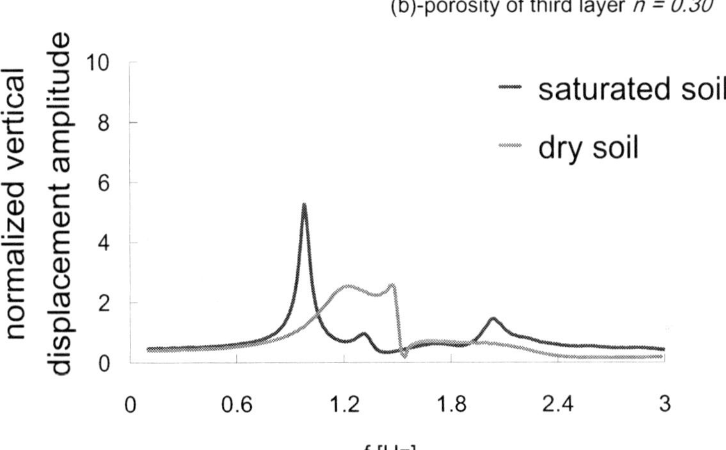

Figure 6: (**a, b**) Normalized amplitude of displacement versus frequency at site S2(−30,0) for dry and saturated soil for the geological profile shown in Figure 4

The effect of the saturation in soil can be seen in Figure 8, for the earthquake source 27/4/2006 with M = 6.5 and for the Sofia districts Reduta and Mladost 4 where the RSRs of acceleration for both dry and saturated soils are compared. The geological profile with a flat free surface shown in Figure 5 is used for this parametric study. Considering the results shown in Figure 8, we

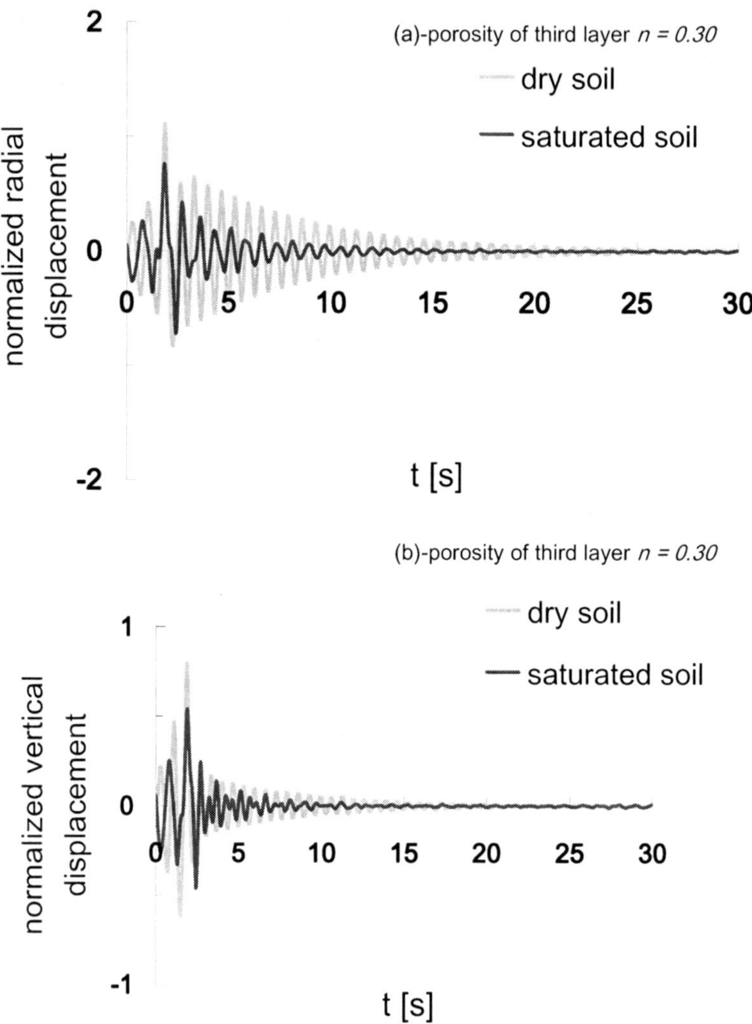

Figure 7: (**a, b**) Normalized amplitude of displacement versus time at site S2(−30.0) for dry and saturated soil for the geological profile shown in Figure 4

can conclude that (a) there is a difference in the dynamic responses at different sites, i.e., there are site effects due to the mixed influence of lateral inhomogeneity and soil saturation; (b) there is a shift in the RSR peak to the right, i.e. towards the higher periods, in saturated soils when compared with dry soil and (c) the shape of the RSR curves does not change dramatically for saturated soil in comparison with the RSR curves for dry soil. In most cases, the main peak of the RSR of saturated soils is smaller than the main peak of the RSR in dry soils.

DRY SOIL
"Reduta"

Horizontal component of RSR

SATURATED SOIL
"Reduta"

Horizontal component of RSR

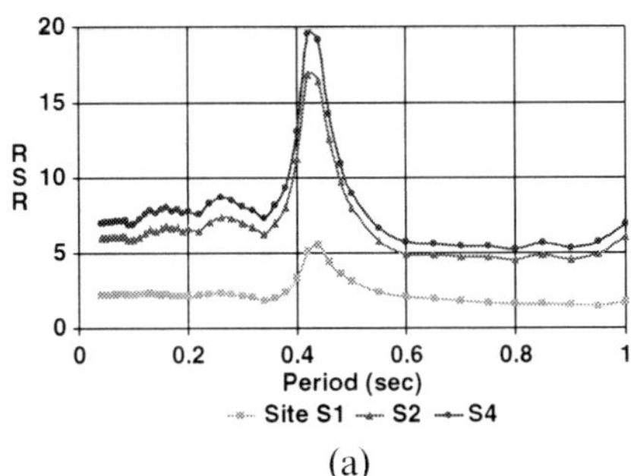

(a)

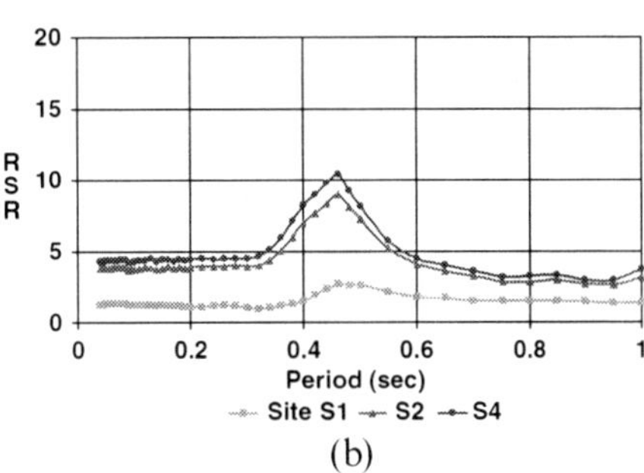

(b)

Vertical component of RSR Vertical component of RSR

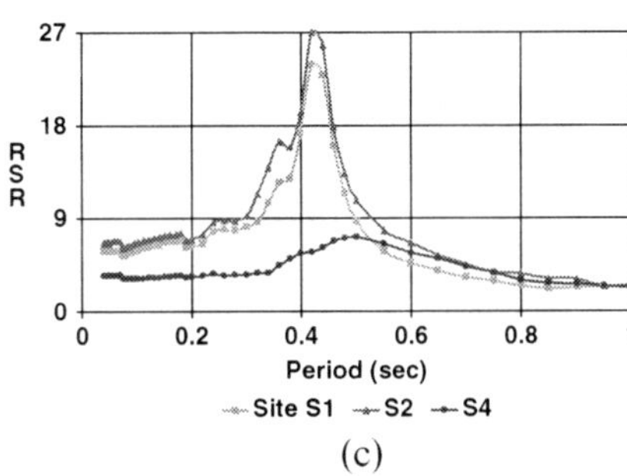

(c)

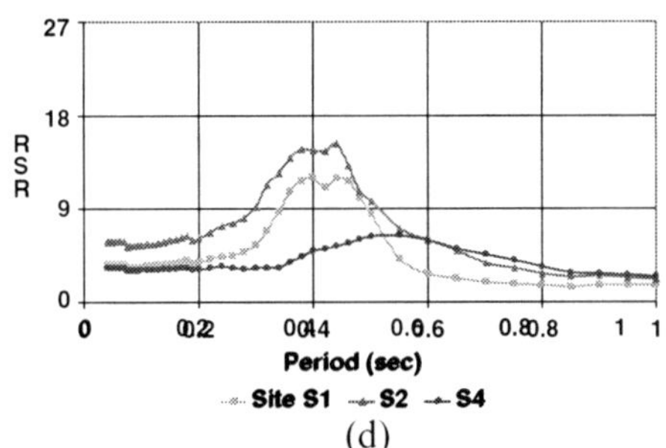

(d)

"Mladost 4"

"Mladost 4"

Horizontal component of RSR Horizontal component of RSR

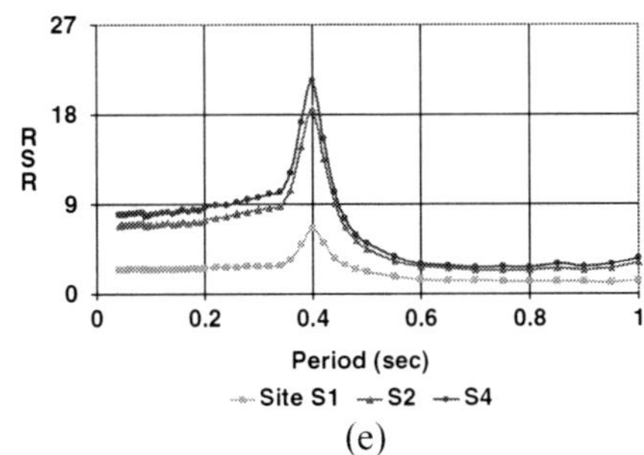

(e)

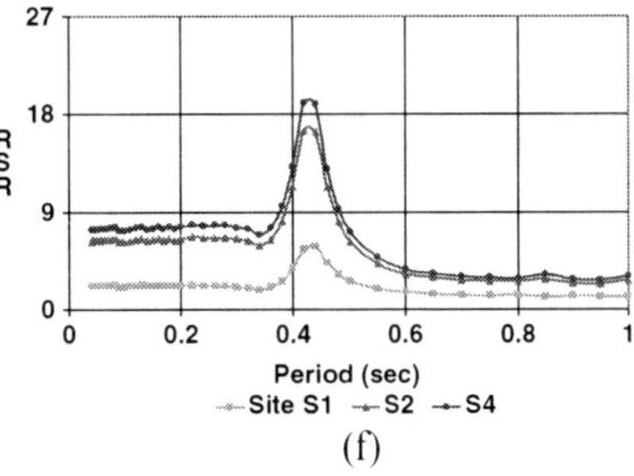

(f)

Vertical component of RSR Vertical component of RSR

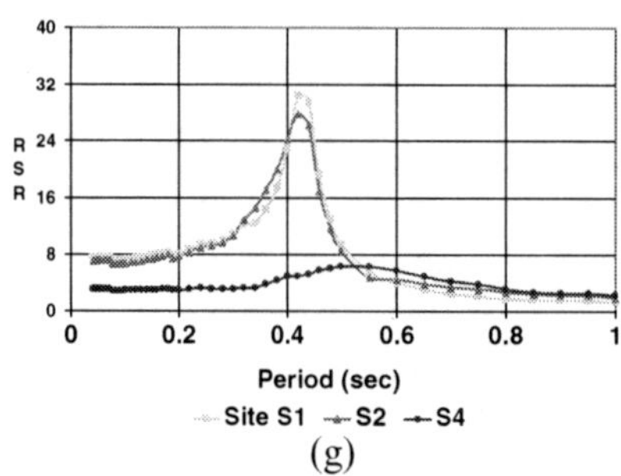

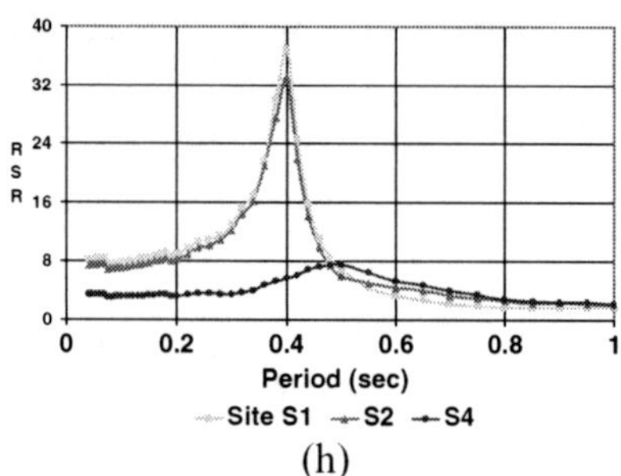

Figure 8: (**a–h**) Influence of soil saturation on horizontal and vertical components of RSRs at the sites S1, S2 and S4 for the geological profile shown in Figure 5 (variant "flat") with mechanical properties of Reduta and Mladost 4 residential districts. Earthquake scenario of 27/4/2006: simulation with magnitude M = 6.5

The presence of fluid in the soil acts as a stiffener of the material; the pore pressure induced by the seismic load resists compression and stiffens the rock. The shear strength of Biot's porous material is provided by the solid framework (matrix), and it is not affected by saturation with fluid, or $\mu_{sat} = \mu_{dry}$, when the Lame coefficient λ_{sat} changes to

$$\lambda_{sat} = \lambda_{dry} + \frac{Q^2}{R}$$

where Biot's coefficients Q and R (see Equation (5)) are functions of the porosity n, solid framework, K_{dry}, solid grain, K_g, and fluid, K_f, bulk moduli. So, the saturated material has a lower compressibility if compared with the dry one (Lin et al., 2005; Degrande et al., 1998; Aznarez et al., 2006) and this fact may explain the lower values of the RSR for the saturated soil in Figure 8.

Fluid in the soil acts also as a viscous damping element because it is possible for it to flow between regions of higher and lower pressure, i.e. diffusion can be provoked which can lead to a damping mechanism because the fluid's motion is dominated by viscous effects.

Figure 9 demonstrates the influence of topography on the site effects computed for the earthquake scenario of 27/4/2006 simulated with magnitude M = 6.5 called (b) for the geologic profiles called variant canyon (Figure 4) and variant flat (Figure 5) situated in the district of Reduta. In both variants, all other mechanical parameters are kept the same. The existence of a free surface relief changes the amplification factors at different sites both qualitatively and quantitatively.

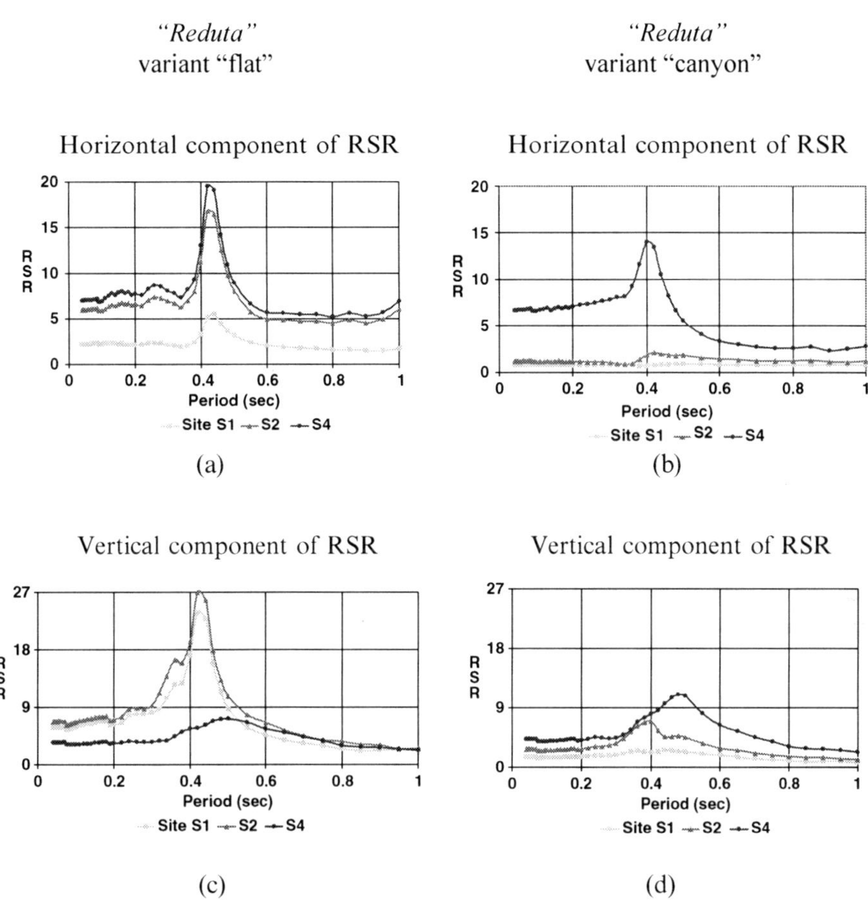

Figure 9: (**a–d**) Influence of the topography on the horizontal and vertical components of RSR at sites S1, S2 and S4 for the geological profiles shown in Figures 4 and 5 with mechanical properties of Reduta district: dry soils. Earthquake scenario of 27/4/2006: simulation with magnitude M = 6.5

6. Conclusions

The study formulates and solves the 2-D elastodynamic problem for in-plane seismic wave propagation in laterally inhomogeneous geological structures imbedded in a vertically inhomogeneous half-space where the earthquake source is buried. For this purpose, an efficient hybrid MS-BIEM in combination with the viscoelastic isomorphism of Biot's dynamic poroelasticity is developed. Each of the two computational techniques is applied in the part of the model in which it works more efficiently.

The proposed mathematical model and the hybrid tool are applied to investigate the site effects phenomena due to the impedance contrasts of soil layers, to surface topography and lateral inhomogeneity, to the influence of earthquake source properties, to the existence of water saturation in soils and to the conversion of body waves into surface waves. The simulations were performed considering the regional structural model of Sofia; the geotechnical parameters and the soil's mechanical properties for the different districts considered here are, in fact, those more common in the central part of the city.

The results obtained show that in realistic structural models, the wave pattern at a site is the complex result of different physical properties like non-parallel layering, topography and water saturation in the soil stratum.

Acknowledgements

The authors acknowledge financial support through the CEI-ICTP Fellowship Programme, 2006, DST Trieste University, Italy and INTAS ref. Nr. 05-104-7584 "Numerical analysis of 3-D seismic wave propagation using modal summation, finite element and finite difference methods."

References

Aznarez J.J., Maeso O., Dominguez J. (2006) BE analysis of bottom sediments in dynamic fluid-structure interaction problems, *Eng. Anal. Bound. Elem.*, 30, 205–223.

Bardet J.P. (1992) A viscoelastic model for the dynamic behaviour of saturated poroelastic soils, *Trans. ASME*, 59, 128–135.

Biot M.A. (1956) Theory of propagation of elastic waves in a fluid-saturated porous solid: I. Low frequency range, *J. Acoust. Soc. Am.*, 28, 168–178.

Degrande G., De Roeck G., Den Broeck V.P. (1998) Wave propagation in layered dry, saturated and unsaturated poroelastic media, *Int. J. Solids Struct.*, 35(N34–35), 4753–4778.

Dineva P., Vaccari F., Panza G.F. (2003) Hybrid modal summation – BIE method for site effect estimation of a seismic region in a laterally varying media, *J. Theor. Appl. Mech. Bulg. Acad. Sci.*, 33(4), 55–88.

Dineva P., Datcheva M., Schanz T. (2006) BIEM for seismic wave propagation in fluid saturated multilayered media. In: *Numerical Methods in Geotechnical Engineering* (H.F. Schweiger, Ed.), *Proc. 6th European Conference on Numerical Methods in Geotechnical Engineering*, 6–8 September 2006-Graz, Austria, published by Taylor & Francis/Balkema, pp. 257–265.

Dominguez J. (1993) *Boundary Elements in Dynamics*, Computational Mechanics/ Elsevier, Southampton/Amsterdam.

Fah D., Suhadolc P., Panza G.F. (1993a) Variability of seismic ground motion in complex media: The Fruili area (Italy). In: *Geophysical Exploration in Areas of Complex Geology, II* (R. Cassinis, K. Helbig, G.F. Panza, Eds.), *Journal of Applied Geophysics*, 30, 131–148.

Fah D., Iodice C., Suhadolc P., Panza G.F. (1993b) A new method for realistic estimation of seismic ground motion in mega cities: The case of Roma, *Earthq. Spectra*, 9, 643–668.

Lin C.H., Lee V.W., Trifunac M.D. (2005) The reflection of plane waves in a poroelastic half-space saturated with inviscid fluid. *Soil Dyn. Earthq. Eng.*, 25, 205–223.

Morochnik V., Bardet J.P. (1996) Viscoelastic approximation of poroelastic media for wave scattering problems, *Soil Dyn. Earthq. Eng.*, 15, 337–346.

Panza G.F., Romanelli F., Vaccari F. (2001) Seismic wave propagation in laterally heterogeneous anelastic media: theory and application to seismic zonation. In: *Advances in Geophysics* (R. Dmowska, B. Saltzman, Eds.), Academic Press, San Diego, CA, Vol. 43, pp. 1–95.

Paskaleva I. (2002) A contribution to the seismic risk assessment of the Sofia City. Report on CNR-NATO program, 65, Annocement, 219.33, May–October 2002, 100p.

Paskaleva I., Panza G.F., Vaccari F., Rajgelj S., Ivanov P. (2003) Deterministic modelling for microzonation of Sofia: An expected earthquake scenario, *Proceedings of the International Conference in Earthquake Engineering, SE 40EEE*, 26–29 August, 2003, Skopje.

Simon B.R., Zienkiewicz O.C., Paul D.K. (1984) An analytical solution for the transient response of saturated porous elastic solids, *Int. J. Numer. Anal. Methods Geomech.*, 8, 381–398.

Slavov S., Paskaleva I., Kouteva M., Vaccari F., Panza G.F. (2004) Deterministic earthquake scenarios for the City of Sofia, *Pure Appl. Geophys. (PAGEOPH)*, Birkhauser Verlag, Basel, 161, 1221–1237.

SEISMIC MONITORING: A CONTRIBUTION TO THE SUSTAINABLE DEVELOPMENT OF THE RUSSE REGION OF BULGARIA

M. KOUTEVA[*,1,2], I. PASKALEVA[1]
[1]*Central Laboratory for Seismic Mechanics and Earthquake Engineering, Bulgarian Academy of Sciences, 3 Acad. G. Bonchev str., 1113 Sofia, Bulgaria*
[2]*ESP-SAND ICTP, Trieste, Italy*

Abstract. Recently published data on seismic hazards due to strong, intermediate-depth Vrancea earthquakes in the area around the city of Russe is summarized including a review of analyses of available data, both macroseismic and instrument, and of recent theoretical seismic hazard estimates. A snapshot of the development of a strong motion registration system is offered, and the benefits of seismic monitoring in the region are discussed. Transforming data produced by seismic monitoring instruments into information for decision makers (e.g. emergency managers, earthquake engineers) are summarized including analytical steps that differ depending on the type of data and the nature of the decision.

Keywords: Dynamic monitoring, seismic hazard, strong intermediate-depth earthquakes, Vrancea, strong motion registration

1. Introduction

The sustainable development of society depends not only on a reasonable policy for economic growth but also on improved management of natural and man-made risks. Over the centuries, Bulgaria has experienced strong earthquakes. The seismicity of neighboring areas in Greece, Turkey, the former Yugoslavia and Romania, especially Vrancea intermediate-depth earthquakes,

---

[*]Central Laboratory Seismic Mechanics and Earthquake Engineering (CLSMEE), Bulgarian Academy of Sciences (BAS), Acad. G. Bonchev block 3, 1113 Sofia, Bulgaria, e-mail: mkouteva@geophys.bas.bg

A. Zaicenco et al. (eds.), *Harmonization of Seismic Hazard in Vrancea Zone,* 183
© Springer Science + Business Media B.V. 2008

contributes significantly to the seismic hazard in Bulgaria. The total area affected by intermediate-depth Vrancea earthquakes comprises 300,000 square kilometers (km^2) with a population of about 25 million people (Zaicenco and Alkaz, 2005). The maximum observed macroseismic intensity in the Vrancea intermediate zone has reached intensity (I)~VIII-IX MSK-64. A special feature of the Vrancea earthquake hazard in Bulgaria is its irregular macroseismic effects. In all cases, southwest to northeast bands of maximum seismic effect were traced (Glavcheva, 1983). The area that has suffered damage of severity (I)~VI includes Romania, Moldova, a large part of Bulgaria and southwestern Ukraine. A number of important cities, numerous industrial and energy facilities (including two nuclear power plants), industrial chemical facilities, lifelines of crucial importance (railways, gas and fuel pipelines, bridges and dams) are situated in this earthquake-prone area. The four strongest earthquakes (M_w > 6) that occurred in the Vrancea zone during the last century (in 1908, 1940, 1977, 1986 and 1990) caused significant damage over a wide part of Europe including distant, long-period elements in the built environment. The quake of March 4, 1977, (M_w = 7.5) caused significant damage in Bulgaria and was felt in Central Europe at distances of about 1,000 km.

The strong, irregular but not infrequent manifestations of the Vrancea focus (70–170 km) have always affected Bulgaria. They have caused a high death and injury toll and extensive damage over the last several centuries. The wave field radiated by Vrancea intermediate-depth earthquakes mainly at long periods (T > 1 s) attenuates with distance less rapidly than the wave fields of earthquakes in other seismically active zones in Bulgaria. Vrancea intermediate-depth earthquakes are thus the dominant hazard in large parts of northern Bulgaria; an appropriate understanding of these earthquakes is of great importance for adequate hazard assessments (Paskaleva 2004; Simeonova et al., 2006a). A review of the macroseismic record of strong, intermediate-depth Vrancea quakes indicates the region of maximum damage is in the northeast from Orjahovo to Silistra on the Danube and to the south to the towns of Razgrad and Dulovo.

Russe lies on the Danube between Orjahovo and Silistra. Russe is the fifth largest city in Bulgaria with a population of 178,000. Among the main Bulgarian cities, Russe is therefore the most exposed to Vrancea intermediate-depth earthquakes. Approximately 200 buildings in Russe are listed in the architectural and historical heritage of the country. Several ancient places of interest including the Rock Monastery of Ivanovo (1979 UNESCO World Heritage Site) are located nearby in the valley of the Russenski Lom River. Russe is the most significant Bulgarian river port and is thus an important element of the international trade of the country. The Russe-Giurgiu Friendship Bridge, the only one in the shared Bulgarian-Romanian section of the Danube, crosses the river near the city. In addition, industries of national importance are located there, e.g., chemical and electrical plants.

This paper makes a case for the necessity of locating strong-motion seismic instrumentation at Russe based on recent seismic hazard assessments, and points out the applicability of scientific results and monitoring data for practical earthquake hazard assessment and risk management.

2. The Vrancea Earthquake Hazard at Russe

The built environment in Russe is subject to earthquakes from the main earthquake zones in the region: Vrancea in Romania and Gorna Oriahovitza, Shabla and Dulovo (Dobrodja) in Bulgaria. The documented macroseismic intensity (MSK-64) at Russe of 32 recent strong and moderate earthquakes is in Table 1 (Glavcheva, 1983). One can see that the maximum intensities correspond to the strong intermediate-depth Vrancea earthquakes.

TABLE 1: Documented macroseismic intensity (MSK-64) at Russe

No.	Earthquake	M	I	H	No.	Earthquake	M	I	H
1	23/01/1838/Vrancea	6,9	*VII*	150	17	24/06/1940/Vrancea	5,8	*III*	120
2	17/08/1893/Vrancea	6,1	*V*	100	18	22/10/1940/Vrancea	6,2	*V*	90
3	10/09/1893/Vrancea	6,1	*V*	100	19	13/11/1940/Vrancea	7,3	*VII*	150
4	31/08/1894/Vrancea	6,5	*V*	150	20	07/09/1945/Vrancea	6,5	*IV*	100
5	31/03/1901/Shabla	7,1	*VII*	25	21	09/10/1945/Vrancea	6,2	*V*	90
6	08/06/1903/Vrancea	5,4	*IV*	150	22	17/05/1953/Vrancea	5,0	*IV*	140
7	13/09/1903/Vrancea	6,2	*IV*	50	23	13/04/1954/Vrancea	4,9	*III*	120
8	06/02/1904/Vrancea	5,9	*IV*	60	24	01/05/1955/Vrancea	5,4	*III*	135
9	06/10/1908/Vrancea	6,8	*VI*	150	25	30/06/1956/Shabla	5,0	*III*	12
10	08/09/1911/Shabla	4,6	*III*	10	26	31/05/1959/Vrancea	4,7	*IV*	20
11	25/05/1912/Vrancea	6,3	*IV*	40	27	04/01/1960/Vrancea	5,3	*IV*	60
12	14/06/1913/G.Orjahovitza	7,0	*V*	25	28	14/01/1963/Vrancea	5,9	*III*	118
13	26/10/1914/Vrancea	5,0	*IV*	120	29	17/07/1974/Vrancea	5,7	*IV*	140
14	01/11/1929/Vrancea	6,6	*V*	150	30	04/03/1977/Vrancea	7,2	*VIII*	110
15	29/03/1934/Vrancea	6,9	*V*	150	31	30/08/1986/Vrancea	7,2	*VI*	133
16	13/07/1938/Vrancea	5,7	*IV*	150	32	30/05/1990/Vrancea	6,9	*V*	90

M = magnitude; I = intensity; H = focal depth.

A map of macroseismic damage observed at Russe after the 1977 Vrancea earthquake (Simeonova et al., 2006b) is shown in Figure 1. The higher intensities correspond to weaker soil classes according Eurocode 8 (EC 8) soil classifications.

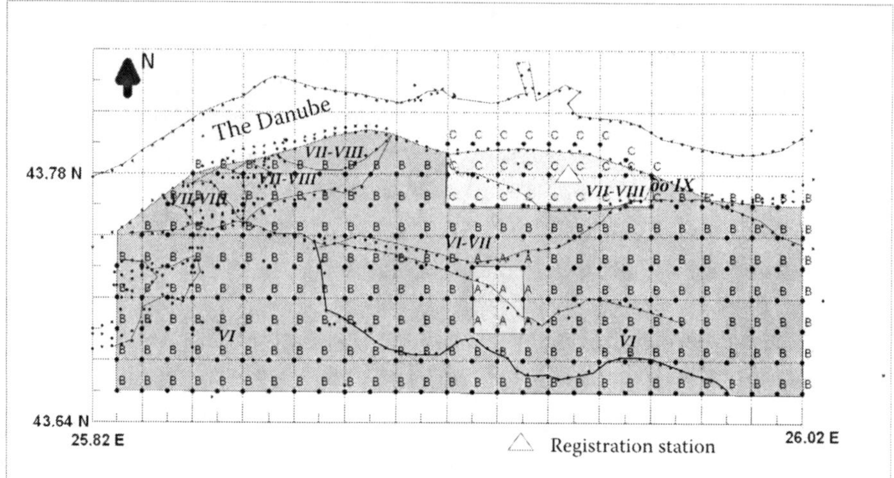

Figure 1: Observed macroseismic intensity, I, MSK-64, due to the March 4, 1977, Vrancea Earthquake (Simeonova et al., 2006b) overlapped with the schematic geological zonation of the city of Russe following the EC8 soil classification (Paskaleva and Kouteva, 2007)

Accelerograms due to Vrancea intermediate-depth earthquakes have been recorded on three seismic networks in Romania as well as on the seismic networks of the Academy of Science in Moldova and the Academy of Science in Bulgaria. A recent, detailed analysis of Bulgarian data using the moving windows technique, a new data processing technique in the frequency-time domain, and tables with the main characteristics of the accelerograms recorded were provided by Paskaleva (2004). For the Bulgarian records, the values of the control periods T_c are $T_c = 0.34–1.47$ s. In fact, even though the Vrancea 1977 event motivated changes in the Bulgarian Code for Design and Construction in Seismic Regions of 1987, the seismic excitation from the 1986 and 1990 events exceeded the seismic loading prescribed. The damage reported in northeastern Bulgaria due the 1986 and 1990 Vrancea events requires community policy makers and stakeholders to pay attention to the design, construction and effective protection of the area's historical heritage.

A comparison of the dynamic coefficients and spectral displacements computed from the accelerograms recorded at the Russe registration station during the August 31, 1986 (VR86) and May 30, 1990 (VR901) Vrancea quakes with the values given in the Bulgarian Code for Design and Construction in Seismic Regions 1987 (soil class III) and Eurocode 8 is shown in Figure 2.

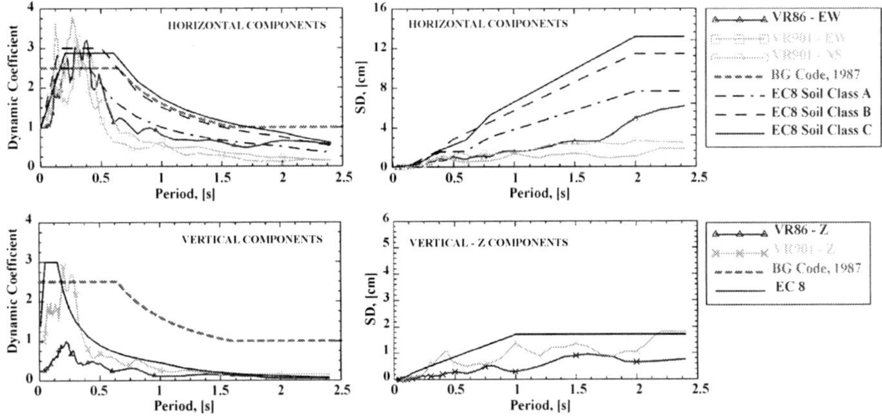

Figure 2: Comparison of the dynamic coefficients (*on the left*) and the spectral displacements computed for 5% damping (*on the right*) for the Vrancea VR86 and VR901 quakes with the values given in the Bulgarian Code for Design and Construction in Seismic Regions 1987 and EC8

One can see that the dynamic coefficients for the horizontal components (upper left) are higher than the recommended values for periods up to 0.6 s. The dynamic coefficient for the vertical component (VR901) is also higher than the recommended values for periods $0.2 < T < 0.5$ s. The spectral displacements for the horizontal components for both earthquakes are lower than the recommended EC 8 values.

2.1. PROBABILISTIC SEISMIC HAZARD ASSESSMENT

Probabilistic and deterministic seismic hazard estimates for Russe were made and published by Todorovska et al. in 1995 and by Paskaleva et al. in 2001. The attenuation laws used were determined from a regression analysis of the peak ground acceleration (PGA) observations available in Romania and Bulgaria due to the Vrancea earthquakes of 1977, 1986 and 1990 grouped by epicentral distance, azimuth, magnitude range and focal depth (Shebalin et al., 1999; Romplus catalogue). These attenuation laws were considered to model the seismic input due to two earthquakes with different magnitude M and focal depth H: (1) M = 7.1, H = 133 km and (2) M = 7.0, H = 90 km. The chosen sets of focal depth and magnitude (H, M) reproduce the Vrancea events of 1986 (VR86) and 1990 (VR90), respectively. For VR86, the observed PGA at Russe was lower than the probabilistic estimate by about 30% (Paskaleva et al., 2001). For VR90, the recorded PGA at Russe and Bozvelijsko exceeded the estimated values by approximately 20% (Paskaleva et al., 2001). For the same seismic events, the deterministic results obtained for Russe (Paskaleva and Kouteva, 2000) underestimated the recorded PGA by 30%. In addition to the probabilistic calculation, mean values of PGA plus one standard deviation for a Vrancea earthquake of M = 7.8, H = 90 km and a return period suggested by EC 8 in 1994 of T = 475 are calculated. The PGA at Russe was about 0.3 g. The results are shown in Figure 3 (Paskaleva et al., 2001).

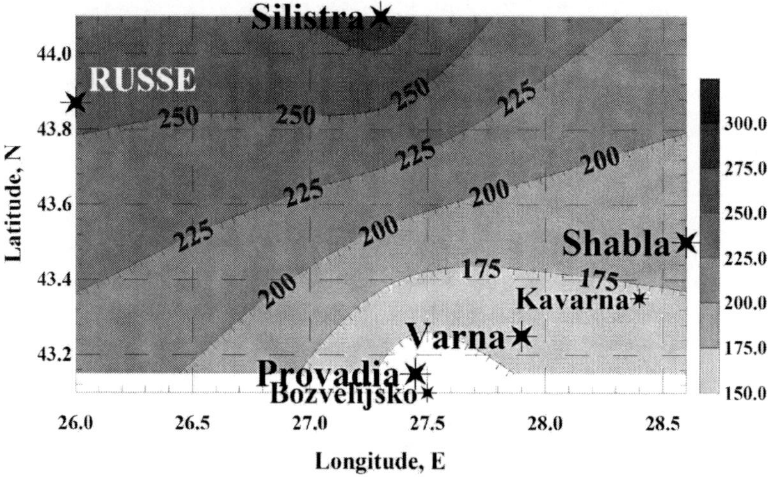

Figure 3: Probabilistic seismic hazard assessment for NE Bulgaria – mean values of PGA plus one standard deviation for a Vrancea earthquake of M = 7.8, H = 90 km and return period T = 475 (Paskaleva et al., 2001)

In 2006, a seismic hazard map for Bulgaria was published by Simeonova et al. (2006a) as a basis for a new building code. The Vrancea earthquake hazard was treated separately for shallow and the intermediate-depth zones. Maps obtained for a return period of 475 years showed that the seismic hazard at Russe is controlled mainly by the Vrancea intermediate-depth seismic zone (VrIdSZ) with expectations of a macroseismic intensity of (I)~VIII MSK in both cases analyzed with exposure only to the VrIdSZ and exposure to all source zones considered (Simeonova et al., 2006a).

Seismic hazard evaluations based on a traditional probabilistic seismic hazard analysis (PSHA) rely on the probabilistic analysis of earthquake catalogues and of ground motion, on macroseismic observations and on instrument records. Recently, PSHA's usefulness in providing a reliable seismic hazard assessment has been limited possibly due to insufficient information about historical seismicity which can introduce relevant errors in a purely statistical approach based mainly on seismic history (Klügel et al., 2006; Klügel, 2007a, b). The standard error of empirical attenuation equations (or the difference between observations and estimates of a physical predictive model) is one of the main sources of errors in traditional PSHA. The macroseismic and instrument data available are very limited, and the statistical determination of the coefficients of frequency-magnitude relationships (FMR) for the zones controlling the seismic hazard in northeastern Bulgaria is affected by significant uncertainties (Molchan et al., 1997).

2.2. NEO-DETERMINISTIC SEISMIC HAZARD ASSESSMENT

In 2000, finalizing the UNESCO-IGCP Project 414 "Seismic Ground Motion in Large Urban Areas" and taking advantage of the existing Central European Initiative (CEI) Network, maps of neo-deterministic seismic hazards including design ground acceleration (DGA) peak ground velocities (PGV) and peak ground displacements (PGD) for some European countries were published by Panza and Vaccari. An extraction of these maps, concerning the Bulgarian territory is given in Figure 4.

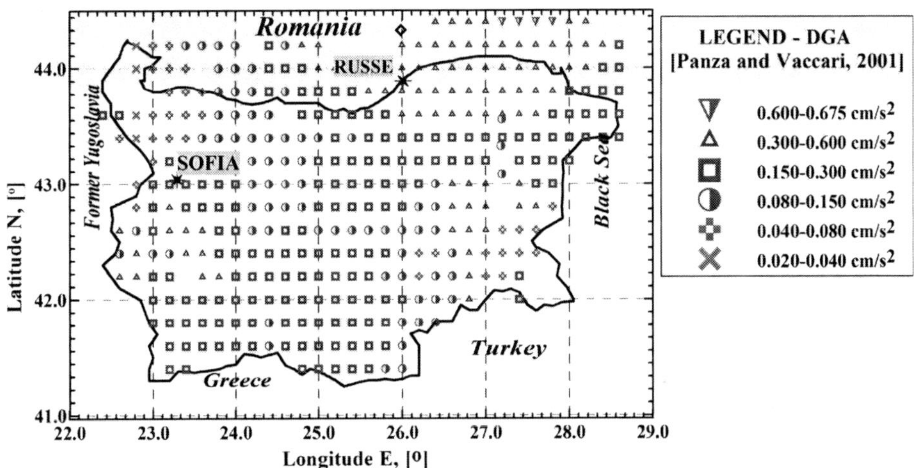

Figure 4: Design ground acceleration (DGA) in Bulgaria according Panza and Vaccari (2000)

To construct these maps, a circle with a 350 km radius centered on a Vrancea epicenter cell was considered. Shallow seismicity was assumed limiting computations to epicentral distances shorter than 90 km. The hypocentral depth was 10 km for events with magnitude $M_w < 7$ and 15 km for larger events. In the case of Vrancea intermediate-depth events, spectral properties specifically determined for Romanian intermediate-depth earthquakes were considered and computations were made for Romania, northeastern Croatia and Hungary within this circle. The hypocentral depths considered were 90 km for magnitudes less than 7.4 and 150 km for larger quakes. According to these maps, DGA up to 0.675 g can be expected in that region. The deterministic results of an M = 7.7, H = 150 km event that occurred in 1940

(Panza and Vaccari, 2000), give a DGA for Russe within the range 0.30–0.60 cm/s^2, maximum displacement of 15–30 cm and maximum velocity of 30–60 cm/s. These acceleration values significantly exceed the prescribed seismic loading in the Bulgarian code of 1987 (seismic coefficient K_c = 0.15 g) and the recorded PGA during both VR86 and VR90 which caused significant damage at Russe. These results indicate that detailed site response investigations that make use of realistic, advanced ground motion modelling methods must be performed.

Theoretical modeling of the seismic loading at Russe due to Vrancea intermediate-depth events considering different frequency intervals was done by Kouteva, Paskaleva, Panza and Romanelli (Kouteva et al., 2000, 2004a, b, c; Paskaleva et al., 2001; Kouteva, 2002; Panza and Vaccari, 2002) applying an analytical deterministic procedure (Panza et al., 2001). The results showed that ground motion at a given site was dependent not only on the elastic and non-elastic characteristics of the medium but also on the parameters describing the geometry and the movement of the seismic source. All ground motion components – transverse, radial and vertical – contribute significantly to seismic input. It has been shown that the site amplifications at Russe due to the different earthquakes show rather different amplification patterns for the various ground motion components (Panza et al., 2002). A parametric/sensitivity analysis of the ground motion at Russe varying the seismic source location, the parameters describing the movement in the source and the Vp/Vs ratio of the superficial layer at the local site (Kouteva et al., 2000, 2004a, b, c; Kouteva, 2002) was also performed. The results showed that all the tensorial properties of the earthquake sources influence the shape and the amplitude of the seismic signal, and among and among the selected parameters, the focal depth contributes most strongly to these changes. Superficial soil conditions also contribute to both the seismic input and to the damage potential of the ground motion (Kouteva et al., 2000, 2007).

3. Seismic Monitoring Systems

Protecting people, the built environment, cultural heritage and the nation's economy from the effects of devastating earthquakes has been the focus of scientific and engineering endeavors for more than 100 years. The development of successful risk management strategies for earthquake hazards requires that the benefits from reduced uncertainty provided by improved seismic monitoring are integrated with the factors that influence risk perception and choice. The extent to which information from seismic monitoring networks can be used to reduce losses from future earthquakes depends, to a large degree, on providing this information to decision makers and other end users in an appropriate form and on the extent to which these individuals and groups understand and make use of the information. An integrated, dynamic monitoring system for earthquake engineering purposes will contribute in general to (i) site effect assessment (the denser the network is, the better the assessment is), (ii) design parameter assessment, (iii) the dynamic behavior of structures and soil-structure interaction analyses, (iv) calibration of the models used for seismic wave propagation and (v) ground motion modeling. Such an integrated seismic network should contain three types of instrumentation: (i) free-field, (ii) building and infrastructure and (iii) free-field and boreholes (e.g. Aldea et al., 2006a, b). For completeness of monitoring, the particular historical heritage of the region has to be included at the monitored sites.

In Bulgaria, real-time seismological monitoring is done by the Geophysical Institute of the Bulgarian Academy of Sciences' National Operative Telemetric System for Seismological Information (NOTSSI) as shown in Figure 5. Since 1981, the Central Laboratory of Seismic Mechanics and Earthquake Engineering (CLSMEE) has operated a strong ground motion network with SMA-1 instruments (analogue registration stations equipped by accelerometers SMA-1), and one three-component SMA 1 accelerometer has been installed at Russe. In 2006, after a long period of difficulties with the maintenance of the system, it was renovated with digital instruments. All the instrumentation is installed at free-field registration sites. Figure 5 also shows the strong motion registration stations maintained by CLSMEE. In the framework of NATO SfP Project 980468 another four GeoSIG accelerometers were purchased and were installed in the cities of Sofia, Russe, Silistra and Musomishte indicated by white rectangles in Figure 5.

4. Uses of Seismic Monitoring

Transforming data produced by seismic monitoring instruments into information for decision makers (e.g. emergency managers, earthquake engineers) requires analytical steps that differ depending on the type of data and the nature of the decision. Various pathways are summarized in Figure 6. Seismic monitoring networks provide the basis for hazard analysis and quantification; the density of network coverage and instrument type determines how much will be learned from each damaging earthquake. Recent earthquakes have shown that the existing seismic monitoring network in northeastern Bulgaria was too sparse to provide the essential information required to understand what happened during these earthquakes.

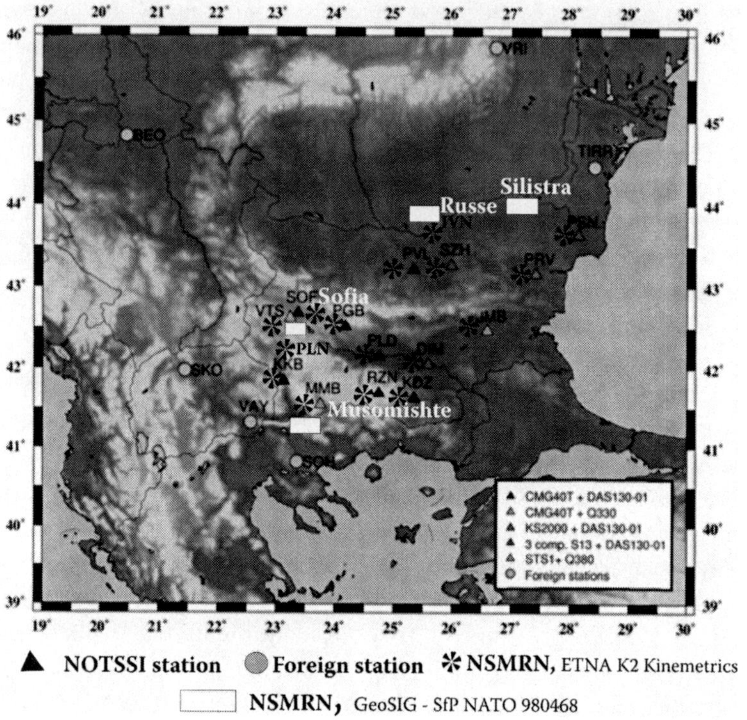

▲ NOTSSI station   ● Foreign station   ✳ NSMRN, ETNA K2 Kinemetrics

☐ NSMRN, GeoSIG - SfP NATO 980468

Figure 5: Configuration of the National Earthquake Registration Systems and foreign stations used in seismic network data processor (SNDP) for real-time data processing (Paskaleva and Kouteva, 2007; Solakov et al., 2007). NSMRN: National Strong Motion Recording Network.

The essential role of seismic information is to reduce uncertainty in risk assessment over time and thereby increase its usefulness for emergency preparedness, loss-avoidance regulations, private risk financing and insurance and/or earthquake predictions. Improved monitoring provides more information; therefore, more complete understanding of geophysical processes, more realistic models and better-informed risk assessments are possible.

Seismic networks provide both parametric (e.g., earthquake origin times, locations and magnitudes) and waveform or seismogram data. These data are used as the basis for both public safety decisions and for scientific and engineering research. Collecting high-quality data at different epicentral distances is critical for validating and verifying the earthquake engineering models that are used in estimating the vulnerability of any environment. Measuring the response of selected representative structural systems during actual earthquakes will provide empirical information on their seismic performance and can also provide information for evaluating the efficacy of current mitigating engineering practices. Immediately after a significant earthquake, there is a need to assess the extent and severity of damage and to identify where emergency actions are needed.

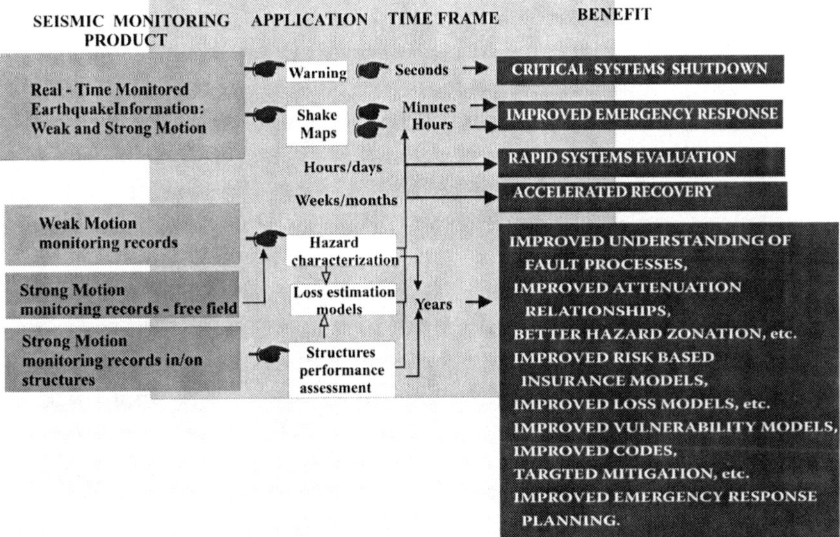

Figure 6: A general flowchart summarizing the information path from seismic monitoring data to the provision of benefits

With appropriate information, decision makers – including public sector agencies, lenders, engineers, builders and insurers – can develop effective mitigation strategies to reduce the impacts of low-probability, high-loss earthquakes. Reducing uncertainties in both hazard and risk assessments through improved monitoring and research and enabling policy makers to gain an improved understanding of these uncertainties also has considerable potential for improving the efficiency of emergency response planning and mitigation activities. The potential benefits from improved seismic monitoring are quite varied. A strong motion seismic monitoring system can supply data that might be used for creating policy instruments (e.g., economic incentives, insurance, building codes) and regulations that can be used to reduce future earthquake damage and the loss of lives while providing financial relief after a disaster. The capability to reliably predict the pattern of ground motion amplification in the monitored region and thus to identify locations that are especially vulnerable as well as ones that are not has the potential to reduce earthquake losses significantly and to guide rational regional development.

5. Conclusions

The total area affected by intermediate-depth Vrancea earthquakes comprises 300,000 km^2 and has a population of about 25 million people. Earthquakes coming from this source are of significant scientific and practical interest because of their social and economic consequences.

Bulgaria regularly experiences strong Vrancea events. The regional seismic hazard in northeastern Bulgaria is controlled mainly by Vrancea intermediate-depth events. In addition to near-field sites, urban areas located at rather large distances from earthquake sources may also thus be prone to severe earthquake hazards.

In areas exposed to significant earthquake hazards where there is a lack of data, there are two general ways to contribute to making reliable seismic hazard and risk assessments and to managing earthquakes for engineering purposes: (i) develop a strong motion registration network and (ii) implement reliable theoretical approaches for making numerically modeled estimates that are validated by recently accumulated monitoring data and updated databases.

The broad range of potential benefits of seismic monitoring can be summarized in three categories: (i) benefits that flow from information that is available immediately following a single earthquake or a swarm of earthquakes (e.g., levels of ground shaking); (ii) intermediate to long-term benefits that occur as society responds to the information provided by monitoring data and (iii) knowledge benefits that will accrue as a result of an improved fundamental understanding of earthquake processes and the distribution of earthquake risk.

Seismic instrumentation is essential to properly establish input ground motion for designing and seismically evaluating and retrofitting existing buildings. The potential benefits from improved seismic monitoring are quite varied. An important role of seismic information is to improve the accuracy (i.e., to reduce the uncertainty) of building damage predictions and loss estimates as the basis for more effective loss-avoidance regulations, as well as to enable more effective emergency preparedness activities and improved earthquake forecasting capabilities. Each earthquake provides a unique opportunity to learn. Improved monitoring of future earthquakes will lead to a more complete understanding of geophysical processes, to more effective hazard mitigation strategies and to improved emergency response and recovery.

Acknowledgements

This study is a contribution to both the NATO SfP 980468 and INTAS 104-7584 projects. The Central European Initiative University Network is gratefully acknowledged.

References

Aldea A., Okawa I., Koyama S., Poiata N. (2006a) Dense Urban Instrumentation for Site Effects Assessment in Bucharest, Romania, First European Conference on Earthquake Engineering and Seismology (13th ECEE & 30th General Assembly of the ESC), Geneva, Switzerland, 3–8 Sept. 2006, Paper Number: 518.

Aldea A., Kashima T., Poiata N., Kajiwara T. (2006b) A New Digital Seismic Network in Romania with Dense Instrumentation in Bucharest, First European Conference on Earthquake Engineering and Seismology (13th ECEE & 30th General Assembly of the ESC), Geneva, Switzerland, 3–8 Sept. 2006, Paper Number: 515.

Glavcheva R. (1983) Parameters of the Macroseismic Field and of the Attenuation of the Intensity with the Distance. In: The Vrancea 1977 Earthquake. Consequences in the Republic of Bulgaria (Brankov G., Bonchev E., Ignatiev N., Boncheva H., Christoskov L., Paskaleva I., Eds.), BAS, Sofia, pp. 151–153 (in Bulgarian).

Klügel J.U. (2007a) Error Inflation in Probabilistic Seismic Hazard Analysis, Engineering Geology, 90(2007), 186–192, available online at www.sciencedirect.com.

Klügel J.U. (2007b) Comment on "Why Do Modern Probabilistic Seismic-Hazard Analyses Often Lead to Increased Hazard Estimates" by Julian J. Bommer and Norman A. Abrahamson, BSSA, 97(6), comment, doi: 10.1785/0120070018.

Klügel J.U., Mualchin L., Panza G.F. (2006) A Scenario-Based Procedure for Seismic Risk Analysis, Engineering Geology, 88(2006), 1–22, available online at www. sciencedirect.com.

Kouteva M. (2002) Estimates of Ground Motion and Hazard Assessment Through Earthquake Scenarios, Marie Curie Fellowship final report, Project EVK2-CT-2000-5700, DST, Univ. of Trieste.

Kouteva M., Panza G.F., Paskaleva I. (2000) An Example for Ground Motions in Connection with Vrancea Earthquakes (Case Study in EN Bulgaria, Russe site), 12 WCEE 2000, Auckland, New Zealand, 30 Jan.–4 Feb., Ref.2185, on CD.

Kouteva M., Panza G.F., Romanelli F., Paskaleva I. (2004a) Modelling of the Ground Motion at Russe Site (NE Bulgaria) Due To the Vrancea Earthquakes, Journal of Earthquake Engineering, 8(2), 209–229.

Kouteva M., Panza G.F., Paskaleva I., Romanelli F. (2004b) Deterministic Earthquake Scenarios Associated with the Intermediate-Depth Vrancea Earthquakes: Case Study of Russe, NE Bulgaria, 13 WCEE, Vancouver, Canada, Aug. 1–6, 2004.

Kouteva M., Panza G.F., Paskaleva I., Romanelli F. (2004c) A View to the Intermediate-Depth Vrancea Earthquake of May 30, 1990: Case Study in NE Bulgaria, Journal of Hungarian Academy of Sciences, 39(2–3), 223–231, April–May.

Molchan G., Kronrod T., Panza G.F. (1997) Multi-scale Seismicity Model for Seismic Risk, BSSA, 87(5), 1220–1229.

Panza G.F., Vaccari F. (2000) Introduction in Seismic Hazard of the Circum-Pannonian Region (Panza G.F., Radulian M., Trifu C., Eds.), Pageoph Topical Volumes, Birkhauser Verlag, pp. 5–10.

Panza G.F., Romanelli F., Vaccari F. (2001) Seismic Wave Propagation in Laterally Heterogeneous Anelastic Media: Theory and Applications to the Seismic Zonation, Advances in Geophysics, 43, 1–95.

Paskaleva I. (2004) Assessment of the Design Parameters for North-East Bulgaria Based on the Recorded Accelerograms During 1986 and 1990 Vrancea Earthquakes, ESC General Assembly Papers, Potsdam.

Paskaleva I., Kouteva M. (2000) An Approach of Microzonation of the Town of Russe in Connection of Recent Vrancea Earthquakes and Shabla Zone, Report under bilateral cooperation with UNIVTS-DST, Trieste, Italy, pp. 200.

Paskaleva I., Kouteva M. (2007) Intermediate Progress Report, NATO SfP Project 980468, CLSMEE – Bulgarian Acad. Sci.

Paskaleva I., Kouteva M., Panza G.F., Evlogiev J., Koleva N., Ranguelov N. (2001) Deterministic Approach of Seismic Hazard Assessment in Bulgaria: Case Study in NE Bulgariia – the Town of Russe, AJNTS, 2001 (1), Albanian Academy of Sciences, Maluka, Tirana, pp. 51–73.

Romplus Earthquake Catalogue: http://www.infp.ro/Labs/catal.html.

Shebalin N., Leydecker G., Mokrushina N., Tatevossian R., Erteleva O., Vassiliev V. (1999) Earthquake Catalogue for Central and Southeastern Europe 342 BC – 1990 AD, European Commission, Report No. ETNU CT 93-0087.

Simeonova S., Solakov D., Leydecker G., Busche H., Schmitt T., Kaiser D. (2006a) Probabilistic Seismic Hazard Map for Bulgaria as a Basis for a New Building Code, Natural Hazards and Earth System Sciences, 6, 881–887.

Simeonova S., Alexandrova I., Solakov D., Popova I. (2006b) Macroseismic Observations Due To the Intermediate-Focus Vrancea Earthquakes (1940 and 1977), Romania, on the Territory of the Town of Russe, Nat. Conf. "Geosciences" 2006, Book of Abstracts, ISBN-10:954-91606-4-5, pp. 323–325 (in Bulgarian).

Solakov D., Nikolova S., Miloshev N., Simeonova S., Dimitrova L. (2007) Bulgarian Seismological Network – Current Status, Orfeus Newsletter, 7(1), http://www.orfeus-eu.org/Organization/Newsletter/vol7no1/BSN/BSN.htm

Todorovska M., Paskaleva I., Glavcheva R. (1994) Earthquake Source Parameters for Seismic Hazard Assessment Examples in Bulgaria, Proc. Xth European Conference on Earthquake Engineering, Vienna, Sept. 1994, Vol. 4, pp. 2573–2578.

Zaicenco A., Alkaz V. (2005) Design of the Strong-Motion Database for Vrancea Earthquakes, Science and Technology for Safe Development of Lifeline Systems, Bratislava, Slovak Republic, Oct. 24–25, 2005.

Codes

Code for Design of Structures in Seismic Regions, Sofia, Bulgaria, 1987.

EUROCODE 8 Basis of Design and Actions on Structures, CEN 1994.

Improved Seismic Monitoring – Improved Decision-Making: Assessing the Value of Reduced Uncertainty, Committee on the Economic Benefits of Improved Seismic Monitoring, Committee on Seismology and Geodynamics, National Research Council, ISBN: 0-309-55178-1, 196 pp.,2006.

AN ASSESSMENT OF THE PARAMETERS CONTROLLING SEISMIC INPUT FOR THE DESIGN AND CONSTRUCTION OF A HIGH-RISE BUILDING: A CASE STUDY FOR THE CITY OF SOFIA

I. PASKALEVA[*] SV. SIMEONOV, N. KOLEVA,
M. KOUTEVA, K. HADJIISKI
*Central Laboratory for Seismic Mechanics and Earthquake
Engineering, Bulgarian Academy of Sciences, 3 Acad. G.
Bonchev str., 1113 Sofia, Bulgaria*

Abstract. The earthquake record and the code for designing and constructing buildings in seismic regions in Bulgaria have shown that the country is exposed to a high seismic risk due to local shallow and regional strong intermediate-depth seismic sources and that the available strong-motion database is quite limited and therefore not at all representative of the real hazard. The problem of seismic microzonation is of crucial importance for Sofia due to rapid urbanization at the national level and to the significant growth of the city that has led to the construction of a large volume of new residential and high-rise administrative and business buildings. Sofia is exposed to a high seismic risk. Macroseismic intensities in the range of VIII–X (MSK) can be expected in the city. The earthquakes that influence the hazard in Sofia originate either beneath the city or are caused by seismic sources located in a radius of 40–300 km. The city is also affected by remote seismic zones located in Turkey, Greece and Romania. The built environment is particularly vulnerable to the long-period elements of the Vrancea seismic zone in Romania. The high seismic risk and the lack of instrument records of regional seismicity require the development of a viable seismic microzonation procedure that supplies seismic input consistent with current codes (norms) and considers local site geology in detail as well. In this study, the results of applying a seismic microzonation procedure for designing and constructing a 34-storey administrative building in Sofia are shown. Both regional (20–300 km zone) and far-distant seismicity (zone with radius larger than 300 km) were analyzed in order to define characteristic

*Central Laboratory for Seismic Mechanics and Earthquake Engineering, Bulgarian Academy of Sciences, 3 Acad. G. Bonchev str., 1113 Sofia, Bulgaria. e-mail: paskalev_2002@yahoo.com, paskalev@geophys .bas.bg

A. Zaicenco et al. (eds.), *Harmonization of Seismic Hazard in Vrancea Zone,*
© Springer Science + Business Media B.V. 2008

parameters for probable earthquake scenarios. Both deterministic and probabilistic seismic hazard estimates were made. On the basis of the analyses, accelerograms were generated considering one-dimensional linear models and making use of the geomechanical and geophysical characteristics of the soil at the construction site that were obtained by *in situ* measurements. Using the results obtained, seismic risk estimates and seismic monitoring prescriptions for the planned building are provided.

Keywords: Seismic hazard, magnitude, ground motion modeling

1. Introduction

Bulgaria lies in the Alpo-Himalayan seismic belt which is characterized by a high level of seismic activity. Comprehensive seismological studies of the region (radius of 300 km) were performed in 1982 in connection with the seismic zonation of Bulgaria (1982) and later, after 1992, to assess the seismic safety of the Kozloduy and Belene nuclear power plants. Recurrence rates of earthquakes in the mega seismic regions containing the structure were taken from investigations that were performed for other important sites (Kozloduy, Belene, Medet in the Phodops Mountains) and from other available publications.

The seismicity of the city of Sofia has been formed mainly by local earthquakes with magnitudes M = 5.5 and M = 7.0. According to the map of possible seismic zones, the building site falls in the Sofia Zone (hypocentral distances R ~ 20 km) with expected magnitude M = 6.7–7.0 and focal depth interval H = 5–15 kilometers (km). The site region can be also affected by the zones with the strongest earthquakes in the country, e.g. the Struma Zone (M = 7.6–6.8). In this sense, the building site is located in the vicinity of the most dangerous seismogenic part of Bulgaria within hypocentral distances 50–80 km.

At present, there is not a uniform methodology for analyzing expected seismic excitations which is a function of the complex, multi-factor nature of the problem. For this reason and to avoid flagrant errors, several analyses that made use of different independent approaches were performed. Following the currently active Code for Design and Construction in Bulgaria (1987), the seismic coefficient 0.27 and intensity I = IX (MSK) have been assigned to Sofia. In fact, this intensity could increase to X in a limited part of the epicentral area of a possible future earthquake with magnitude M > 6.5.

2. Assessing the Parameters Controlling the Seismic Hazard in the Target Region and the Seismic Zone Distribution

Geological outline. Sofia City is situated in the central southern part of the Sofia Kettle. Regarding its structure, the Sofia Kettle represents a complex, asymmetrical graben of the block structure of the western Srednogorie region. This graben is a post-Alpine structural unit of the Neogene-Quaternary age oriented southeast. Longitudinal, transversal and oblique faults cut the graben and the surrounding horsts. During the last 25–30 years, quite a lot of geological and geophysical data have been collected. A total of 180 boreholes that reached the Preneogene basement and 354 km of seismic profiles have been analyzed by Ilieva and Jossifov (1998). The detailed precise map of the Preneogene basement, scale 1:25,000, compiled by Ilieva and Josifov (1998) is a major result (Figure 1). An analysis of the Preneogene basement map demonstrates that some of its most important elements – the faults and faulting zones – are spread out in all parts of the Sofia depression. The geometric

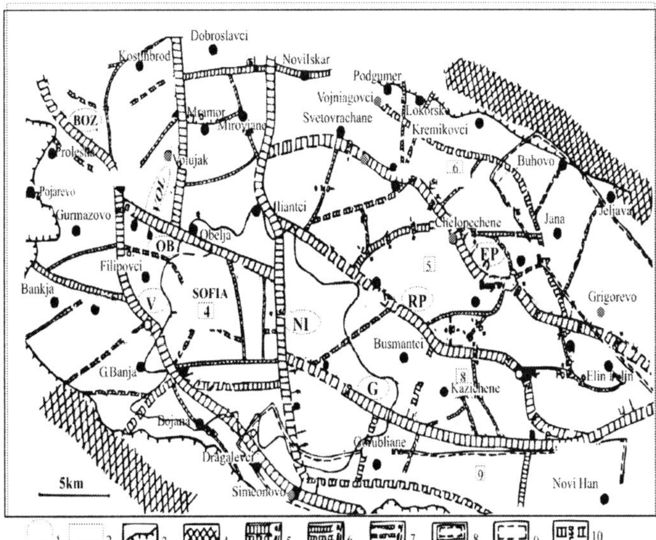

Figure 1: Scheme of the fault and block structures of the Preneogene basement in the Sofia depression (Josifov et al., 1998; Ilieva and Josifov, 1998) (redrawn by Paskaleva, 2002): 1. naming the block structures of 1st rank; (V – *Vitosha* North fault, NI – *Novi Iskar* fault, G – *Gorubljane* fault, RP – *Ravno Pole* fault, VOL – *Volujak* fault, Boz – *Bojhurishte* fault; 2. blocks; 3. boundary of the basement outcrop; 4. deep fault; 5–7. faults of the basement of 1st, 2nd and 3rd rank: (a) confirmed, (b) non confirmed; 8. zones of possible listric faulting; 9. naming the block structures of 1st rank; 10. vertical amplitude of the fault

parameters of the faults (length, width, slope, vertical and horizontal amplitude) vary in a significant range. The lengths vary from a few to more than 50 km, and the widths from several to 100–150 m. Two basic classes of faults have been differentiated (Ilieva and Jossifov, 1998): deep fault zones and faults in the Preneogene basement. The deep fault zones are structures with a large length, width and amplitude that outline areas of significant size.

The faults of the Preneogene basement are divided into three groups according to their length, vertical amplitude and width: (a) first rank faults with lengths over 20 km and varying widths that can surpass 250 m and amplitude over 100 m and that can reach more than 300–400 m in separate sectors; (b) second rank faults with lengths within the interval 5–20 km, predominating widths up to 50 m and amplitudes up to 100 m and (c) third rank faults with lengths up to 5 km, predominating widths up to 10–20 m and amplitudes up to 50 m. Due to the faulting, the Sofia depression is divided into two mega blocks – western and eastern – each of which contains first- and second-rank blocks.

2.1. SEISMIC REGIME

The seismically active faults can be divided into the following three groups based on the depth of their mobilization during earthquakes: (a) faults with seismic activation of up to 10 km (the Bozhurishte, the Elin Pelin, the Draganovo faults), (b) faults with seismic activation of up to 20 km (the Bankya fault), (c) faults with seismic activation of up to 30 km (the Vitosha, the Sub-Vitosha fault zones and the Chepitsi, the Vladaya, the Gnilyane faults). Fault activity is historically documented. There is a mobile sector of the Vitosha faults in the vicinity of the Boyana quarter of Sofia City whose occurrence is related to the 1858 Sofia earthquake (M = 6.5–7.0) (Solakov et al., 2001a, b). The same earthquake caused the activation of old landslides and rock falls and the appearance of new ones along the Vitosha fault zone between the Boyana and Dragalevtsi quarters of the city. A considerable number of these landslides and rock falls is still active, an indication of the long-term mobility of the Vitosha fault zone.

During the last 10 years, instrument registration and the publication of seismic data (Botev et al., 1999; Matova, 2001) have enabled an analysis of the spatial distribution of low seismicity in the area investigated. The epicenters of numerous weak earthquakes (M < 3.5) are localized mainly along the Negushevo and the Vitosha fault zones of the northern and the southern boundaries of the Sofia graben. Some epicenters follow the Bozhurishte, the Bankya, the Gnilyane, the Ravno Pole and the Elin Pelin faults and the Sub-Vitosha fault zone.

Strong and moderate earthquake epicenters are concentrated along faults and also in fault crossing points and blocks of the middle and the southern parts of the Sofia graben (Figure 2). Weak earthquake epicenters have very impressive localizations along the faults as well as in the horsts to the north, east and south of the Sofia graben. The epicenters of the weak earthquakes with M < 4.0 are localized generally along the Negushevo fault zone which is located in the northern periphery of the Sofia graben. The mobility of the Vitosha fault zone of the southern graben periphery is also significant but smaller than that at *Negushevo*. The influence of the low seismicity is limited. It is detected mainly in the southern and the northern periphery of Sofia City. Local earthquakes have their sources in the crust. Shallow earthquakes emerge from hypocenters mainly at a depth of up to 10 km (Matova, 2001). Earthquake hypocenters at depths of 10–30 km are concentrated near the Vitosha fault zone where it crosses the Chepintsi and the Vladaya faults. The hypocenters are in

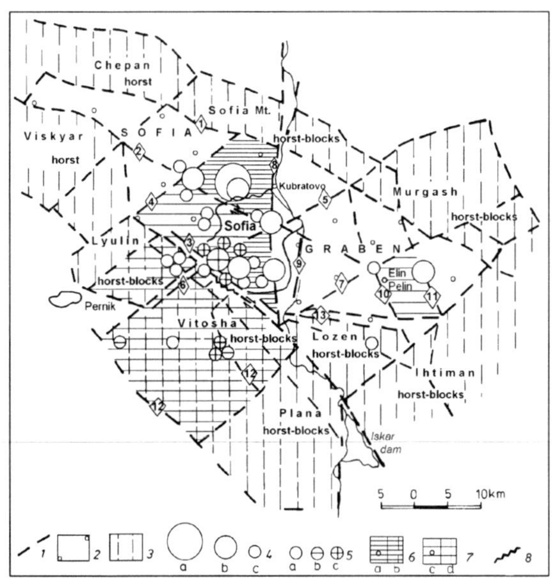

Figure 2: Matova (2001) Seismic events with magnitude M = 4.00–7.00 in the blocks of the Sofia graben and the adjacent horsts. 1. faults: (a) of the block boundary (the names are indicated in Figure 1.3), (b) sector of the *Vitosha* fault zone activated during the 1858 Sofia earthquake (M = 6.5–7.00); 2. block of the Sofia graben; 3. block of the adjacent horsts; 4. epicenters of earthquakes with magnitude: (a) M = 6.0–7.0; (b) M = 5.0–5.9, (c) M = 4.0–4.9; 5. depths of earthquake hypocenters: (a) up to 10 km, (b) 11–20 km, (c) 21–30 km; 6. blocks of considerable seismic mobility: (a) of the graben, (b) of the horsts; 7. blocks of moderate seismic mobility: (a) of the graben, (b) of the horsts; 8. seismic active sector of the Vitosha fault during the 1858 Sofia earthquake

the near vicinity of the contact point of the blocks of the Sofia graben and the Vitosha horst. The relatively deep seismic sources create a significant danger for large areas of Sofia. Strong, moderate and weak seismicity are related specifically to the faults investigated and to the blocks of the Sofa graben and the surrounding horsts. As a product of the investigations carried out about the recent geodynamics of the Sofia complex graben (Shanov et al., 1998), a geological and geophysical three-dimensional model of the Earth's crust for a part of the graben (Shanov et al., 1998) was constructed. The analyses of the seismotectonic situation and structure of the basement of the Sofia depression, the reference structural model, were used to construct a regional model.

2.2. GEOLOGICAL CONDITIONS AND LITHOLOGY OF SEDIMENTS IN THE SOFIA KETTLE

The Sofia Kettle is one of the numerous continental basins in southern Bulgaria filled in with Miocene-Pliocene sediments. The nature of these sediments is heterogeneous in composition and they are present in rocks of different ages that are found along the edges of the depression. The Sofia Kettle is filled with Neogene and Quaternary sediments that are loose and more or less compact. The total thickness of the Neogene and Quaternary sediments reaches 1,200 m (near the town of Elin Pelin).

The layer that characterizes local geological conditions is mainly Quaternary sediments that represent the uppermost part of the geological sequence in the Sofia Kettle. These deposits are widespread and lie over older rocks and soils. The Quaternary cover is thick at 3–100 m or more (Frangov and Ivanov, 1999). Quaternary deposits are characterized generally by wide areas of spreading and by great diversity in composition and properties. In general, Quaternary soils possess relatively high pore volume and low deformation modules. The permeability of the sediments by water predetermines high ground water levels as well as the specificity of suffusion and liquefaction of fine water-saturated sands. The thickness of the Quaternary aquifer is provisionally accepted to be in the 20–30 m range in the western part of the Sofia graben and up to 50–70 m in the eastern part. Analyses of the available data from shallow boreholes (30–50 m) show an abrupt change in velocities from 300–400 to 1,700–1,800 m/s, coinciding with the level of ground water (Petkov et al., 1971). Three velocity layers can be distinguished in the sedimentary complex (Neogene and Quaternary sediments) in Sofia (Demirev et al., 1971; Ivanov, 1997; Ilieva and Josifov, 1998) (i) a top layer with a depth of 3–4 m (depending on the ground water level) with a velocity of the longitudinal waves of 300–400 m/s; (ii) an intermediate layer at a depth of 15–90 m and a velocity of 1,300–1,800 m/s and (iii) a lower one with a velocity of 1,850–2,100 m/s. The velocities of the longitudinal seismic waves in the Neogene sediments are of the order of 1,500–2,000 m/s and in the basement are in the range of 4,000–5,600 m/s. An analysis of vertical graphs shows a very great variety of values of the velocities of the propagation of longitudinal seismic waves. In the depth interval under consideration, they change from 300 to 4,000 m/s. The presence of significant lateral and vertical velocity heterogeneities has been recognized, and in some boreholes, an inversion of the velocities in depth has been observed.

The three types of local geological conditions recognized from available data and from the geological map correspond to a certain degree to the four groups of soil classes accepted in the Bulgarian code for Design and Construction in Seismic Regions (1987) (Figure 3). The culture layer is composed of old structures, technogenic soils and industrial and household waste. The first layer of old structure remnants is the thickest in the central parts of the city and can be as much as 10 m. Technogenic soils are distributed around open pits, big apartment buildings and power stations. They are composed of reworked rocks and soils and industrial waste. Neogene sediments are represented to depths in the 25–30 m range by sands of various grain size distributions. The important feature of their granulometry is the high silty fraction (42–82%) which to a great degree increases the peculiarity of their physical and mechanical properties, i.e., low bulk density, high pore coefficient, high moisture content and plastic consistency (Frangov and Ivanov, 1999).

The bulk density of the Quaternary sediments is about 1.84–2.07 g/cm^3. The non-consolidated Neogene sediments (sands, clays, claysands, alevrolites, etc.) are characterized by a density not surpassing 2,000 kg/m^3. The rocks constructing the basement (limestone, marls, sandstone, andesites, etc.) have densities varying from 2,500 to 2,600 kg/m^3.

Field measurements on the site for the 34-storey administrative building in Sofia were made. The field investigation was carried out by (a) the downhole shooting (DSh) method for compression (P) waves and (b) the DSh method for shear (SH and SV) waves. The DSh method was applied by using the 24-channel PS logging system McSeis-170f for seismic data acquisition, processing and analysis and the borehole Pick BP-3040 of OYO Corp., Japan.

Field measurements were accomplished with specialized equipment at the building site applying three schemes of excitation, one vertical and two horizontal (Figure 4). DSh was carried out in depth with a step of 1 m for compression (P) waves. For the SV-wave measurements, the step was 2 m since they are performed to control the SH-wave travel-time measurement. The DSh records acquired had 500 μs resolution and additional amplification selected according the respective ambient noise level.

The Vp/Vs ratio and the Poisson coefficient (v) from the surface of the site down to 42 m are computed by the experimental data acquired from the DSh deep boreholes. The geological materials from the surface of the site (central part of Sofia) down to 42 m may be classified in three groups (by the ratio

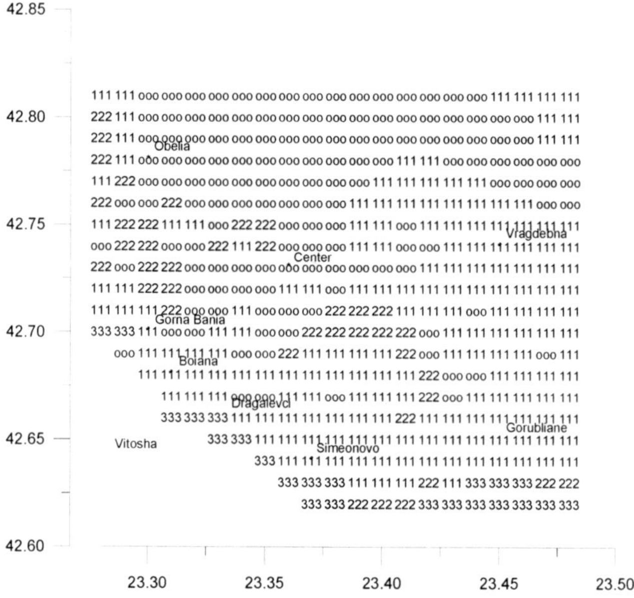

Figure 3: Map of geological conditions in Sofia: 3. hard rock; 2. hard sedimentary rock; 1. intermediate soft Upper Neogene deposits; 0. Soft Quaterenry deposits (Paskaleva et al., 2004a, b)

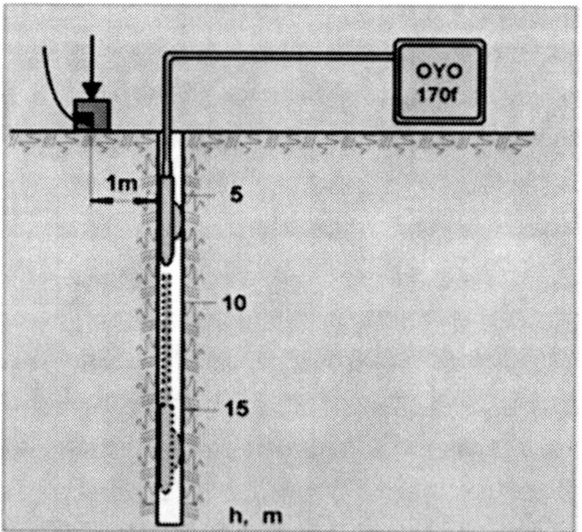

Figure 4: Scheme for excitation and recording the P-, SH- and SV-waves by DSh

Vp/Vs and the Poisson coefficient) as follows: (a) slightly damped ground materials (down to 3 m) characterized with Vp/Vs ratios of about 1.71 and Poisson coefficient of about 0.190; (b) damped ground materials with Vp/Vs from 2.72 to 4.79 and Poisson coefficient from 0.392 to 0.446 and (c) ground materials below the underground water level with Vp/Vs from 4.52 to 8.10 and Poisson coefficient from 0.462 to 0.488. The underground water level determined for the investigated site was about 11 m.

3. Recent Seismic Hazard Assessments

3.1. EARTHQUAKE RECORD

During the past centuries, the territory of Bulgaria has suffered strong earthquakes. Strong seismic manifestations in the Sofia region date from the 15th century. The historical record shows that strong earthquakes occurred in 1450, (M ~ 6) and in 1,557 (M ~ 5.5), and the quakes of 1818 (VIII–IX MSK) and 1858 (M $\approx$ 6.3, I_0 = IX MSK) occurred in the vicinity of Sofia. The 1818 earthquake caused significant damage in Sofia; as a consequence of this quake, thermal springs appeared in the western part of the city. The biggest documented seismic event in Sofia was the earthquake of September 18/20 1858 with epicentral intensity Io = IX (Petkov and Christoskov, 1965) According to the empirical relationship of epicentral intensity versus magnitude (Glavcheva, 1990) I_0 = 1.45M–3.24 Lg (h) + 2.82, if we accept Io = IX and focal depth h = 10 km, we obtain M = 6.5. Some other investigations of the same event have shown that the magnitude of this 1858 earthquake may even have reached M = 7.0 (Christoskov et al., 1986).

The earthquakes that occurred in the 20th century were documented more accurately in the publications of Watzov (1902, 1908). The strongest earthquake in Sofia to date was the earthquake of October 18, 1917 with epicentral intensity Io = VII–VIII (Petkov and Christoskov, 1965) and M = 5.2 (Christoskov and Grigorova, 1968; Ivanov, 1931; Kirov, 1952; Petkov and Christoskov, 1965). This earthquake caused a change in the level of some mineral springs in the Sofia Zone. There is information (Ivanov, 1931) that the ground motion amplitudes reached 5–10 cm. This quake was followed by a year-long series of aftershocks. Since that earthquake, the Sofia Zone has been relatively quiet as it is at present. From 1918 to 1997 (80 years) no earthquakes of M > 5.0 have been documented in this zone.

Two thirds of the earthquakes occurred in seismic sources at depths not more than 10 km below the Earth's surface. In this depth range, the entire energetic variety of seismic sources is present including the strongest known earthquake (1858) with epicentral intensity Io = IX and magnitude M = 6.5.

About one sixth of the earthquakes lineout the second depth range of 10 < H < 25 km where it seems that earthquakes with magnitudes not larger than M = 5 occurred at these depths. The deepest seismic sources have been localized at depths of 30 km, where again mainly earthquakes with magnitudes less than five occur. There is only one historical event (M = 6.0, H = 25 km) associated with this depth interval, but the information source has not been considered as a very reliable one.

3.2. SEISMIC MICROZONATION STUDIES

Existing publications concerning the seismic microzonation of the city are based on engineering, geological, hydrological and geophysical investigations and macroseismic data (Petkov and Christoskov, 1965; Petrov and Iliev, 1970; Petkov et al., 1971; Demirev et al., 1971). The coefficient of variation of seismic intensity has been calculated according to Medvedev's formula (1977). The results have shown that the values of the incremental intensities can be 3° and for the most unfavorable ground conditions even 3.5.

A hybrid neo-deterministic procedure (Panza et al., 2001) was applied to compute the peak ground acceleration (PGA) distribution, and a set of synthetic accelerograms corresponding to receiver sites with 100 m between two consecutive locations was generated. Epicentral distances of 10–20 km (Slavov, 2000; Paskaleva, 2003; Paskaleva et al., 2003; Slavov et al., 2004) were taken into account in the computations performed. The PGA distribution is shown in Figure 5, and the boundaries of horizontal and vertical acceleration according to the current Bulgarian code for designing buildings and structures in seismic regions are shown by straight lines. The distribution of the vector of the horizontal components is shown in Figure 6. According to these figures, for epicentral distances less than 15 km, the PGA does not exceed the standard

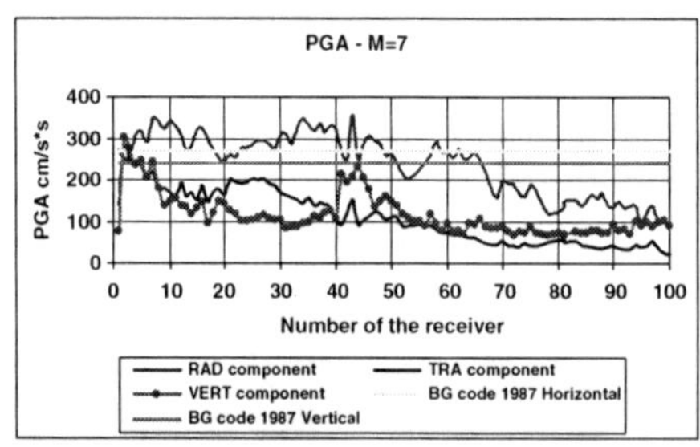

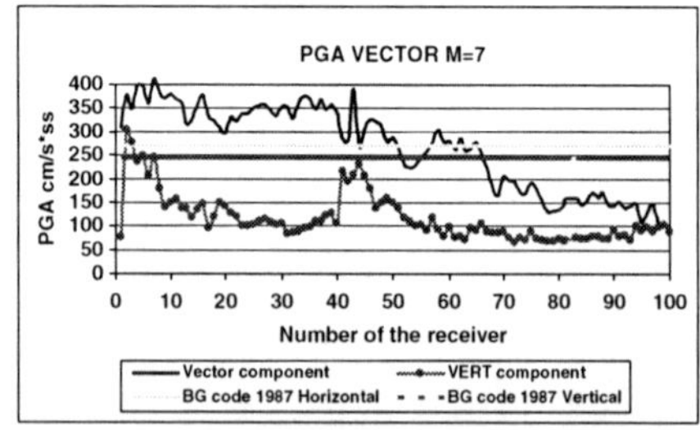

Figure 5: PGA distribution for M = 7

Figure 6: Vector distribution of the horizontal PGA for M = 7

accelerations values Kc = 0.27 g for horizontal components. The PGA of horizontal and vertical components attenuates with an increase in the epicentral distance. For epicentral distances greater than 16 km, the PGA values, mean and mean +1 standard deviation exceed the acceleration prescribed in the Bulgarian Code (1987) (standard values). For the vertical component, the PGA values, mean and mean +1 standard deviation become smaller than the standard ones for epicentral distances greater than 12 km.

4. Seismic Hazard Computations

4.1. ZONES WITH MACRO INTENSITY I > VI, MSK AND LOCAL SEISMOGENIC STRUCTURES IN A RADIUS OF 30–100 KM

The region covering the vicinity of the target site includes the central part of the eastern Balkans and a large part of the territory of Bulgaria. As stated previously, thorough seismological investigations of the region (radius 300 km) were performed for the seismic zonation of Bulgaria (Bonchev et al., 1982) and later for assessing the seismic safety of the Kozloduy and Belene nuclear power plants (Sokerova et al., 1992; Dachev et al., 1995). The seismicity of Bulgaria is related to seismic zones (Sokerova et al., 1992) defined on the basis of the spatial seismicity distribution and the potential seismogenic zones (Bonchev et al., 1982) as the following: Rhodopes, Kresna, Maritza, Sofia and Gorna Orjahovitza. Seismicity outside of Bulgaria has been connected to the following seismic zones: northern Greece, the Aegean Sea, Edrine, the Anatolian fault and the Vrancea Zone.

4.2. ZONES CONSIDERED AS POTENTIAL SEISMOGENIC SOURCES: MODELS AND CORRESPONDING SOLUTIONS OF THE PEAK GROUND ACCELERATION FOR THE TARGET CONSTRUCTION AREA

Kresna seismic zone: Two of the strongest earthquakes in Europe during the 20th century occurred in this zone in April 1904 ($M_S = 7.1$ and $M_S = 7.8$). These earthquakes occurred in the Kresna-Simitly region where the active Krupnik and Struma faults cross. The deepest earthquakes have also occurred in the Kresna Zone where the thickness of the Earth's crust is significant (it is 45–40 km beneath the Rhodopes Massive). The hypocenters of the earthquakes were distributed mainly in the uppermost 30 km of the crust with a maximum concentration between 5 and 20 km. The maximum focal depth in this zone has been up to 50 km (Sokerova et al., 1992).

Maritza seismic zone: The strongest earthquakes that have occurred in this zone were the Chirpan earthquake on April 14, 1928 ($M_S = 6.8$) and the Plovdiv earthquake on April 18, 1928 ($M_S = 7.0$). This seismic zone has developed in the upper 20 km of the crust, but some single events have been detected at depths up to 45 km. The highest hypocenter density has been observed at depths of 5–10 km (Sokerova et al. 1992).

Sofia seismic zone: The seismic activity in this zone is mainly connected with the marginal neotectonic fault in the Sofia graben. Available historical documents show evidence for destructive earthquakes that occurred in this zone in the 15th and 18th centuries (Watzov, 1902). The first well-documented strong event ($M_s = 6.0$) occurred in 1818 near Sofia (Christoskov et al., 1979). The strongest earthquakes that occurred in this zone were the ones in 1858 (in the vicinity of Sofia) and in 1905 (near the town of Tran in the western part of the zone). The earthquakes in this zone occur in the upper 20 km of the crust with observed hypocentral concentrations at depths of 5 and 15 km (Sokerova et al., 1992). In the Sofia Zone, significant damage can be expected due to earthquakes generated by local sources since external sources can cause damage of macroseismic intensity up to $I = VII$ (MSK) (Josifov et al., 1998).

Gorna Orjahovitza seismic zone: The strongest known earthquakes in this zone was the quake of 1813 ($M_S = 7.0$) followed by a lack of seismic activity up to 1986 when two moderate earthquakes ($M_S = 5.3$ on February 21 and $M_S = 5.7$ on December 7) struck near Stragitza. Seismic activity has been considered to be concentrated mainly in the upper 15 km of the crust, but single events occurred at depths up to 25–30 km (Sokerova et al., 1992).

Effects of earthquakes in Macedonia: After 1900, three strong earthquakes, particularly the one on March 8, 1931, caused damage corresponding to intensity $I = V$ (MSK). It is interesting to mention that the first event with the lowest magnitude ($M = 6.0$) caused macroseismic intensity $I = V$ on the east bank of the Iskar River (Elin Pelin).

Effects of earthquakes in northern Greece: Considering the strongest earthquakes in this zone, the Sofia Kettle falls into the $I = IV$ range of the generalized macroseismic field. Some local effects of $I = V$ have been observed too, e.g. the earthquake of Sept. 26, 1932 ($M = 6.9$) caused intensity $I = V$ at Bogurishte, Elin Pelin, Pernik, Kremikovtzi and Bankja.

Effects of earthquakes in northwest Turkey: This is the most distant region of investigation with respect to the Sofia Kettle. It generates shallow, crustal earthquakes. According to available data, the strongest one was on August 9, 1912 and was $M = 7.3$; the intensity of the generalized macroseismic field in the Sofia Kettle was $I = IV$. Local effects of $I = IV$ were also observed in Sofia and its vicinity due to the earthquakes of March 18, 1953 (M = 7.2) and October 6, 1964 (M = 6.8).

Effects of intermediate-depth (H > 60 km) earthquakes in Romania (Vrancea): Compared with all the other seismic sources outside of Bulgaria, these earthquakes have the greatest effects on the Sofia Kettle, and this zone has been the most active macroseismic one. During the last century, nine earthquakes with magnitudes M > 6.0 occurred in the Vrancea Zone and were felt in varying degrees in Sofia. The maximum observed effects were I = V–VI in the October 10, 1940 quake (M = 7.3) and in the March 4, 1977 quake (M = 7.2). In addition, the earthquakes in Macedonia on March 7, 1931 (M = 6.0), March 8, 1931 (M = 6.7) and July 26, 1963 (M = 6.1) caused damage in Bulgaria with intensity I > III MSK. The Sofia Kettle experienced intensity I = IV of the generalized macroseismic field of these three earthquakes.

4.3. SHAKING FIELD OF THE TARGET CONSTRUCTION REGION

Prognostic estimates of expected seismic loads for the region of the target site at maximum magnitude M and focal depth H were carried out by E. Bonchev et al. (1982) and Christoskov et al. (1979) as part of the prognostic seismic zonation of the whole country. These results were extracted from shaking maps constructed for the time periods 100, 1,000 and 10,000 years. These maps were based on complex maps of the potential seismogenic zones, the recurrence rates in these potential seismogenic zones and models of the isoseists of earthquakes with magnitude M > 4.6 where the data on the seismic sources observed and seismic source dimensions were accepted by Bonchev et al. (1982).

Attenuation laws: A crucial point in seismic hazard analysis is calculating the ground motion parameter from magnitude data. Due to the insufficient strong-motion database available for Bulgaria (Nenov et al., 1990), to date no seismic attenuation laws have been derived for the country. It is common engineering practice to use empirical attenuation laws that have been obtained on the basis of sufficiently large, homogeneous strong-motion databanks containing records from regions with similar geological and seismotectonic characteristics. The relationships of Ambraseys et al. (1996) are derived for rock since Ambraseysand Bommer (1991), Ambraseys et al. (2005) are based on mixed soil conditions excluding soft soil. Idriss (1985) pointed out that the PGA values in a wide range of accelerations are almost identical for both soils – rock and stiff. These three relationships were derived using the European strong motion data bank, whereas the other two are based on available strong motion Balkan data. The attenuation laws R1 and R2 published in the report on the investigations of the Balkan Region by IZIIS were used in our hazard calculations. The prognostic

I. PASKALEVA ET AL.

PGA for the target site computed using different attenuation laws is shown in Table 1. One can see in Table 1 that the maximum accelerations are associated with the Sofia Zone bearing in mind the uncertain estimates of the distance between the target construction site and the seismic source.

4.4. PROBABILISTIC SEISMIC HAZARD ASSESSMENTS

The analysis of the seismic hazard for the target site is an assessment of the probability to exceed given levels of ground motions. For this purpose different versions of EQRISK and NEQRISK software (McGuire, 1976, 1978; Lee and Trifunac, 1985) are used most often. The complete procedure for assessing the seismic input for a given system includes three major analyses: (a) defining the seismic models; (b) generating uniform seismic risk spectra and (c) producing synthetic accelerograms.

TABLE 1: Peak ground acceleration (g) for maximum magnitudes with different attenuation relationships

Zone	M_{max}	km	Ambraseys et al. (1996)	Ambraseys and Bommer (1991)	Idriss (1985)	R1	R2	Mean
Struma Mesta	7.5	95	0.059	0.057	0.055	0.075	0.051	0.059
Plovdiv	7.1	50	0.085	0.084	0.095	0.141	0.104	0.102
Sofia	7.0	15	0.244	0.255	0.33	0.311	0.250	0.278
Sofia	7.0	10	0.354	0.381	0.52	0.363	0.299	0.383

The definition of the seismic models concerns both the physical and the observed quantitative recurrence ratio of seismic events and an analytical description of the geological soil conditions. The methodology adopted in most probabilistic seismic hazard analyses is analogous to the methodology applied in a deterministic seismic hazard assessment. The major differences concern the size of the seismic sources since the probabilistic approach allows us to have very small, plain ruptures as well as large seismic seismotectonic provinces, and the procedure assumes constant seismogenic potential for the whole seismic source. This means that the chance for a given earthquake to occur is equal in the framework of the whole seismic source. The seismicity of the region containing the target construction site was determined using seismic sources in three different models: Model D1 (1991 Sofia and 1992 Skopje progress reports by IZIIS on the safety of the Kozloduy nuclear power plant site), Model S3 (Energoproekt, 1989, 2005) and Model M2 (Todorovska et al., 1994) (see Figure 7). These models were made for similar studies on the sites of other important structures in Bulgaria. Available geological, seismological and geophysical information for a radius of 30–40 km around the target site was factored in.

Similar to a deterministic procedure, a probabilistic analysis takes in into account ground motion attenuation but it considers a family of curves, each of which is related to a certain ground motion parameter. For this task, the 1992 results obtained for ground motion attenuation for the Kozloduy nuclear power plant site – namely R1 and R2 – were used with an increase in the epicentral distance for a certain level of earthquake excitations. Using the results supplied in the progress reports on the seismic safety of the Belene nuclear power plant site (1991, 1993) three different combinations of the standard deviation were applied for regional sources and the local zone: (a) a conservative alternative $+\sigma = 0.7$; (b) a mean alternative $+\sigma = 0.6$ and (c) $+\sigma = 0.5$. The probabilistic approach integrated the influence of all possible earthquakes of different severities that might occur at different places with different probabilities. The result obtained is the probability of exceeding different levels of selected ground motion parameters (intensity, maximum expected ground motion,

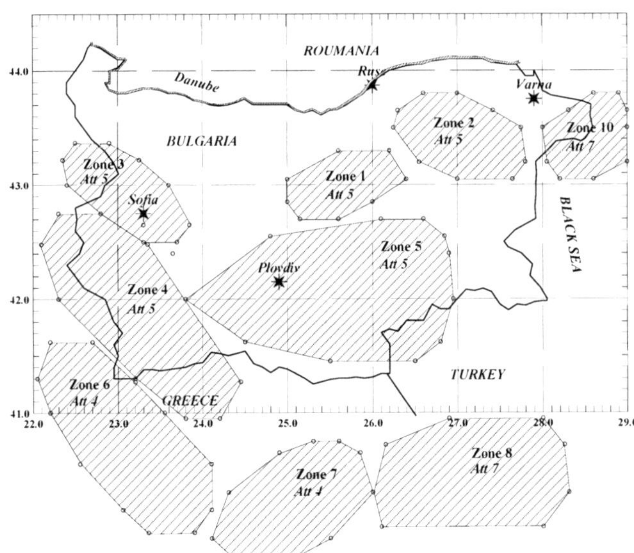

Figure 7: Seismogenic zones Model M2 (Todorovska et al., 1994)

spectra, etc.) at the site for a given time period. This part of the seismic hazard assessment (McGuire, 1976, 1978) does not deal with the frequency range within which the maximum ground acceleration could occur.

A comparison of the hazard curves generated using models D1, M2 and S3 considering the R1 attenuation law is shown in Figure 8. PGA values computed for different return period and attenuation laws considering the three seismic models are shown in Table 2.

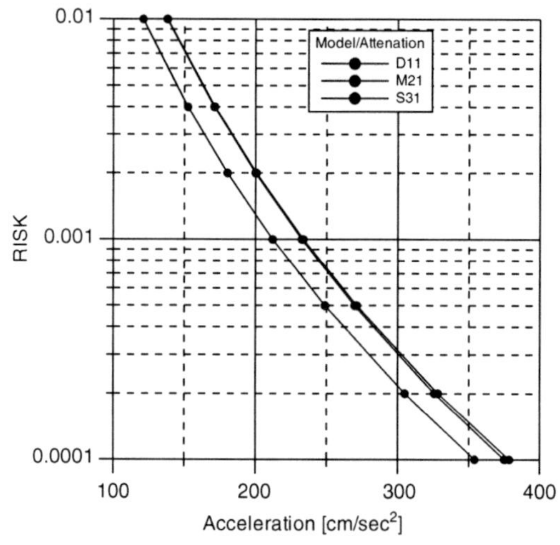

Figure 8: Hazard curves following models D11, M21, S31* and attenuation law R1

TABLE 2: Alternative peak ground accelerations (cm/s^2) by model and attenuation laws

Return period (years)	D11[a]	M21	S31	D12[a]	M22	S32
100	138	121	138	138	100	117
250	171	153	171	171	123	143
500	200	180	201	200	144	166
1,000	233	212	234	232	168	191
2,000	270	249	271	270	195	220
5,000	326	305	328	326	235	263
10,000	375	354	379	375	270	300

[a]Note: D11, M21 and S31 are models D1, M21 and S31 with attenuation law R1; D12, M22 and S32 are models D1, M2 and S3 with attenuation law R2

4.5. CONSTRUCTING A UNIFORM SEISMIC RISK SPECTRA

Considering the spectral nature of ground motion, constructing a uniform seismic risk spectra involves a certain return period. Taking into account the frequency range in which maximum accelerations occur, the procedure used for generating uniform seismic risk spectra aims to define a frequency-dependent risk function as follows (Lee and Trifunac, 1985).

(a) The geometry of the seismic sources, i.e., point, area, deep sit or volume, is defined. In each zone, the estimated number of seismic events N (Mj) for each magnitude interval dM is then defined. Uncertainties in the seismicity estimates and assessments of the severity of earthquakes are also defined by analyses of historical seismicity and considerations of the geology and the tectonics of the region accompanied by expert and statistical estimates.

(b) Each seismic zone is divided into unit source elements assuming that the epicenter of any earthquake with severity M_j could be located with equal probability in each unit. Seismicity is thus distributed among all elements.

(c) The frequency-dependent description of the attenuation of ground motion amplitudes is determined, and the type of distribution of observed ground motion amplitudes $S_{ij}(w)$ around the estimated mean value $S(w)$ is chosen. Thus we define the function $Q_{ij}[S(w)]$ describing the probability to exceed $S(w)$ at the site for an event with severity M_j that occurs in the unit source of the seismic zone.

(d) The expected number of cases $Ne[S(w)]$ for which $S(w)$ will be exceeded at the site in all seismic units in each seismic zone due to all probable earthquakes is then estimated. Then the probability $P[S(w)]$ that $S(w)$ would be exceeded at least once at the given site in a given time period is computed.

(e) The frequency-dependent risk function $P[S(w)]$ estimated for a limited number of discrete frequencies is constructed.

Thus it is possible to construct uniform risk spectra for a given probability level (Lee and Trifunac, 1985; Todorovska et al., 1994). The Fourier spectra with the corresponding risk not to exceed an exploitation period of 100 years is shown in Figure 9a. These spectra and the Eurocode 8 spectra for soil conditions "C" were used to generate three component accelerograms (X, Y, and Z).

According to Eurocode 8, the return period of a design earthquake is 475 years which corresponds to a 10% probability of exceedence for a 50-year construction exploitation period. To limit the damage and financial losses due to weaker but more frequent earthquakes, a second seismic hazard level has been defined to correspond to a 95-year return period (10% probability of exceedence for a 10-year exploitation period). Particular structures (e.g., nuclear power plants, dams, tailings, bridges) have special codes. The computed probability of exceedence of different levels of a chosen ground motion parameter (e.g., maximum expected acceleration, spectra) at a site for a given return period and a given exploitation period is shown in Figure 9b.

The intensity (MSK) hazard curve expressed (Simeonova et al., 2006) is shown in Figure 10. One can see on this figure that the design earthquake of 475 years corresponds to intensity I = VIII with a probability of exceedence of 10% in a 50-year exploitation period. For a 100-year period, this probability is 19% for I = VIII, and for a 1,000-year return period there is a 10% probability

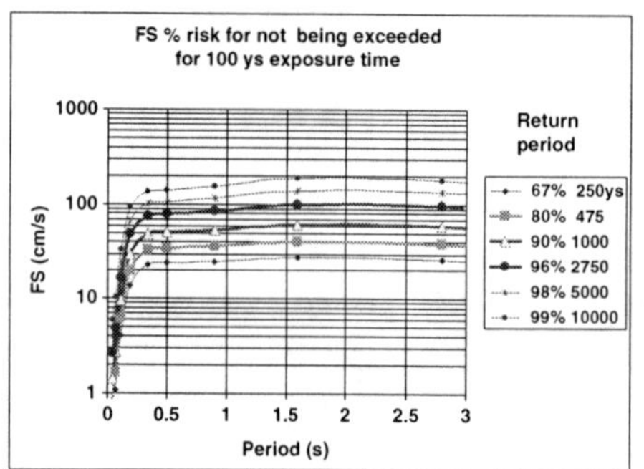

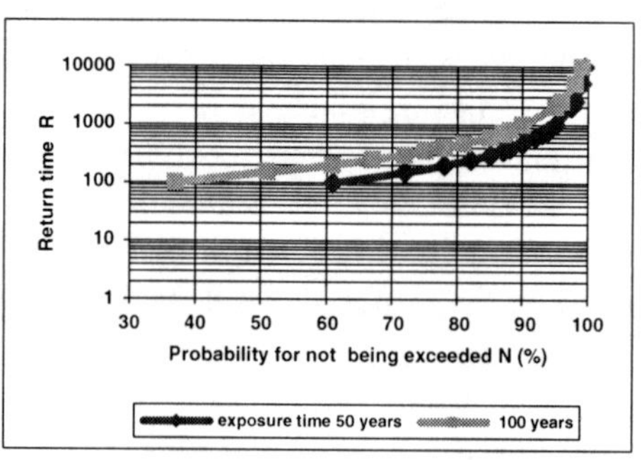

Figure 9a: Fourie spectra with corresponding percentage of risk of non- exceedence in 100-year exploitation period

Figure 9b: Probability of exceedence of different levels of preliminary selected ground motion parameter for a given return period and given exploitation period

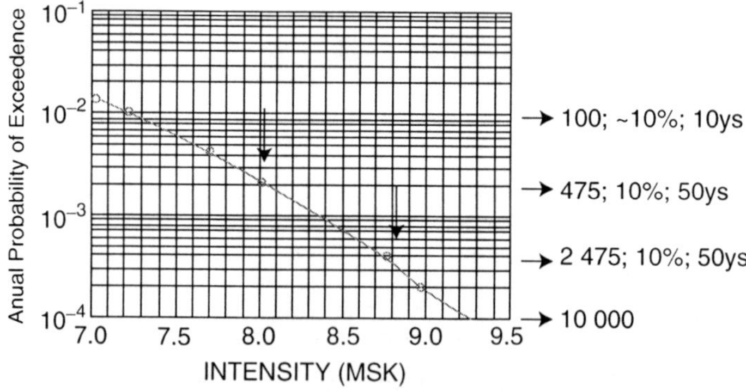

Figure 10: Intensity hazard curve for Sofia City (Simeonova et al., 2006)

of exceedence of I = VIII+ in a 100-year exploitation period. Following Richter's empirical ratio Log (A) = 0.33I − 0.50, we obtain an acceleration amplitude A = 295 cm/s^2.

Synthetic accelerograms. Synthetic accelerograms are a product of the results of the previous parts of this complex study. The analysis of the engineering-geological cross section characterized by a specific geological structure described by the corresponding densities and velocities of the longitudinal and transverse seismic waves with respect to the soil classification provided in Eurocode 8/2001 has shown that: (1) the mean transverse seismic wave velocity at 30 m depth is V_{s30} = 270 m/s, which classifies a construction site as class "C" [(V_s = 180–360 m/s and SPT = 15–50 (bl30 cm)]; (2) the mean transverse seismic wave velocity at 50 m depth is V_{s50} = 375 m/s which classifies a construction site as class "B" [(V_s = 360–800 m/s and SPT ≥ 50 (bl30 cm)] and (3) the corresponding free vibration periods at depths 30 and 50 m are T_{30} = 0.44 and T_{50} = 0.53 s.

In this analysis, accelerograms at an epicentral distance of 15 km due to an earthquake with expected M = 7.0 and I = VIII and an assumed attenuation law by Todorovska et al. (1994) are computed. The diffuse seismic model M2 (Figure 7) was adopted for the description of all seismic zones in spite of the fact that remote zones such as the Aegean Sea and the Kresna Zone can also be modeled as point sources. The frequency-dependent risk function obtained is used to generate accelerations at confidence levels of 0.8 and 0.9 applying two simplified models – "rock soil conditions" and "intermediate soil conditions" – as shown in Figure 3. Synthetic accelerograms with different probabilities of exceedence/non-exceedence of selected ground motion parameters were generated considering different site-to-source and site soil conditions, and different hypocentral distances were computed. The maximum expected accelerations extracted from the accelerograms generated for different seismic zones with different percentages of non-exceedence and "weak" soil conditions (soil conditions for site-to-source S = 0 and s = 2,1,0 for the site – Figure 3), I = VIII and epicentral distance 15 km are shown in Table 3a. The maximum deformations at the site of the structure are given in Table 3b. The maximum accelerations, velocities and displacements of the rotations at the construction site are computed in Table 3c and the maximum accelerations, velocities and displacements of the torsion at the construction site are in Table 3d.

TABLE 3a: Model of "Weak" geological conditions, site to source S = 0, Vs < 350 m/s, site s = 2, 1, 0, I = VIII, hypocentral distance 20 km

Shakability	Axes	A_{max} cm\s^2	V_{max} cm\s	D_{max} cm	Site-to source	Site	% of non-exceedence
Sofia	x	228	17	11	0	2	80
	y	197	23	14	0	2	80
	z	132	10	7	0	2	80
475 years	x	231	20	12	0	1	80
20%	y	202	25	15	0	1	80
100 years	z	131	12	8	0	1	80
	x	330	16	10	0	0	80
	y	281	21	13	0	0	80
	z	193	9	7	0	0	80
Sofia	x	323	24	16	0	2	90
	y	277	31	19	0	2	90
	z	131	12	8	0	2	90
475 years	x	324	28	17	0	1	90
10%	y	284	35	21	0	1	90
50 years	z	186	17	11	0	1	90
	x	468	23	14	0	0	90
	y	395	30	17	0	0	90
	z	274	13	9	0	0	90

TABLE 3b: Maximum deformations at the site of the structure

Shakability	Magnitude/ intensity M/I	Epic. distance R (km)	Axes	Site-to-source	Site	% of non-exceed-ence	Deformation X $\times E^{-5}$	Y $\times E^{-5}$	Z $\times E^{-5}$
1	2	3	4	5	6	7	8	9	10
	M = 7	15	x	0	0	50	10	4	3
		15	x	0	0	50	9	3	2
		15	x	1	1	50	12	4	3
		15	x	2	2	50	12	4	3
		20		0	0	90	26	11	7
		20		2	2	90	34	15	10
		20		2	1	90	34	15	9
		20		2	0	90	33	15	9
475 years/20%/100years exp. period	I = VIII	20	x	2	2	80	19	9	5
			x	1	1	80	1	1	0.4
475 years/10%/50 years expl. period	I = VIII	20	x	2	2	90	27	12	8
	I = VIII	20	x	1	1	90	41	22	11

TABLE 3c: Maximum accelerations, velocities and displacements of the rotations at the construction site

Shakability	Magnitude/ intensity M/I	Epicentr. Distance R [km]	Axes	Site-to-source	Site	% of non-exceedence	Rotation		
							a_{max} rad/s²	V_{max} rad/s	d_{max} rad
1	2	3	4	5	6	7	11	12	13
	M = 7	15	x	0	0	50	3311	53	7.6
		15	x	0	0	50	3012	48.6	6.4
		15	x	1	1	50	1832	41	8.9
		15	x	2	2	50	1822	40	8.9
		20		0	0	90	5331	172	15
		20		2	2	90	3941	163	26
		20		2	1	90	3926	163	25
		20		2	0	90	5694	172	17
475 years/20%/100 years	I = VIII	20	x	2	2	80	3225	116	15
	I = VIII	20	x	1	1	80	2773	142	20
475 years/10%/50 years	I = VIII	20	x	2	2	90	4875	164	22
	I = VIII	20	x	1	1	90	4121	196	27

TABLE 3d: Maximum accelerations, velocities and displacements of the torsion at the construction site

Shakability	Magnitude/ intensity M/I	Epicentr distance R [km]	Axes	Site-to-source	Site	% of non exceedence	Rotation		
							A_{max} x E^{-5} rad/s²	V_{max} x E^{-5} rad/s	D_{max} x E^{-5} rad
1	2	3	4	5	6	7	14	15	16
	M = 7	15	x	0	0	50	1417	43	4
		15	x	0	0	50	1266	40	3
		15	x	1	1	50	874	38	4
		15	x	2	2	50	878	38	4
		20	x	0	0	90	3283	129	11
		20	x	2	2	90	2828	126	15
		20	x	2	1	90	2959	130	15
		20	x	2	0	90	3572	138	11
475 years/20%/100 years expl. prd	I = VIII	20	x	2	2	80	2205	87	9
	I = VIII	20	x	1	1	80	2529	116	16
475 years/10%/50 years expl. prd	I = VIII	20	x	2	2	90	3017	123	12
	I = VIII	20	x	1	1	90	3567	163	22

5. Seismic Risk Assessment

The operating basis earthquake (OBE) represents the level of ground motion at a site at which only minor damage is acceptable. OBE is determined by probabilistic methods as a minimum 50% probability that a specific level of motion will not be exceeded in the lifetime of the structure. For a structure with a life time of 100 years, a 60% probability that the level will not be exceeded requires estimating evaluation parameters for a return period of approximately 200 years. The maximum design earthquake (MDE) is the largest, reasonably conceivable earthquake that appears to be possible within a tectonic province. An MDE will produce the maximum level of ground motion for which the structure should be designed. Actually, damage to buildings greatly depends on local site conditions. In addition, evidence from many earthquakes repeatedly illustrates that damage in the near field is quite different from that in the far field. Also, damage is highly selective in terms of both the natural frequency of structures and the frequency content of ground shaking. Although the effects of local site conditions and epicentral distance on building damage have been confirmed by many earthquakes and have been investigated extensively (Aki, 1993; Finn, 1991; Bard, 1995; IAEE, 1996), these local site effects have not been incorporated using a damage probability matrix (DPM) into the vulnerability analysis of buildings. To incorporate local site effects into damage estimates for existing buildings, subjective expert opinions are often used. Medvedev (1962) first systematically summarized the effect of site conditions on building damage.

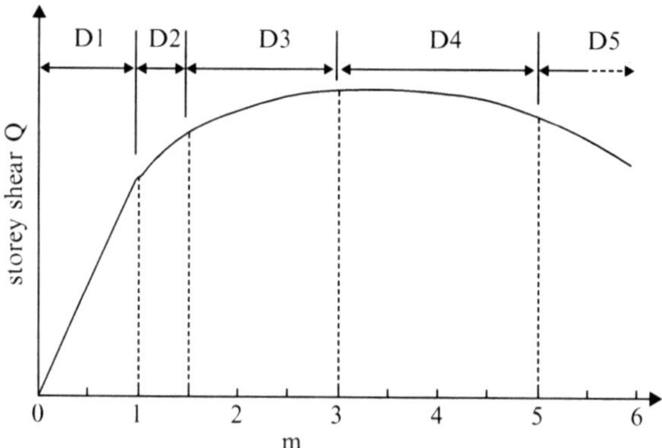

Figure 11: A schematic diagram illustrating a typical plot of the base shear Q against the ductility factor μ. The segments of the curve representing various damage states D1, D2, D3, D4 and D5 (corresponding to no damage, slight damage, moderate damage, extensive damage and complete damage) are indicated

In reality, local site conditions and epicentral distance may influence the amplitudes of the seismic action as well as the frequency content of strong ground motions. Wena et al. (2002) attempted to analyze the combined effects of both soil conditions and epicentral distance on the vulnerability of a typical 21-storey reinforced concrete building frame/shear wall building and thus proposed DPM. The maximum storey ductility factor is a key parameter for indicating building damage. The threshold ductility factors μ for the onset of slightly damaged, moderately damaged, extensively damaged and completely damaged states are 1.0, 1.5, 3.0 and 5 respectively. Figure 11 (Wena et al., 2002) shows a typical plot of the base shear Q versus the ductility factor. Five damage states – D1, D2, D3, D4 and D5 – are assumed corresponding to undamaged, slightly damaged, moderately damaged, extensively damaged and completely damaged, respectively. These values are empirical constants and may vary from one type of building to another.

A popular way to present the DPM is to plot the probability for exceeding a particular damage state versus seismic intensity input as vulnerability or fragility curves. DPM for probability of damage (P) versus intensity for soil (site) conditions SCII (40 m of fill, alluvium) are shown in Figures 12a–b (Near = near-field earthquakes; Far = far-field earthquakes). For the selected building type (high-rise of more than 21 stories) for the target site which consists of 40 m of fill, alluvium (SCII), the probabilities of structures suffering moderate damage to complete collapse are 1, 18, 70 and 96% for near-field earthquakes and 9, 26, 85 and 98% for far-field earthquakes of I = VII, VIII, IX, X respectively. Therefore, site effect is very important and not to be neglected. In short, a high-rise reinforced concrete building resting on a soft site is more conducive to damage than the same structure on a rock site. In addition, high-rise buildings are more susceptible to damage from far-field earthquakes than from near-field quakes because far-field seismic ground motions are richer in higher period content than near-field seismic motions as per field observations during large earthquakes.

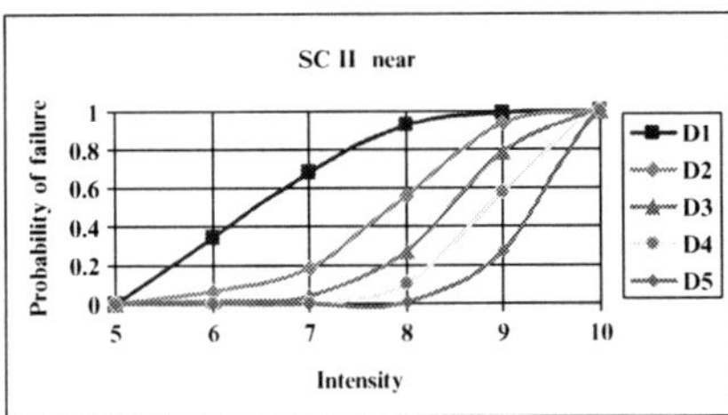

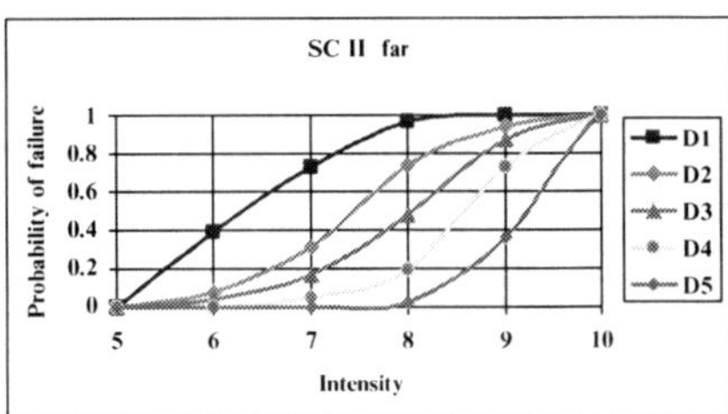

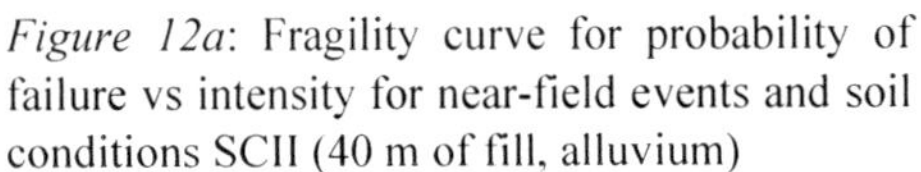

Figure 12a: Fragility curve for probability of failure vs intensity for near-field events and soil conditions SCII (40 m of fill, alluvium)

Figure 12b: Fragility curve for probability of failure vs intensity for far-field events and soil conditions SCII (40 m of fill, alluvium)

6. Conclusions

The results discussed in this work address the need for reliable procedures, capable of predicting realistic seismic input. The results obtained from different approaches were validated against corresponding specifications recommended in Eurocode 8 and in the current Bulgarian construction code. The seismic environment plays a significant role in the appropriateness of seismic hazard assessments. In highly seismic regions where the largest earthquakes may occur every 100–300 years, 475-year shaking could be considered as design ground motion. A deterministic scenario for such an event would allow examining the ground motion effects caused by rupture propagation that might not be available from probabilistic analyses. This would lead to insights into the risk for particular structures or could be used for seismic microzonation. Regional assessments often benefit most from deterministic models in which the probability of occurrence of the scenario in one city, for example, is small but the probability for occurrence is large for the region. The concept of multiple deterministic scenarios allows for rational preparations, even if the details of the earthquake predicted may be wrong.

A seismic hazard evaluation based on the traditional probabilistic seismic hazard analysis (PSHA) relies on the probabilistic analysis of earthquake catalogues and of ground motion, on macroseismic observations and on instrument records. Recently, PSHA has been found to have limitations in providing a reliable assessment possibly due to insufficient information about historical seismicity which can introduce relevant errors in a purely statistical approach based mainly on seismic history.

For areas where very limited numbers of or no instrument earthquake records are available, a synthetic time series can be generated to estimate expected ground motion to define a pre-disaster seismic load for the built environment and for seismic microzonation.

Acknowledgements

The support from the NATO SfP 980468 and INTAS 104-7584 projects is gratefully acknowledged.

References

Aki, K. (1993) Local site effects on weak and strong ground motion. Tectonophysics, 218(1), 93–111.

Ambraseys, N. and Bommer, J. (1991) The attenuation of ground acceleration in Europe Earthq. Eng. Struct. Dyn., 20, 1179–1202.

Ambraseys, N., Simpson, K., Bommer, J. (1996) Prediction of horizontal response spectra in Euro. Earthq. Eng. Struct. Dyn., 25(4), 371–400.

Ambraseys, N., Douglas, J., Sarma, S., Smid, P. (2005) Equations for the estimations of strong motions from shallow crustal earthquakes using data from Europe and the Middle East: horizontal peak ground acceleration and spectral acceleration, Bull. Earthq. Eng., 3, 1–53.

Bard, P. (1995) Effects of surface geology on ground motion: recent results and remaining issues. Proceedings of the 10th European Conference on Earthquake Engineering, Vienna, pp. 305–323.

Bonchev, E., Bune, V., Christoskov, L., Karaguleva, J., Kostadinov, V., Reisner, G., Rizhikova, S., Shebalin, N., Sholpo, V., Sokerova, D. (1982) A method for compilation of seismic zoning prognostic maps for the territory of Bulgaria, Geol. Balc., 12(2, 3), 3–48.

Botev, E., Babachkova, B., Dimitrov, B., Velichkova, S., Tzoncheva, I., Donkova, K., Toteva, T. (1999) Preliminary data on the seismic events recorded by NOTSSI in July-December 1997, J. Bulg. Geophys, 25(1–4) 215–224.

Christoskov, L., Grigorova, Ek. (1968) Energetic and time-spatial characteristics of the destructive earthquakes in Bulgaria after 1900. News Geoph. Inst., BAS, XII, 79.

Christoskov, L., Sokerova, D., Rijikova, Sn. (1979) New catalogue of the earthquakes in Bulgaria for period V B.C. – XIX A.D., GFI Founds, BAS, pp. 1–54.

Christoskov, L., Georgiev, Tz., Deneva, D., Babachkova, B., Dachev, H., Velev, A., Dolbokina, I., Christov, D., Ivanchev, G. (1986) On the study of the seismic activity of the Sofia zone as a base for organization of complex prognostic studies, Bulg. Geoph. J., XII(2), BAS, Sofia (in Bulgarian), 73–80.

Dachev, H., Vaptzarov, I., Filipov, L., Solakov, D., Simeonova, S., Nikolova, S., Sokolova, P., Botev, E., Georgiev, Tz. (1995) Seismology, geology, neotectonics, seismotectonics and seismic hazard assessment for the PNPP Belene site, Rep. of Proj.: Investigations and activities for increasing of the seismic safety of the PNPP Belene site, Geoph. Inst., BAS, S, I, 250.

Demirev, A., Stefanov, G., Petkov, I. (1971) Microseismic characteristic of the central city part of Sofia. Proceedings of the Third European Symposium on Earthquake Engineering, pp. 115–120.

Glavcheva, R. (1990) Parametrization of the isoseismal from Bulgarian earthquakes, Bulg. Geoph. J. XVI, 4, 38–44 (in Bulgarian).

Finn, W.D.L. (1991) Geotechnical engineering aspects of microzonation. Proceedings of the Fourth International Conference on Seismic Zonation, Vol. I., Earthquake Engineering Research Institute, Oakland, CA, pp. 199–259.

Frangov, G., Ivanov, Pl. (1999) Engineering geological modeling of the conditions in the Sofia graben and land subsidence prognoses. Proceedings of the 4th WG, Dec. 3–6, Sofia, pp. 17–13.

Ivanov, P. (1931) A contrubution to the Sofia earthquakes of October 18, 1917. Min. Agricult., Sofia, pp. 1–33 (in Bulgarian).

Idriss, I.M. (1985) Evaluating seismic risk in engineering practice. Proceedings of the eleventh International Conference on Soil Mechanics and Foundation Engineering, San Francisco, 12–16 August, A.A. Balkema, Rotterdam.

International Association for Earthquake Engineering (IAEE) (1996) A word list Earthquake Resistant Regulations, Tokyo, Japan.

Ivanov, P. (1997) Assessment of the geological conditions in the Sofia kettle under seismic impact. Proceedings of the International IAEG Conference, Athens, Balkema, Rotterdam, pp. 1265–1270.

Ivanov, P., Frangov, G. (1997) Zoning of the territory of Sofia kettle according to the deformation properties of the sediments. Proceedings of the Second Working Group Meeting "Expert Assessment of Land Subsidence Related to Hydrogeological and Engineering Geological Condition in the Regions of Sofia, Skopje and Tirana", Skopje October, pp. 62–66.

Ilieva, M., Jossifov, D. (1998) Structure of the preneogene basement in the Sofia depression. BAS Bulgarian Geophys. J., XXIV(1–2) Sofia, 109–120.

Kirov, K. (1952) A contribution to the study of the earthquakes in the Sofia Kettle. Annual of Gen, Direction for Geology and Mine Exploration, Vol. 5, p. 407 (in Bulgarian).

Lee, V., Trifunac, M. (1985) Uniform risk spectra of strong earthquake ground motion: NEQRISK, Depart. Of Civil Eng. Rep. No 85-05, Univ. of Southern California, Los Angeles.

Matova, M. (2001) Recent manifestations of seismotectonic activity in Sofia region and their land subsidence potential. Proceedings of the Final Conference of UNESCO-BAS Project on land subsidence, June 27–30, Sofia, pp. 93–98.

McGuire, R.K. (1976) EQRISK Evaluation of Earthquake Risk to Site Open-file Report R76-67.

McGuire, R.K. (1978) FRISK Seismic Risk Analysis Using Faults as Earthquake Sources Open-file Report R78-1007.

Medvedev, S.V. (1962) Engineering seismology. Moscow: National Publisher of Literatures on Structure, Architecture and Architectural Materials; 1962 (in Russian).

Medvedev, S.V. (1977) Seismic Intensity Scale MSK – 76, Publ. Inst. Geophys. Pol. Acad. Sci. 117, 95–102.

Nenov, D., Georgiev, G., Paskaleva, I., Lee, V.W., Trifunac, M.D. (1990) Strong earthquake ground motion data in EQINFOS: Accelerograms recorded in Bulgaria between 1981 and 1987, Dept. of Civil Engineering Joint Report No CE 90-02, Bulgarian Academy of Sciences, Central Lab. for Seismic Mechanics & Earthquake Engineering (C.L.S.M.E.E.), Bulgaria, and University of Southern California (U.S.C.), Los Angeles, U.S.A 55 pages.

Panza, G.F., Romanelli, F., Vaccari, F. (2001) Seismic wave propagation in laterally heterogeneous anelastic media: theory and applications to the seismic zonation, Advances in Geophysics, Academic Press, Vol. 43, pp. 1–95.

Paskaleva, I. (2002) Visitor's Report NATO Project, DST, Triest, Italy.

Paskaleva, I. (2003) A note on peak velocities and surface strains associated with strong earthquakes on the territory of Bulgaria and case study for Sofia, CD – ROM First Int. Conf. Science and Technology for safe development of life line systems, 4–5 November, Sofia.

Paskaleva, I., Panza, G.F., Vaccari, F., Rajgelj, S., Ivanov, P. (2003) Deterministic modelling for microzonation of Sofia: An expected earthquake scenario. Proceedings of the International Conference in Earthquake Eng. SE 40EEE, 26–29 August, Skopje.

Paskaleva, I., Panza, G.F., Vaccari, F., Ivanov, P. (2004a) Deterministic modelling for microzonation of Sofia – an Expected Earthquake Scenario. Acta Geod. Geoph. Hung.,Vol. 39(2–3), 275–295

Paskaleva, I., Matova, M., Frangov, G. (2004b) Expert assessment of the displacements provoked by seismic events: Case study for the Sofia metropolitan area. Pure and Applied Geophysics, Birkhauser Verlag, Basel, No. 161, pp. 1265–1283.

Paskaleva, I., Kouteva, M., Vaccari, F., Panza, G.F. (2008) Application of the neo-deterministic seismic microzonation procedure in Bulgaria and validation of the seismic input against Eurocode 8, http://www.mercea08.org/.

Petkov, I., Christoskov L. (1965) On seismicity in the region of the town of Sofia concerning the macroseismic zoning, Ann. Sofia Univ., 58, 163–179.

Petkov, I., Velchev, Tz., Dachev, Chr. (1971) Complex geophysical investigations in the city of Sofia and its vicinity in connection with the microseismic zoning. Proceedings of the Third European Symposium on Earthquake Engineering, pp. 103–108 (in Russian).

Petrov, P., Iliev, I. (1970) The effect of engineering geological conditions on seismic microzoning in Sofia. Proceedings of the 3rd European Symposium on EE, Sofia, pp. 79–86.

Simeonova, S., Solakov, D., Leydecker, G., Busche, H., Schmitt, T., Kaiser, D. (2006) Probabilistic seismic hazard map for Bulgaria as a basis for a new building code Nat. Hazards Earth Syst. Sci., 6, 881–887, www.nat-hazards-earth-syst-sci.net/6/881/2006/.

Slavov, Sl. (2000) Ground motion modelling in the City of Sofia (TRIL – ICTP Visitor's report).

Slavov, Sl., Paskaleva, I., Kouteva, M., Vacacri, F., Panza, G.F., (2004) Deterministic earthquake scenarios for the City of Sofia, Pure and Applied Geophysics (PAGEOPH) Birhauser Verlag, Basel, Vol. 161, pp. 1221–1239.

Sokerova, D., Simeonova, S., Nikolova, S., Solakov, D., Botev, E., Glavcheva, R., Dineva, S., Babachkova, B., Velichkova, S., Maslinkova, S., Donkova, K., Rizikova, S., M. Arsovski, M., Matova, M., Vaptzarov, I., Filipov, L. (1992) Geomorphology, neotectonic, seismicity and seismotectonic of NPP Kozloduy, Final Report (Summary) on IAEA Mission: Design basis earthquake for seismic upgrading of NPP Kozloduy, Sofia, p. 200.

Solakov, D., Christoskov, L., Simeonova, S. (2001a) Possible consequences from strong Earthquakes in the Territory of Bulgaria. Minno Delo Geologia J., 1, 52–57.

Solakov, D., Simeonova, St., Christoskov, L. (2001b) Seismic hazard assessment for the Sofia area, Annali di Geofisica 44, 3, 541–555.

Shanov, S., Tzankov, Tz., Nikolov, G., Bojkova, A., Kurtev, K. (1998) Character of the resent geodynamics of Sofia complex graben, Review of the Bulg. Geol. Soc., 59(I), 3–12.

Todorovska, M., Paskaleva, I., Glavcheva, R. (1994) Earthquake source parameters for seismic hazard assessment examples in Bulgaria. Proceedings of X-th European Conference on Earthquake Engineering, Vienna, Sept., Vol. 4, pp. 2573–2578.

Watzov, Sp. (1902) The Earthquakes in Bulgaria in XIX, Sofia, pp. 1–93 (in Bulgarian).

Watzov, Sp. (1908) Construction of the seismography in Bulgaria, Bulg. Literary Soc., BAS, LXIX (in Bulgarian).

Wena, Z.P., Hua, Y.X., Chaub, K.T. (2002) Site effect on vulnerability of high-rise shear wall buildings under near and far field earthquakes, Soil Dyn. Earthq. Eng. 22, 1175–1182.

CodeBulgarian code for design in seismic areas 1987, BAS 68.

Funds ENERGOPROEKT 1989 and 2005.

Report 1992 Investigations and Activities for Increasing the NPP "Kozloduy" 1992, IZIIS, Skopije.

Report on Contract on Investigations and activities for increasing the safety of the Kozloduy NPP site, Vol. A – III, Sofia, 1991, IV-213. Shebalin et al. (1974) Catalogue of earthquakes (part I, 1901–1970, part II, prior to 1901) UNDP/UNESCO Survey of the Seismicity of the Balkan Region, UNESCO, Skopje, Yugoslavia).

Funds of the Committee of Geology, Sofia Investigation and activities for increasing the NPP"Belene" 1991–1993.

Funds of the Ministry of the Environment and waters, Sofia (1997–1998) Detailed investigations for the purposes of seismic microzonation of Sofia. (in Bulgarian).

Funds of the Ministry of the Environment and waters, Sofia, (1998) Josifov, D., Paskaleva, I., Botev, E., Ilieva, M. Seismicity of the Sofia kettle and seismic microzonation of Sofia; Structure and geodynamics of the Sofia kettle and seismic microzonation of Sofia (in Bulgarian).

Funds CLSMEE Report Seismic microzonation of the site for Business building – 34 stories, Sofia, 2007.

AN ESTIMATION OF THE MAXIMUM INTERSTORY DRIFT RATIO FOR SHEAR-WALL TYPE STRUCTURES

G. KOLEVA[*,1], I. SANDU[2], S. AKKAR[3]
[1]*Central Laboratory for Seismic Mechanics and Earthquake Engineering, Bulgarian Academy of Sciences, Sofia, Bulgaria*
[2]*Institute of Geology and Seismology, Academy of Science, Chisinau Moldova MD2028*
[3]*Earthquake Engineering Research Center, Department of Civil Engineering, Middle East Technical University 06531 Ankara Turkey*

Abstract. In displacement-based engineering, the maximum interstory drift ratio (MIDR) is one of the most influential parameters for evaluating the seismic performance of existing structural systems. MIDR is also a key parameter in force-based designs satisfying serviceability limits for new structural systems. A set of predictive equations is derived for estimating MIDR on shear-wall systems with fundamental periods ranging from 0.5 to 1.25 s. The equations are derived from a recently compiled ground-motion dataset that consists of 532 accelerograms recorded from 131 strong-motion events in seismically active regions in Europe and neighboring countries. The moment magnitude (M_w) values in the database are within the limits of $5 \leq M_w \leq 7.6$ with recording distances of up to 100 km. The proposed equations estimate the interstory drift ratio for different horizontal component definitions (geometric mean, maximum and random). The equations include focal mechanism and site class as explanatory variables. The quadratic magnitude dependence and magnitude-dependent geometric decay are included in the functional form of the predictive model. To verify the functional form, residual and error analyses and a p-value test were done.

Keywords: MIDR, predictive equation, ground motion, statistical analysis, regression analysis, p-value

[*]Central Laboratory Seismic Mechanics and Earthquake Engineering (CLSMEE), Bulgarian Academy of Sciences (BAS), Acad. G. Bonchev block 3, 1113 Sofia, Bulgaria.

A. Zaicenco et al. (eds.), *Harmonization of Seismic Hazard in Vrancea Zone,*
© Springer Science + Business Media B.V. 2008

1. Introduction

The purpose of this study is to evaluate the reliability of a recently derived set of predictive equations that estimates the maximum interstory drift ratio (MIDR) on shear-wall structures. The equations are capable of estimating MIDR values for shear-wall buildings with fundamental periods of 0.5, 0.75, 1.00 and 1.25 seconds (s). We used robust statistical analyses (including a p-value test) to evaluate the proposed predictive model. This study makes use of the methodology presented in Akkar and Bommer (2007a, b) and in many ways mimics the schemes proposed in those papers. The ground motions used here are listed in Akkar and Bommer (2007b) along with the date, magnitude, focal mechanism, site class and number of records per event. We used the shear-wall MIDR values calculated from the mathematical model proposed by Miranda and Akkar (2006) for deriving and evaluating the proposed predictive model. In other words, the MIDR values computed from the Miranda and Akkar model are accepted as actual deformation demands on shear-wall models. This assumption is consistent because the aforementioned model runs a response history analysis for calculating MIDR for a given structural model.

The following sections first describe the database and predictive model and then present the relevant statistics for evaluating the proposed mode.

2. Dataset and Method of Analysis

2.1. DATA

We used the dataset recently compiled by Akkar and Bommer (2007a, b) that has been used for deriving predictive equations for peak ground velocity and spectral quantities (spectral displacement and pseudo-spectral acceleration). The database contains 532 accelerograms from 131 earthquakes with moment magnitude values ranging from 5.0 to 7.6. The dataset contains digital and analogue records. Refer to Akkar and Boomer (2007b) for more details about the distributions of magnitude, focal mechanism, distance and site class of the dataset.

Comparative statistical techniques for regression analyses of observed and predicted data were conducted for geometric mean (GM), maximum (MAX) and random (RND) components at fundamental periods from 0.50 to 1.25 s.

The predictions for GM, RND and MIDR were obtained from the same functional form used previously by Akkar and Bommer (2007a, b). This form includes magnitude, focal mechanism and site class as explanatory variables. Quadratic magnitude dependence and magnitude-dependent geometric decay are also included in the functional form of the predictive model. Equation (1) shows the functional form

$$\log(MIDR_{xx}) = b_1 + b_2 M_w + b_3 M_w^2 +$$
$$(b_4 + b_5 M_w)\log(\sqrt{R_{jb}^2 + b_6^2}) + b_7 S_S + b_8 S_A + b_9 F_N + b_{10} F_R \tag{1}$$

where M_w is the moment magnitude, R_{jb} is the Joyner–Boore distance in kilometers (km); S_A and S_S are the variables representing the influence of site class taking values of 1 for stiff and soft soil sites respectively and zero otherwise; F_N and F_R are variables for the influence of fault type taking values of 1 for normal and reverse ruptures respectively and zero otherwise.

TABLE 1: Regression coefficients b_1-b_{10} for Equation (1) for different horizontal component definitions (GM, MAX and RND). S1 and S2 denote the intra- and inter- event standard deviations, respectively

Damp	Period	b_1	b_2	b_3	b_4	b_5	b_6	b_7	b_8	b_9	b_{10}	S1	S2
GM													
0.10	0.50	−6.973	1.638	−0.120	−2.231	0.189	6.674	0.240	0.118	−0.052	0.074	0.300	0.125
0.10	0.75	−8.052	1.839	−0.129	−2.002	0.166	4.870	0.272	0.135	−0.045	0.027	0.283	0.157
0.10	1.00	−8.859	2.089	−0.149	−2.128	0.185	5.104	0.306	0.142	−0.025	0.014	0.279	0.137
0.10	1.25	−9.074	1.961	−0.125	−1.625	0.104	4.698	0.300	0.109	−0.034	0.013	0.272	0.131
MAX													
0.10	0.50	−6.318	1.516	−0.113	−2.430	0.215	7.191	0.239	0.118	−0.064	0.098	0.314	0.115
0.10	0.75	−7.857	1.851	−0.131	−2.147	0.179	5.717	0.264	0.128	−0.051	0.040	0.301	0.147
0.10	1.00	−8.571	2.039	−0.145	−2.142	0.181	5.045	0.301	0.145	−0.045	0.008	0.294	0.133
0.10	1.25	−8.858	2.016	−0.136	−1.875	0.142	3.973	0.320	0.135	−0.049	0.010	0.289	0.131
RND													
0.10	0.50	−6.541	1.604	−0.123	−2.539	0.228	7.762	0.240	0.122	−0.062	0.091	0.320	0.121
0.10	0.75	−7.670	1.741	−0.121	−2.046	0.166	5.534	0.267	0.140	−0.059	0.051	0.305	0.162
0.10	1.00	−8.468	1.989	−0.143	−2.201	0.194	5.356	0.287	0.153	−0.026	0.057	0.300	0.148
0.10	1.25	−8.830	1.995	−0.136	−1.899	0.148	4.410	0.297	0.136	−0.035	0.047	0.303	0.137

The regression coefficients b_1–b_{10} of predictive Equation (1) are computed using a maximum likelihood one-stage regression as described in Akkar and Bommer (2007b) and tabulated (Table 1) for each fundamental period: $T = 0.5$, 0.75, 1.00 and 1.25 s. The regressions consider inter- and intra-event variability separately, and a pure error analysis is included as used by Ambraseys et al. (2005) to determine the magnitude of dependence on the standard deviations.

2.2. METHOD OF ANALYSIS

2.2.1. Data Sorting

The major purpose of this study is to discuss the reliability of a new set of predictive equations for MIDR in terms of robust statistical parameters. Previous work (Akkar and Bommer, 2007a, b) has justified the need for a general predictive equation that depends on the parameters of site class, magnitude, distance and focal mechanism. In this way, the predictive model can account for different regions in Europe.

The first step in this investigation was sorting and grouping the data according to the R_{jb} and M_w parameters. Then we adjusted the predictions and observed data for the same site class and fault type conditions (rock and strike-slip fault mechanism) by using Equation (1) for each record in the dataset. The selected distance intervals ($d_1 = 5$–15 km, $d_2 = 15$–25 km, $d_3 = 25$–35 km, $d_4 = 35$–65 km, $d_5 = 65$–85 km) resulted in the following record distribution: 144, 96, 82, 120 and 48 respectively. Note that 42 records were not used in verification procedures because they fell into $0 < d < 5$ km and $85 < d < 100$ km distance ranges.

TABLE 2: The distribution of records according to event magnitude and Joyner–Boore distance Rjb

Nr.	Mw&Rjb	(5–15)km	(16–25)km	(26–35)km	(36–65)km	(66–85)km
1	M(5.0–5.2)	34	22	15	21	4
2	M(5.3–5.5)	37	19	12	15	3
3	M(5.6–5.8)	10	21	21	28	6
4	M(5.9–6.1)	28	15	14	22	5
5	M(6.2–6.4)	10	8	3	4	2
6	M(6.5–6.7)	11	6	9	14	14
7	M(6.8–7.0)	6	3	2	9	3
8	M(7.1–7.3)	5	2	3	4	0
9	M(7.4–7.6)	3	0	3	3	11

The magnitude axis was also divided into following subintervals: $5.0 \leq M_w \leq 5.2$, $5.3 \leq M_w \leq 5.5$, $5.6 \leq M_w \leq 5.8$, $5.9 \leq M_w \leq 6.1$, $6.2 \leq M_w \leq 6.4$, $6.5 \leq M_w \leq 6.7$, $6.8 \leq M_w \leq 7.0$, $7.1 \leq M_w \leq 7.3$, $7.4 \leq M_w \leq 7.6$. Only the mean class interval values $<d_1> = 10$ km, $<d_2> = 20$ km, $<d_3> = 30$ km, $<d_4> = 50$ km, $<d_5> = 75$ km and $<M_{w1}> = 5.1$, $<M_{w2}> = 5.4$, $<M_{w3}> = 5.7$, $<M_{w4}> = 6.0$, $<M_{w5}> = 6.3$, $<M_{w6}> = 6.6$, $<M_{w7}> = 6.9$, $<M_{w8}> = 7.2$, $<M_{w9}> = 7.5$ of the above bins were considered in numerical comparisons. The distribution of data (number of records) for the above magnitude and distance bins is shown in Table 2.

2.2.2. Adjusting Data

The original Akkar and Bommer (2007b) dataset contains records from several site classes—rock, stiff, soft and very soft—and focal mechanisms - normal, reverse and strike-slip. To reduce the inhomogeneous distribution, we made adjustments using the regression coefficients in Equation (1) and following Akkar and Bommer (2006). We accepted one fundamental case: soil type = rock and focal mechanism = strike slip. We then transformed all other records of different site classes and fault types to this fundamental case. We performed the above calculation with the help of expressions similar to Equation (2) which provides an example of changing rock (R) and reverse-fault (RR) records to rock (R) and strike slip (SS) records.

$$\log(MIDR\,_{R,SS}^{Corr}) = \log(MIDR\,_{R,RR}^{Obs}) - b_{10} \tag{2}$$

We used the adjusted dataset throughout the comparative analysis.

3. Evaluation of the Functional Form and Interpretation of the Results

The evaluation is based on the adjusted rock site and strike-slip data as described in the previous section.

3.1. RESIDUAL ANALYSIS

Figure 1 shows the residual scatters (residuals for GM, RND and MAX, components) against magnitude together with a best-fit line to reveal the existence of possible trends in the residuals. The results presented are for a shear-wall building with a fundamental period of 1.0 s.

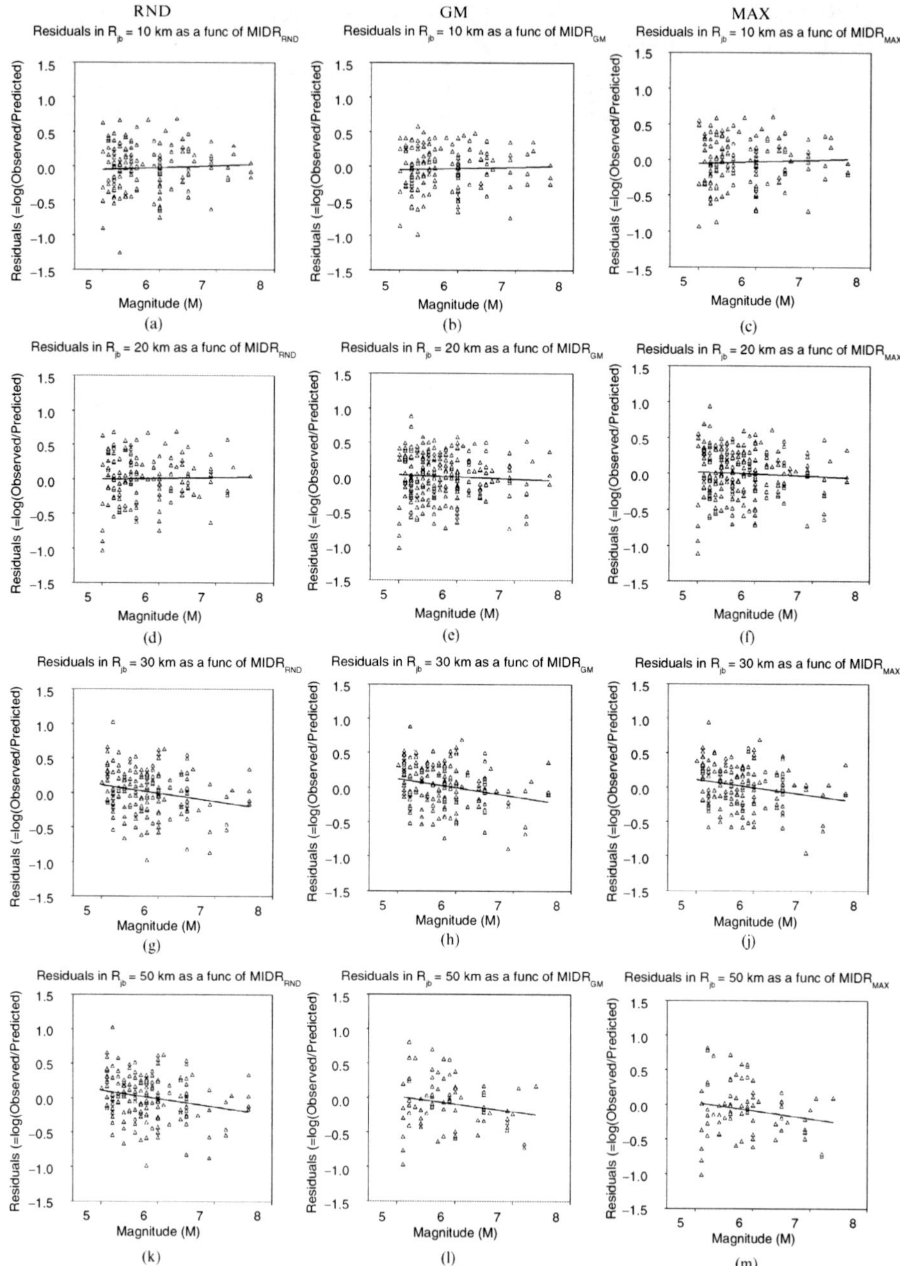

Figure 1: Residuals as functions of magnitude for different MIDR estimations (RND: random component, GM: geometric mean component, MAX: max component from left to right) and distances (R_{jb} = 10, 20, 30, 50 km)

The comparative plots show that the residuals as a function of magnitude do not display a biased trend with the changes in distance intervals. This observation may indicate that the proposed predictive model can adequately address variations in the actual database.

3.2. ACTUAL DATA VARIATION AND PREDICTED TRENDS

Figure 2 presents the observed (actual) data trends and the estimates from the predictive equation. The chosen fundamenal period is T = 0.5 s for this comparative case. The predictive equation seems to capture adequately the actual variation in the data trends.

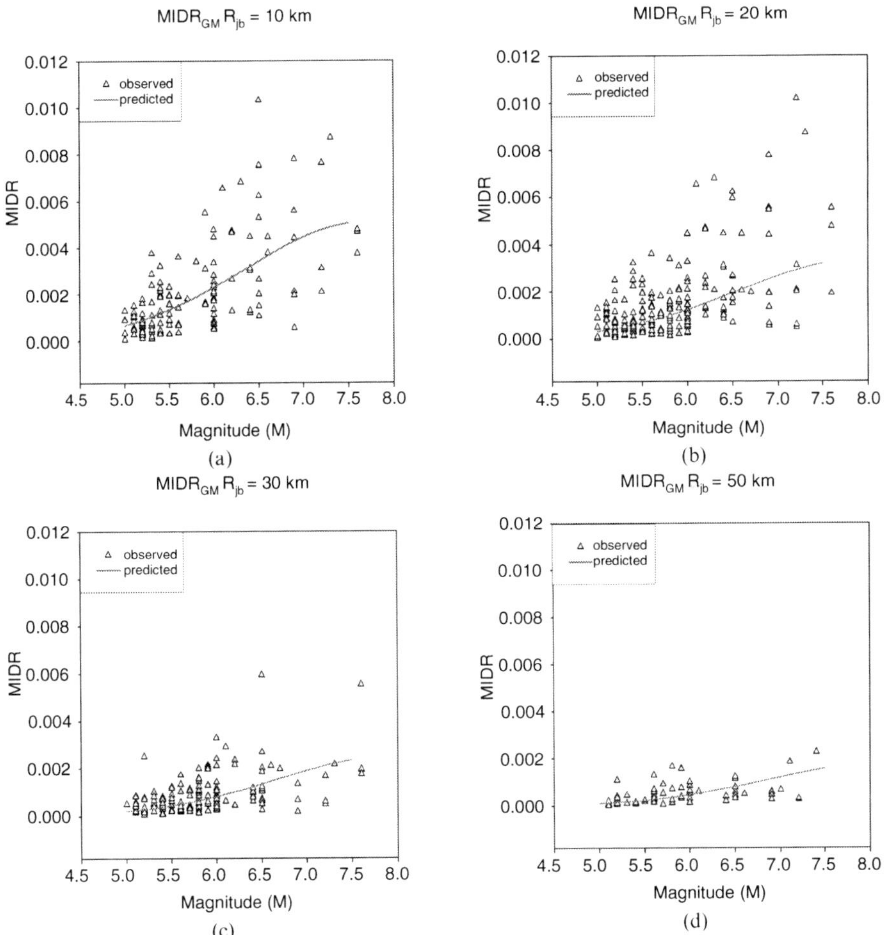

Figure 2: Variation of the ratio MIDR/GM as function of magnitude for different distances (R$_{jb}$ = 10, 20, 30, 50 km)

Figure 3 shows the ratio of MAX to GM as a function of magnitude and different distance values. The ratios are computed for the MIDR values of a shear-wall building with a fundamental period equal to 0.5 s. The scatter plots (ratios from the observed values) and the trend lines obtained from the predictive model once again indicate that the predictive model is capable of capturing actual variations in the data.

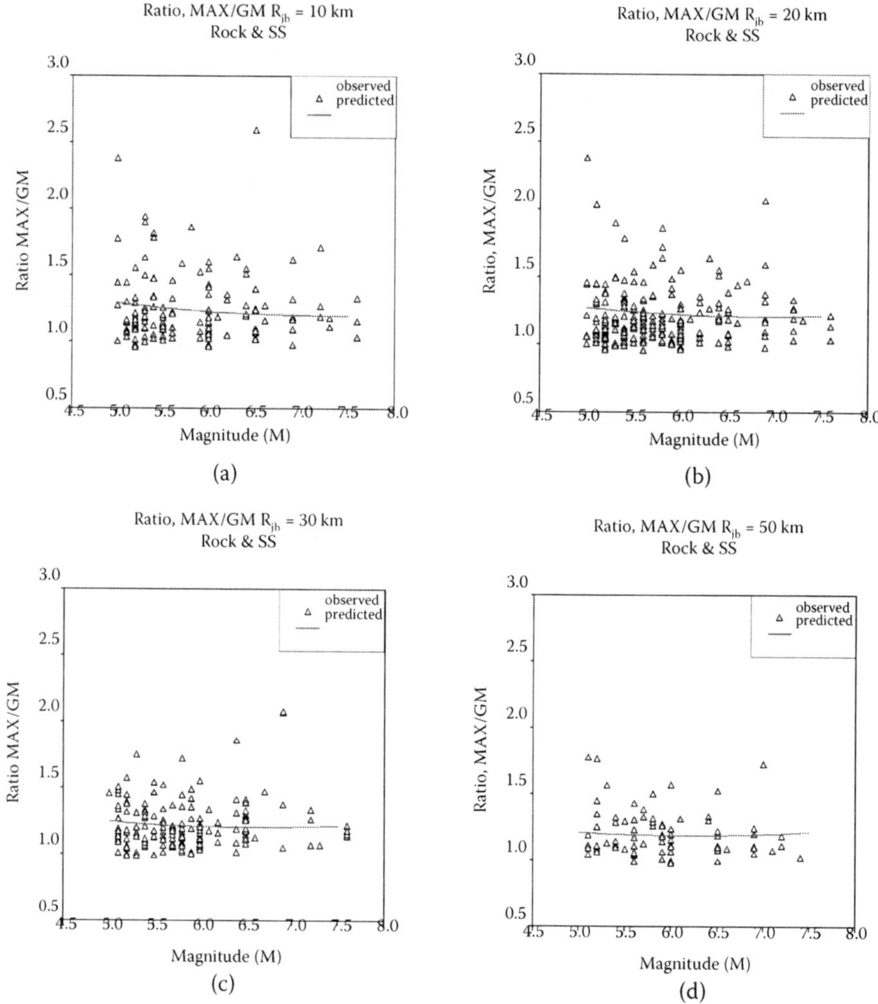

Figure 3: Variation of the ratio between $MIDR_{MAX}$ to $MIDR_{GM}$ as a function of magnitude for different distances (R_{jb} = 10, 20, 30, 50 km)

3.3. ERROR ANALYSIS

Figures 4–6 show the mean and ± sigma estimates for GM, MAX and RND MIDR estimates together with the corresponding observed scatters. The plots also show the error bars computed for the previously described magnitude bins. The plots suggest that the predictive model can describe the overall variation in the actual data fairly.

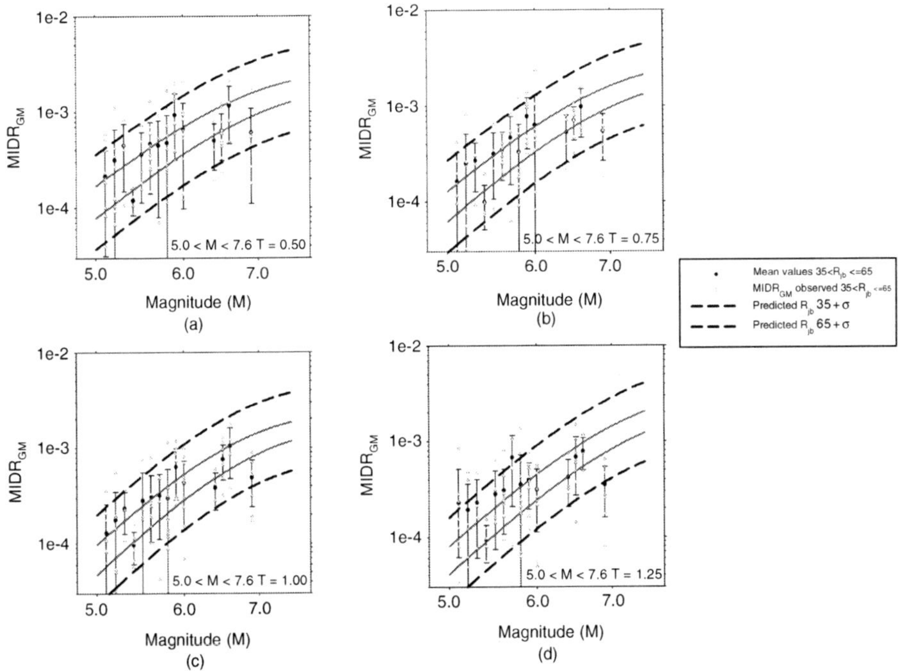

Figure 4: Error analysis of MIDR$_{\mathrm{GM}}$ values as a function of magnitude for different periods: **(a)** T = 0.5 s **(b)** T = 0.75 s **(c)** T = 1.0 s **(d)** T = 1.25 s

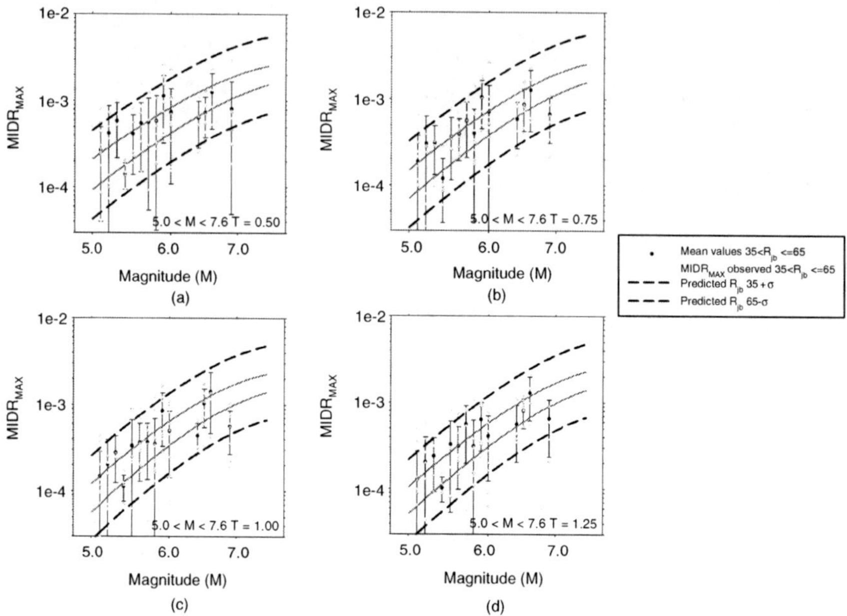

Figure 5: Error analysis of MIDR$_{MAX}$ values as a function of magnitude for different periods: (**a**) T = 0.5 s (**b**) T = 0.75 s (**c**) T = 1.0 s (**d**) T = 1.25 s

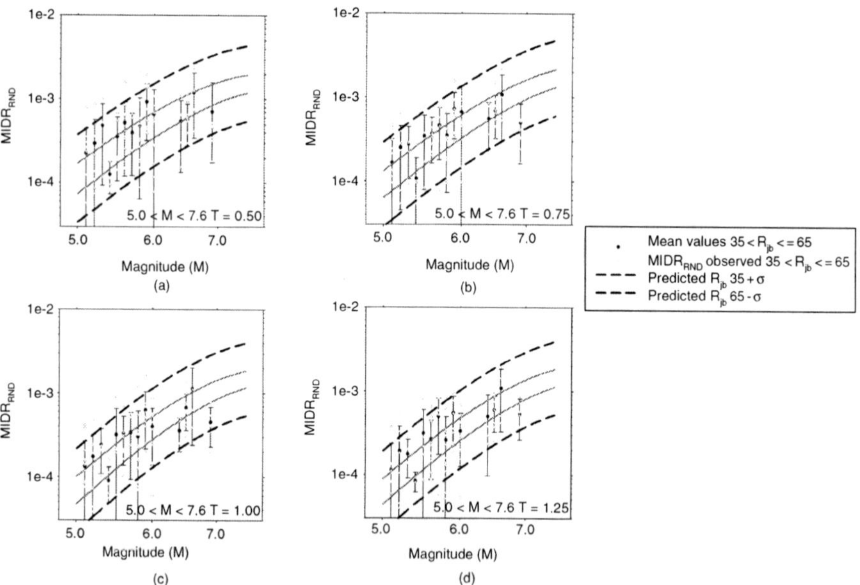

Figure 6: Error analysis of MIDR$_{rnd}$ values as a function of magnitude for different periods, (**a**) T = 0.5 s (**b**) T = 0.75 s (**c**) T = 1.0 s (**d**) T = 1.25 s

4. P-value

Determining the p-value in testing a hypothesis requires a log-normal distribution of residuals suggesting also that the mean values of all residual data in the set should be close to zero. This condition remained valid for all datasets and for a majority of subclasses of datasets investigated after sorting. Subsets that still had no data were considered to be useless to the investigation.

The confidence interval for each subset was computed using the following expression

$$\overline{z} - z_{\alpha/2}\, \sigma / \sqrt{n} \leq \mu \leq \overline{z} + z_{\alpha/2}\, \sigma / \sqrt{n} \tag{3}$$

where n is the size of the subset with a known variance σ^2, z is a standard normal random variable with $\mu = 0$ and $\sigma^2 = 1$ for which $\overline{z}$ and $z_{\alpha/2}$ represent a simple mean and upper $100\alpha/2$ percentage point of the same standard normal distribution with α the significance level and μ the true mean value.

The upper and lower limits of the confidence interval define the critical values as boundaries of two regions: acceptance and rejection. The null hypothesis here requires that μ as the true mean value ($\mu = 0$, for log-normal distribution) should be covered by a confidence interval in the acceptance region. Under the null hypothesis, the probability $(1 - \alpha)$ is the confidence coefficient that the test statistic falls between critical values. Rejecting the true null hypothesis and defining a type I error and probability α as the significance level and the size of the test with the p-value as the smallest level α at which the data are significant, for a two-tailed test we may define it as

$$P\text{-}value = 2\,[\,1 - \Phi(|z|)\,] \tag{4}$$

where $\Phi(|z|)$ is a standard normal cumulative distribution (Kreyszig, 1999).

All p-values obtained for the same components and periods were combined in the same figure by using contour plotting (Figures 7–9).

Grids 9 x 5 were used to construct the contour plots, with a higher density of grids in the lower part of R_{jb} (more data with small epicentral distances). Therefore, the precision of interpolation on each graph decreases with increasing values of distance R_{jb}.

The low p-value (light gray color in Figures 7–9) is due to sparse records at high values in the magnitude domain (Table 2), especially for distances greater than 30 km. The high p-values (dark color in Figures 7–9) are characteristic of small values in the magnitude domain that are kept in a large number of records in local seismic stations. Additional explanations for regions with small p-values comes first from critical points (with no records [Table 2]) and second, from the residual distribution (at $M_w = 6.0$ and $R_{jb} = 10$ km, at $M_w = 5.7$ and $R_{jb} = 75$ km) which shows (for $M_w = 6.0$ and $\alpha = 0.05$) the confidence interval did not include the real value $\mu = 0$ suggesting $p < 0.05$ and rejecting the null hypothesis.

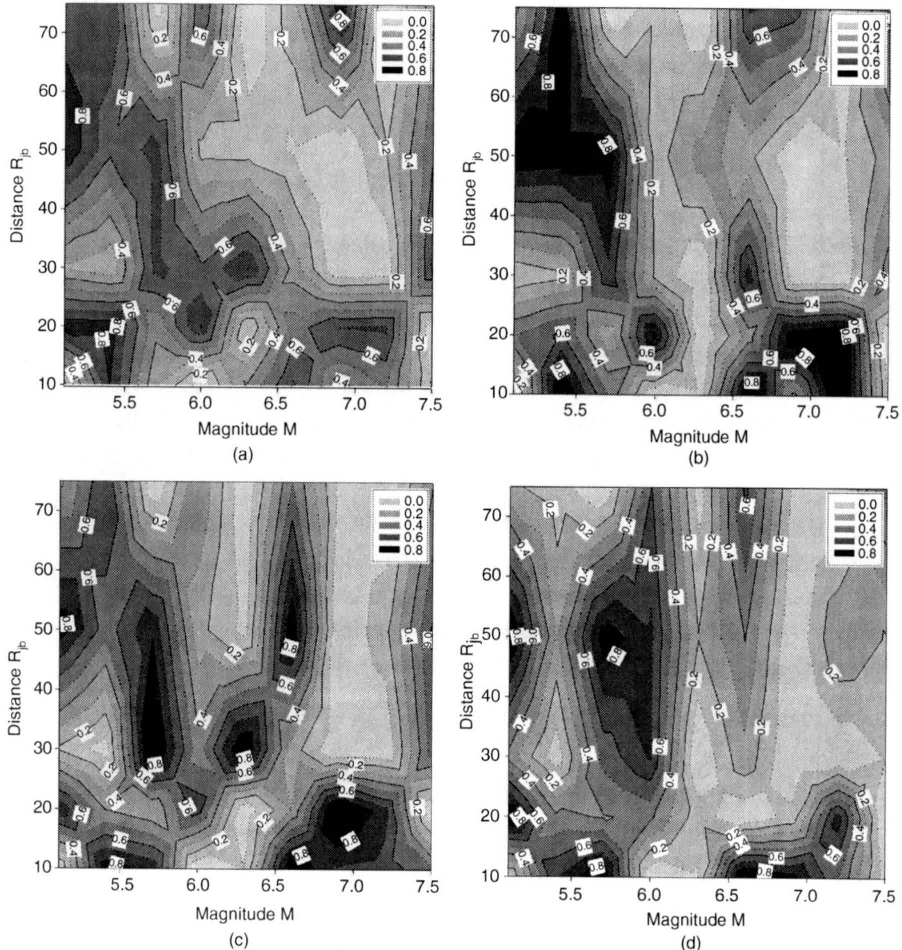

Figure 7: P-values of Maximum Interstory Drift Ratio (MIDR), Geometric Mean (GM) value, for different fundamental periods: (**a**) T = 0.5 s (**b**) T = 0.75 s (**c**) T = 1.00 s and (**d**) T = 1.25 s as a function of: Joyner and Boore distance R_{jb} and magnitude M_w

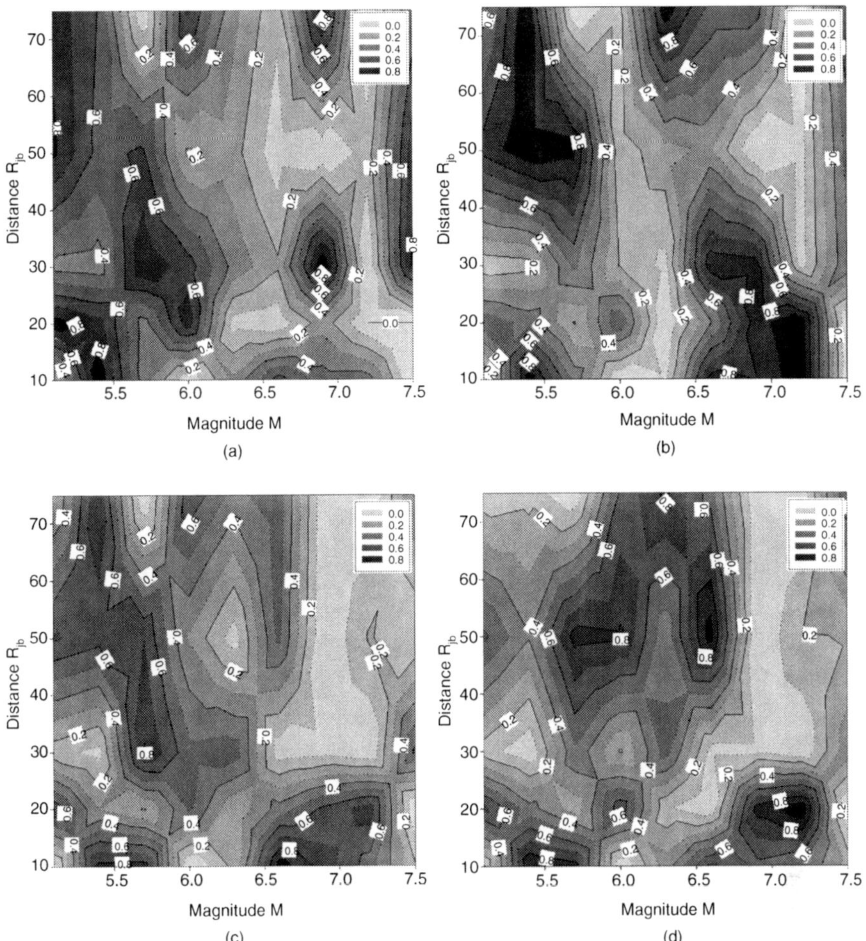

Figure 8: P-values of maximum interstory drift ratio (MIDR), maximum (MAX) component, for different fundamental periods: (**a**) T = 0.5 s (**b**) T = 0.75 s (**c**) T = 1.00 s and (**d**) T = 1.25 s as a function of: Joyner and Boore distance R_{jb} and magnitude $\mathbf{M}_w$.

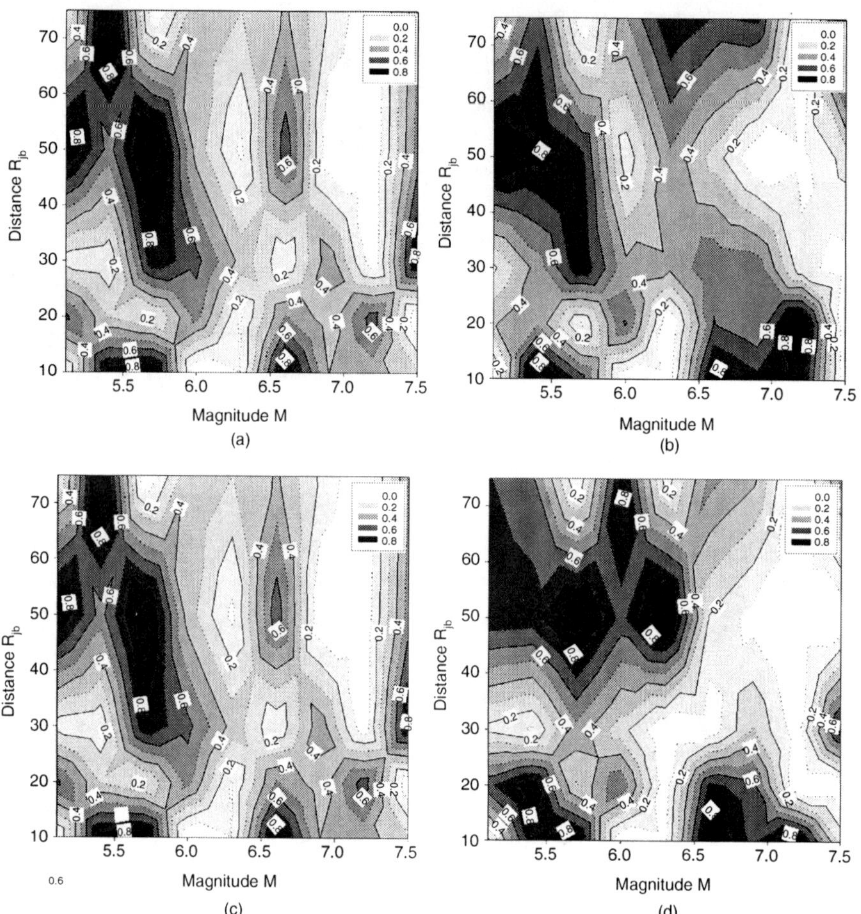

Figure 9: P-values of Maximum Interstory Drift Ratio (MIDR), Random (RND) component, for different fundamental periods: (**a**) T = 0.5 s (**b**) T = 0.75 s (**c**) T = 1.00 s and (**d**) T = 1.25 s as a function of Joyner and Boore distance R_{jb} and magnitude M_w

This investigation shows that the predictive equation is a good fit with the observed dataset. In terms of statistical parameters, Equation (1) is accepted with a significance level $\alpha = 0.05$ for all M_w and R_{jb} except the individual points mentioned before.

5. Conclusions

We evaluated the reliability of a predictive model that can be used for estimating the MIDR values of shear-wall frames as a function of magnitude, distance, site class and focal mechanism. The predictive model was derived by making use of the database and functional form presented by Akkar and Bommer (2007a, b). Our comparative evaluations showed that the proposed model can be used with confidence for estimating shear-wall MIDR values for fundamental periods between 0.5 and 1.25 s.

Acknowledgements

We thank Mr. Emrah Yenier for computing the actual MIDR values and for helping us with the regression analysis. This study is funded by NATO-SfP 980468 project.

References

Akkar S., and Bommer J., 2007a, Empirical Prediction Equations for Peak Ground Velocity Derived from Strong-Motion Records from Europe and Middle East. *Bulletin of the Seismological Society of America*, **97**(2): 511–530, doi: 10.1785/0120060141

Akkar S., and Bommer J., 2007b, Prediction of elastic displacement response spectra in Europe and Middle East. *Earthquake Engng Struct. Dyn.* **36**: 1275–1301. Published online in Wiley InterScience (www.interscience.wiley.com) doi: 10.1002/eqe.679.

Ambraseys N.N., Douglas J., Sarma S.K., Smit P., 2005, Equation for estimation of strong ground motions from shallow crustal earthquakes using data from Europe and Middle East: horizontal peak ground acceleration and spectral acceleration. *Bulletin of Earthquake Engineering*, **3**(1): 1–53.

Kreyszig E., 1999, *Mathematical Statistics (Chapter 23)*, / In: Advanced Engineering Mathematics, 8th Edition, ISBN 047133328-X, Singapore, pp. 1104–1153.

Miranda E., and Akkar S., 2006, Generalized Interstory Drift Spectrum. *Journal of Structural Engineering (ASCE)*, **132**(6): 840–852.

SHAKE MAPS OF STRENGTH AND DISPLACEMENT DEMANDS
FOR ROMANIAN VRANCEA EARTHQUAKES

D. LUNGU[1], I. CRAIFALEANU[*,2]
*Technical University of Civil Engineering, 124, Lacul Tei Blvd.,
020396 Bucharest, Romania*
*National Institute for Historical Monuments, 16, Enachita
Vacarescu St., 040157 Bucharest, Romania*
*[2]Technical University of Civil Engineering, 124, Lacul Tei Blvd.,
020396 Bucharest, Romania*
*National Institute for Building Research, INCERC, 266
Pantelimon St., sector 2, 021652 Bucharest, Romania*

Abstract. The paper presents results from an extensive study involving the mapping of key ground motion characteristics and of linear/nonlinear spectral ordinates for all strong Vrancea earthquakes that occurred in Romania in the last three decades. The scope of the study was to provide a comprehensive image of the spatial distribution of the parameters considered for the specified seismic events. By corroborating the results with data from previous studies by the authors, the study reveals interesting conclusions from the structural engineering point of view.

Keywords: Shake maps, Vrancea earthquakes, strength demands, displacement demands

1. Introduction

Mapping ground motion characteristics and spectral response ordinates represents a valuable technique for assessing the spatial distribution of several key parameters that provide information on earthquakes and on their effects on

[*]Iolanda-Gabriela Craifaleanu, Reinforced Concrete Dept., Technical University of Civil Engineering Bucharest, 124 Lacul Tei Blvd., RO-020396 / Earthquake Engineering Dept., National Institute for Building Research, INCERC, 266 Pantelimon St., Sector 2, RO-021652, Bucharest, Romania; e-mail: i.craifaleanu@gmail.com.

A. Zaicenco et al. (eds.), *Harmonization of Seismic Hazard in Vrancea Zone,*
© Springer Science + Business Media B.V. 2008

buildings. In the last few years, an extensive study was performed by the authors of this paper on Vrancea earthquakes with moment magnitude M_w, larger than 6.0 recorded in Romania in the last three decades (Lungu et al., 1999). In the first phase of the study, maps were generated for peak ground motion characteristics, effective peak ground characteristics and control (corner) periods of response spectra. The second phase of the study focused on the development of maps for linear/nonlinear acceleration and displacement spectra.

The following Vrancea events were considered: August 30, 1986 (moment magnitude $M_w = 7.1$, focal depth $h = 133$ km), May 30, 1990 ($M_w = 6.9$, $h = 91$ km) and May 31, 1990 ($M_w = 6.4$, $h = 79$ km). The originally digitized records from seismic networks in Romania and in neighboring countries were used in the study. Seismic values were mapped for each event. Then based on map ordinates, interpolation surfaces and contours of constant values were determined by using specialized GIS software.

2. Maps of Peak Ground Motion Values

Maps obtained for peak ground acceleration (PGA) for Romania and for Bucharest are shown in Figures 1 and 2.

All the maps of peak ground motion values, both for PGA and for peak ground velocity (PGV) showed significant differences between the spatial distribution patterns for each seismic event considered. At the origin of this phenomenon is mainly the diversity of the source mechanisms as well as the wave propagation pattern of the events, features that were pointed out in previous studies on Vrancea earthquakes.

It should be mentioned that the group of stations providing seismic records was not the same for all the earthquakes considered either due to the expansion of seismic networks in Romania or to the occasional malfunction of some older accelerometers.

As observed in the above maps, the northeast-southwest pattern of map contours that was very marked for the August 30, 1986 earthquake is not present on the maps of the other two seismic events considered. The existence of records in the northern part of the country for these last two events would have provided valuable information for the generation of the interpolation surface.

Another remarkable feature of the maps is the clear difference between the contour patterns of the two earthquakes in 1990. In the first 1990 event, large PGA values compared with the other stations were recorded in the eastern part of the Romanian Plain (Figure 1b), but this was not the case for the second event (Figure 1c). Again, this agrees with conclusions from previous studies on these earthquakes.

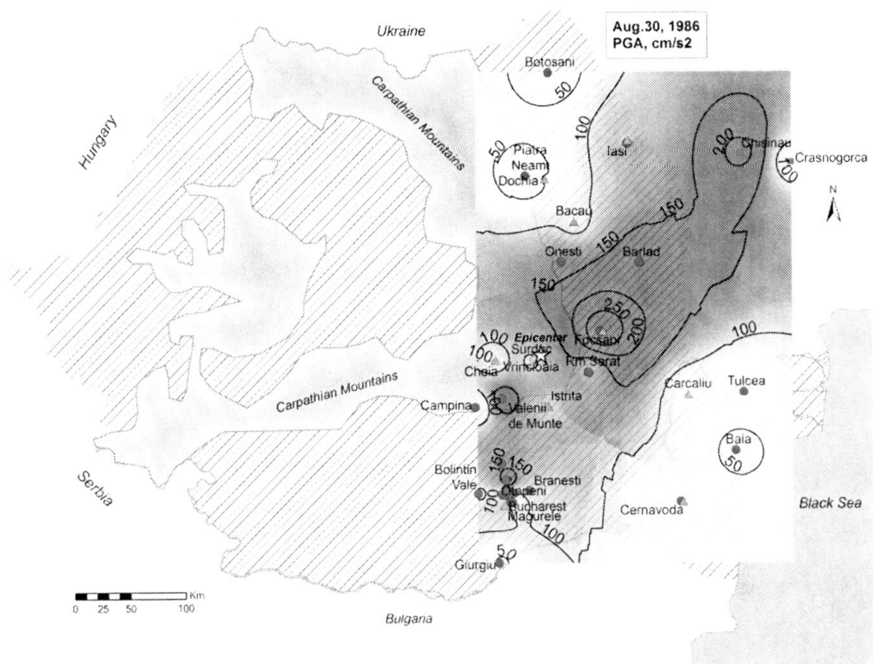

a) August 30, 1986

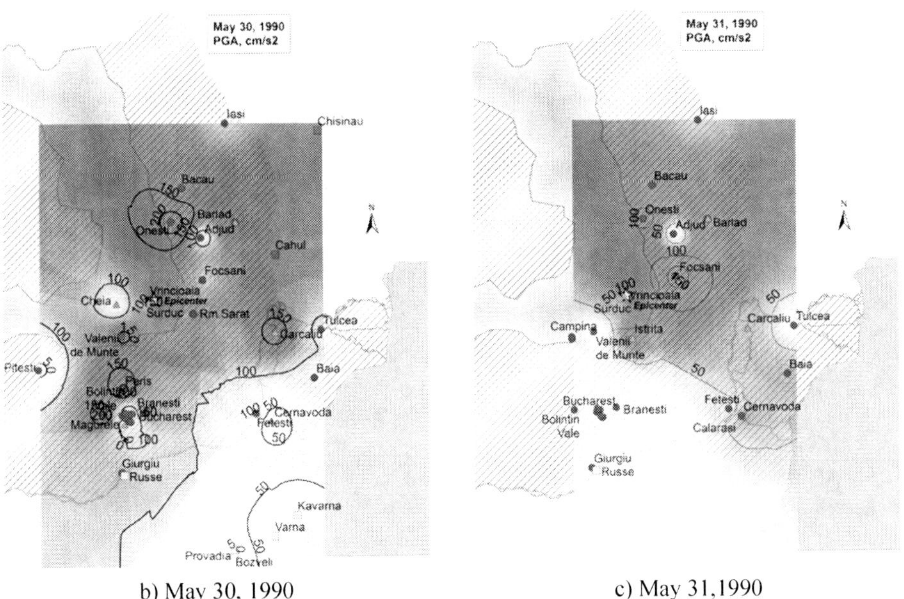

b) May 30, 1990   c) May 31,1990

Figure 1: PGA distribution for three Vrancea seismic events in Romania

    D. LUNGU AND I. CRAIFALEANU

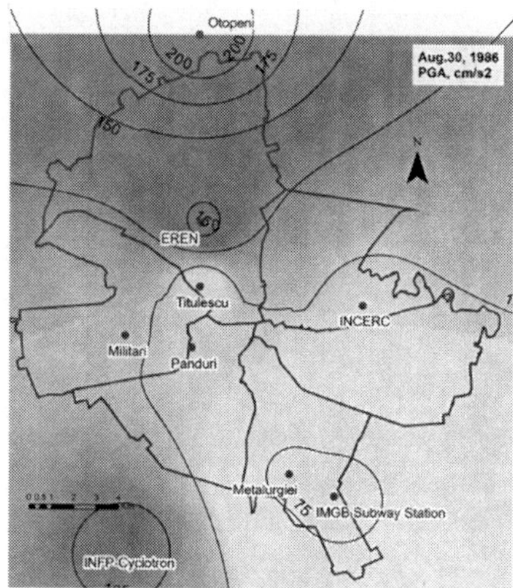

a) August 30, 1986

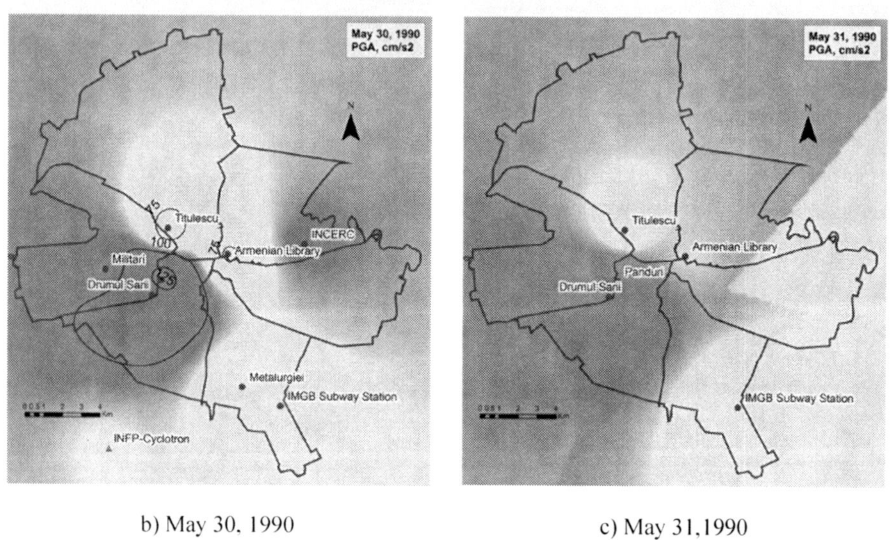

b) May 30, 1990   c) May 31,1990

Figure 2: Bucharest. PGA distribution for three Vrancea seismic events

Detailed views of the spatial distribution of PGA values in Bucharest for the three seismic events considered are shown in Figure 2.

In the case of Bucharest, the contour patterns appear very different from one seismic event to the other. For example, in the 1986 event, the lowest PGA values were recorded in the southeast of the city, but for the 1990 events this pattern was not observed.

3. Maps of Effective Peak Ground Motion Values

The maps obtained for effective peak ground acceleration (EPA) and effective peak ground velocity (EPV) are shown in Figures 3–6.

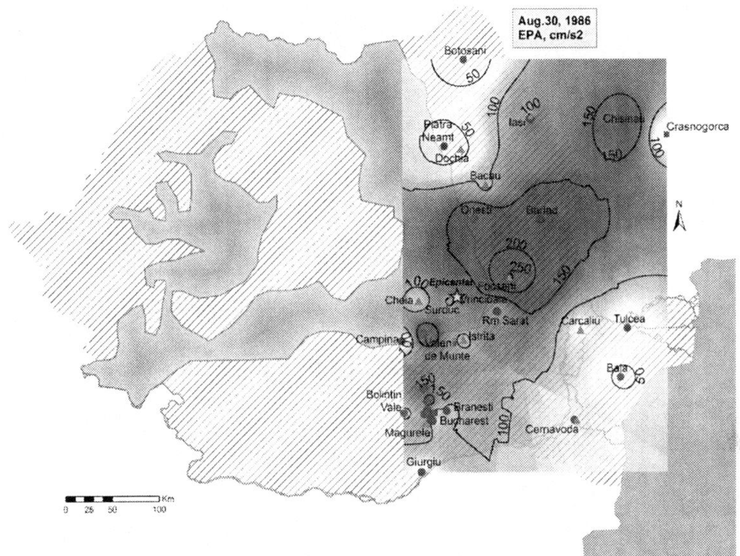

a) August 30, 1986

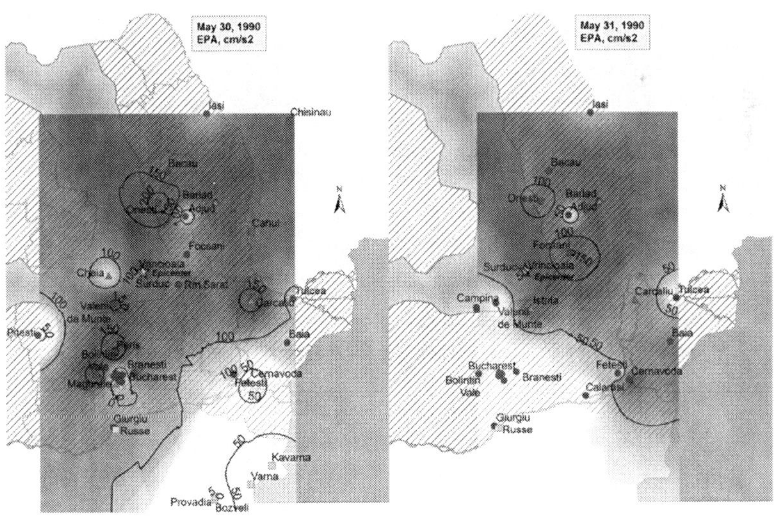

b) May 30, 1990   c) May 31,1990

Figure 3: EPA distribution for three Vrancea seismic events in Romania

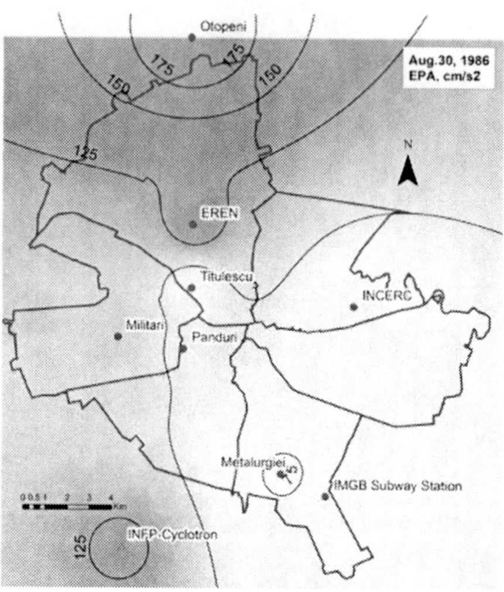

a) August 30, 1986

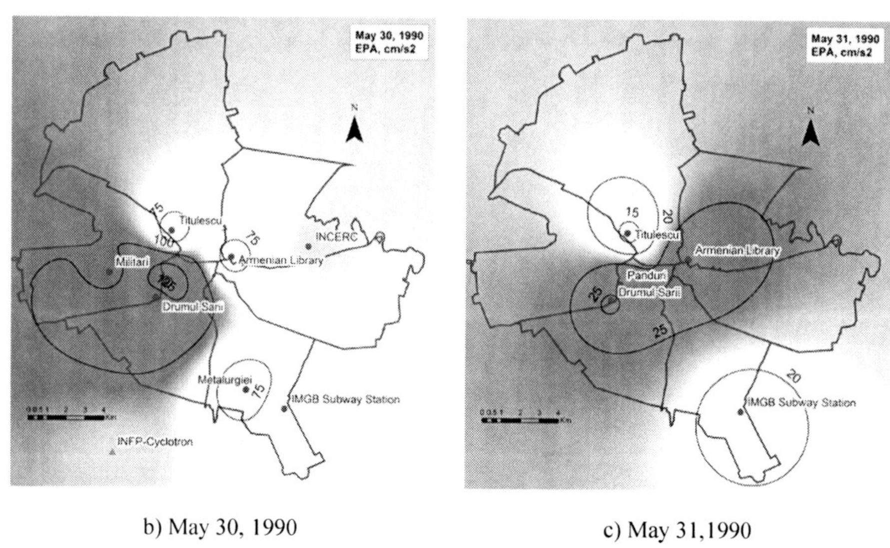

b) May 30, 1990   c) May 31,1990

Figure 4: Bucharest. EPA distribution for three Vrancea seismic events

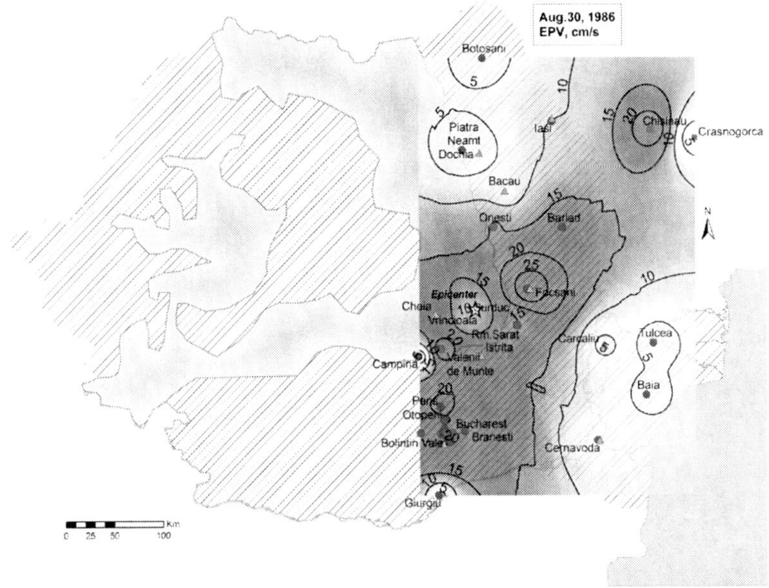

a) August 30, 1986

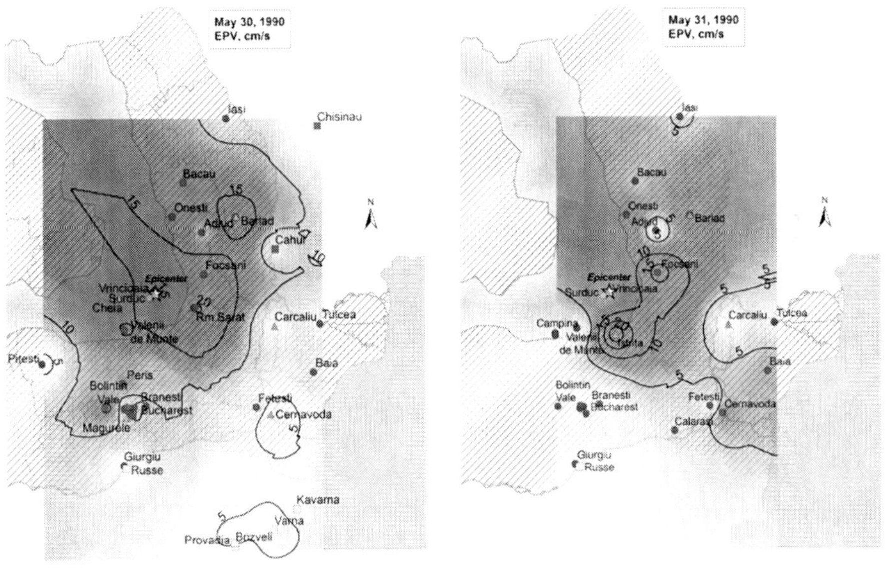

b) May 30, 1990   c) May 31,1990

Figure 5: EPV distribution for three Vrancea seismic events in Romania

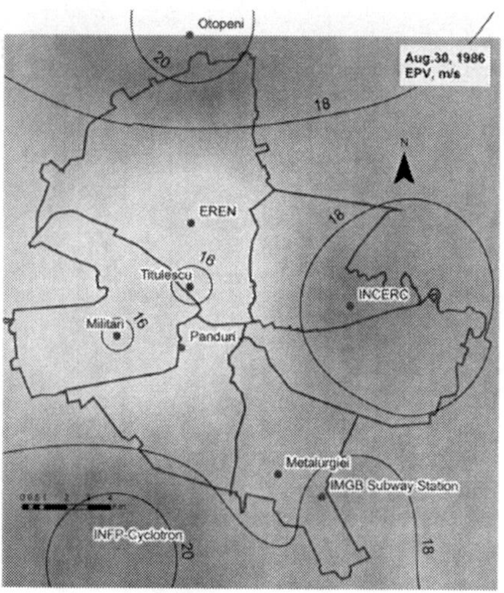

a) August 30, 1986

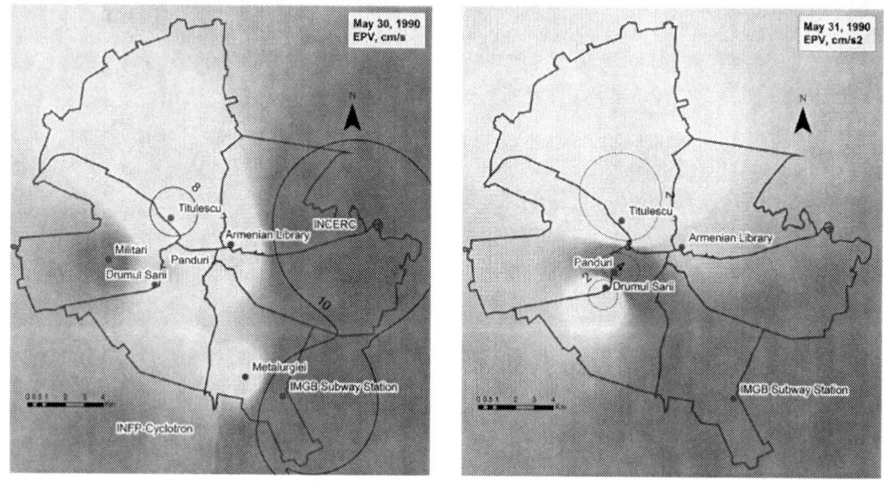

b) May 30, 1990								c) May 31, 1990

Figure 6: Bucharest. EPV distribution for three Vrancea seismic events

The effective peak ground acceleration and effective peak ground velocity were calculated using the following expressions introduced by Lungu et al. (1995).

$$EPA = \left(SA_{\text{averaged on 0.4 s}}\right)_{max} / 2.5 \tag{1}$$

$$EPV = \left(SV_{\text{averaged on 0.4 s}}\right)_{max} / 2.5 \tag{2}$$

where $(SA_{\text{averaged on 0.4 s}})_{max}$ denotes the maximum values of the acceleration response spectra averaged on a 0.4 second (s) period mobile window, and $(SV_{\text{averaged on 0.4 s}})_{max}$ denotes the maximum values of the velocity response spectra averaged on a 0.4 s period mobile window.

The configuration of EPA maps is quite similar to the configuration of PGA maps, both for Romania and for Bucharest. This together with a similar observation made for PGV and EPV in (Craifaleanu et al., 2006), confirms the validity of the expressions (1) and (2) for calculating the effective peak ground motion values.

4. Maps of Corner Period of Response Spectra, T_C

The control (corner) period of the response spectra T_C provides the most reliable and stable information on the frequency contents of ground motion. The control period was calculated by using the following expression:

$$T_C = 2\pi\, EPV / EPA \tag{3}$$

The resulting maps of T_C are shown for Romania and for Bucharest in Figures 7 and 8.

For the August 30, 1986 earthquake (Figure 7a), the largest values of T_C occur in the southwestern part of the zone analyzed whereas for the 1990 earthquakes (Figures 7b and 7c), a very different contour pattern can be observed.

For Bucharest, the general tendency appears to consist in larger values of T_C in the southeastern part of the city compared with the rest of the stations. This feature was underlined in previous studies by Lungu et al. (Lungu et al., 2005) where it was found to be related to local soil conditions.

The values of T_C tend to decrease with seismic magnitude.

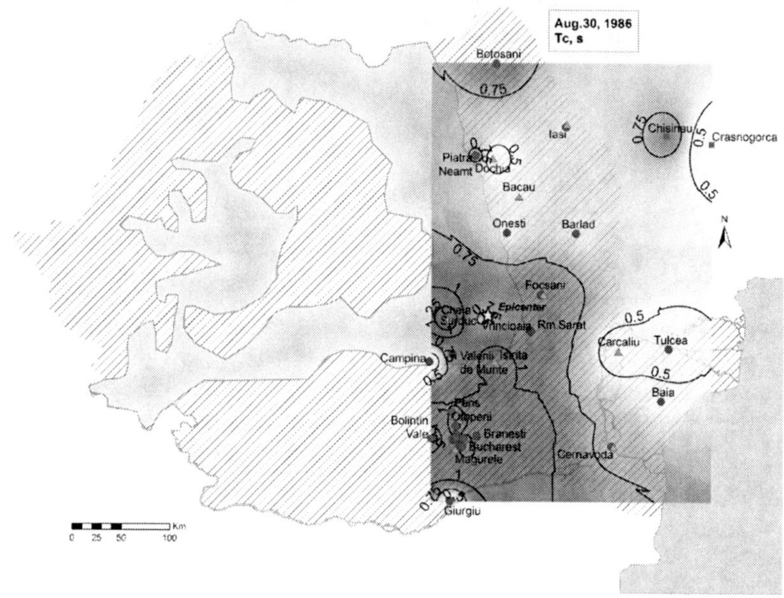

a) August 30, 1986

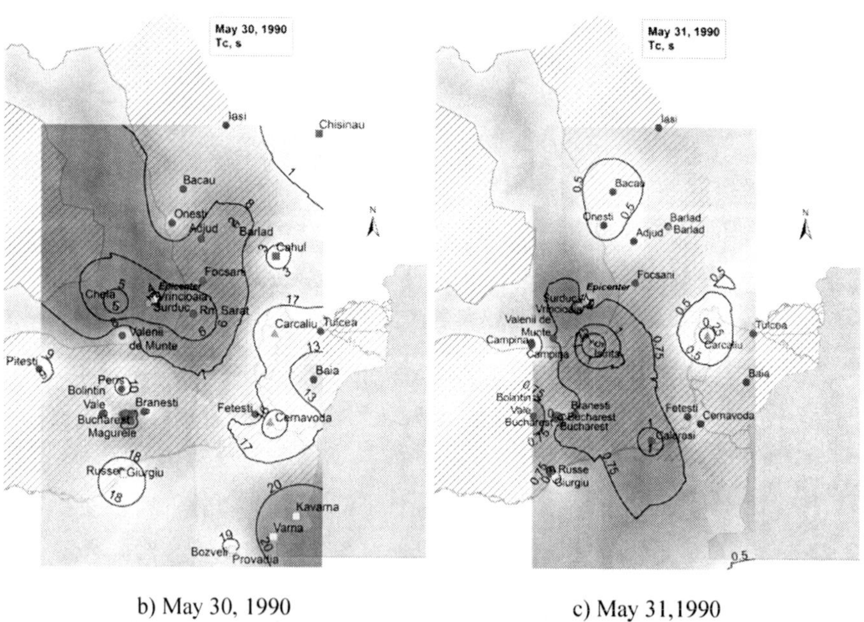

b) May 30, 1990　　　　　　　　　　　c) May 31,1990

Figure 7: T_C distribution for three Vrancea seismic events in Romania

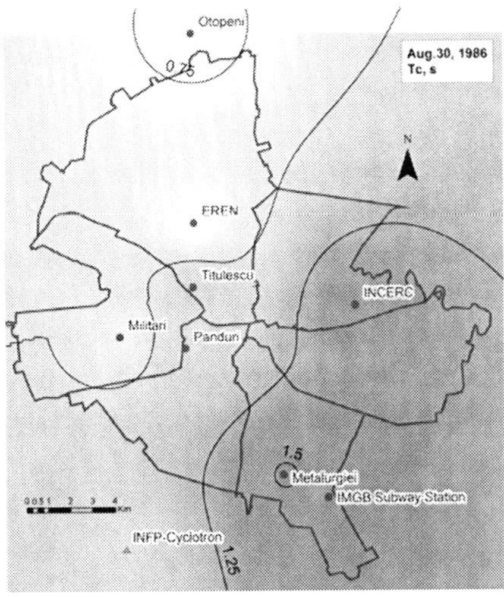

a) August 30, 1986

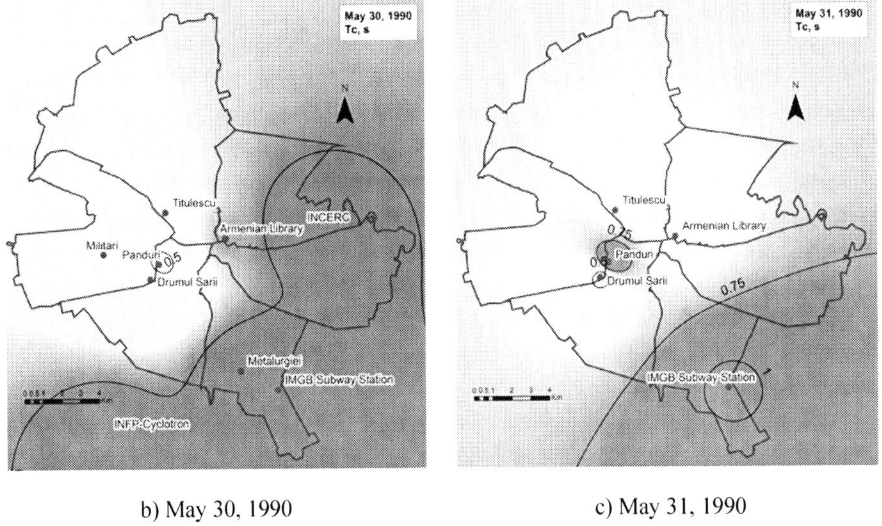

b) May 30, 1990   c) May 31, 1990

Figure 8: Bucharest. T_C distribution for three Vrancea seismic events

5. Maps of Elastic and Inelastic Response Spectrum Ordinates

Response spectrum ordinates used to generate maps were computed for the earthquake of August 30, 1986, the strongest seismic event analyzed in the study. A damping ratio of 5% and three values of structure period ($T = 0.5$, 1.0 and 1.5 s) were considered.

Inelastic behavior was modeled by considering a simple elastic-perfectly plastic hysteretic model. Spectral values were calculated for a set of specified values of ductility, i.e. $\mu = 1$ (elastic behavior) and $\mu = 1.5, 2, 3, 4, 5$ and 6 (inelastic behavior).

The study focused on spectral accelerations and spectral displacements, in order to explore the spatial distribution of seismic demands on building structures with different stiffness and ductility characteristics.

For the sake of space requirements, only parts of the maps are presented.

The strong variability of response spectrum ordinates due to different factors complicates discerning general tendencies; however, the results of previous studies on elastic/inelastic spectra for the ground motions considered provide a reliable source of information.

5.1. SPECTRAL ACCELERATIONS

Maps of elastic and inelastic spectral acceleration ordinates for the earthquake of August 30, 1986, the strongest seismic event analyzed in the study, are shown in Figures 9–12.

By examining the maps, a clear tendency of decreasing spectral ordinates with an increase in ductility can be observed. The spatial distribution of spectral accelerations becomes more and more uniform as ductility increases. This phenomenon can be observed on all maps, irrespective of the structure period for which the spectral ordinates were computed. However, the interpolation surface does not flatten uniformly as the rate of variation of spectral ordinates with ductility is different from one ground motion record to another.

The structure vibration period also does have an important influence on spectral accelerations. However, for $T = 1.5$ s, the amplitude of this variation attenuates considerably as a consequence of the frequency contents of the ground motions analyzed.

Contour maps are also sensitive to factors like the number of stations that provided seismic records and the values of the numerical parameters used to generate the interpolation surface.

One of the most significant consequences of the observations above is that for building structures with inelastic behavior, the spatial distribution of seismic strength demand is more uniform than for structures behaving elastically. As a result, for common structures for which inelastic behavior is allowed for the design earthquake, the influence of the other factors affecting spatial distribution is less important than anticipated.

Detailed analyses of inelastic response spectra for the Vrancea seismic motions considered in the present study can be found in Lungu et al. (1996) and Craifaleanu (1996–2005).

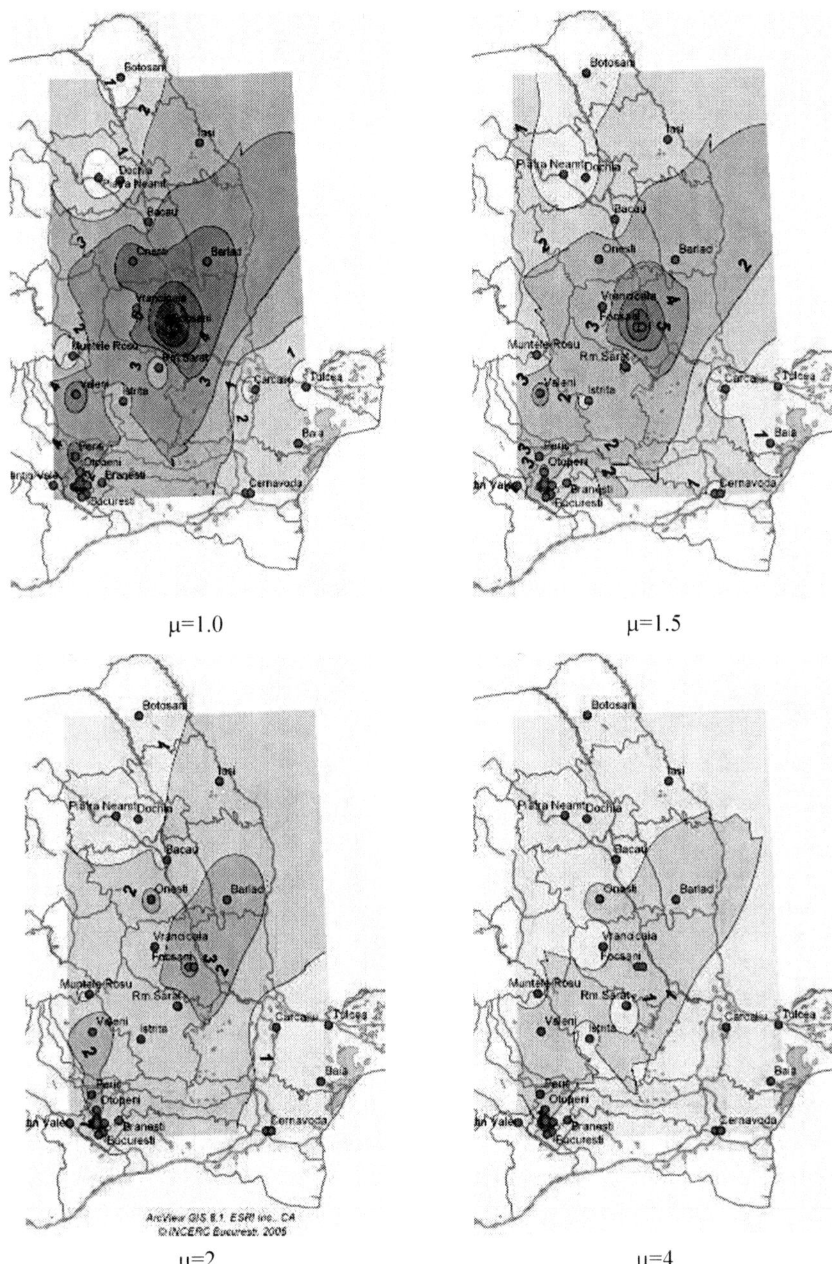

Figure 9: August 30, 1986. Distribution of elastic (μ = 1.0) and inelastic (μ = 1.5, 2.0, 4.0) spectral acceleration for structure period $T = 0.5$ s

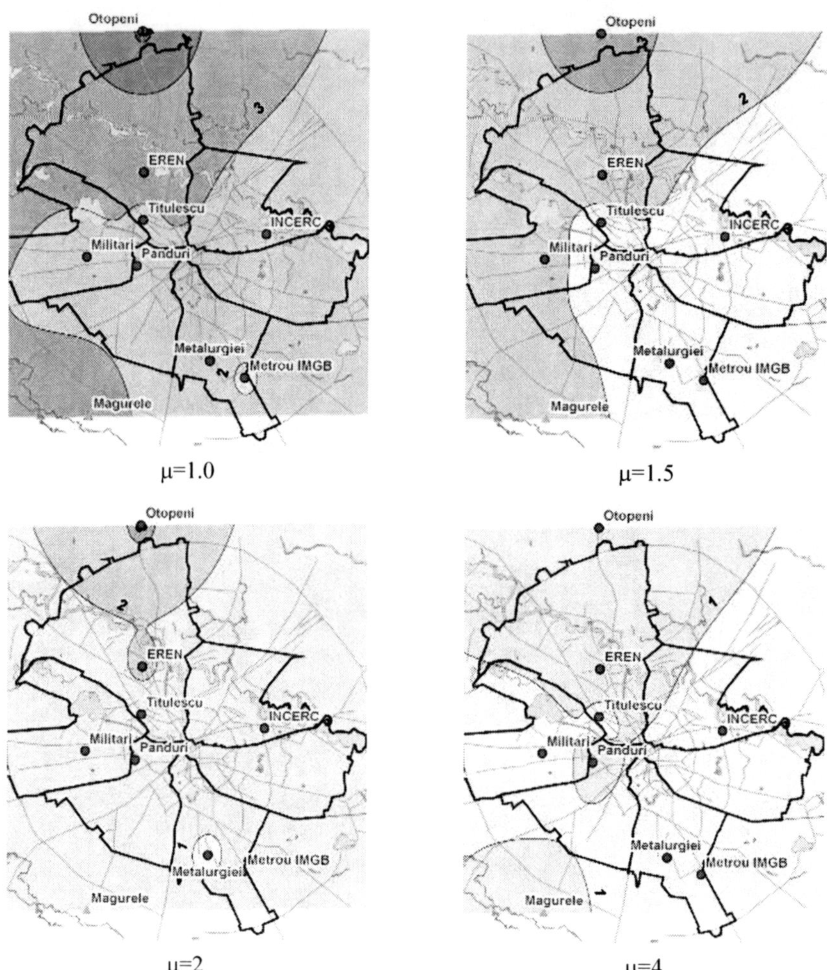

Figure 10: Bucharest, August 30, 1986. Distribution of elastic ($\mu = 1.0$) and inelastic ($\mu = 1.5$, 2.0, 4.0) spectral acceleration for structure period $T = 0.5$ s

Apart from the previous observations, it is also worth noting for the maps of Romania (Figures 9 and 10) that the northeast-southwest orientation of map contours for the 1986 earthquake is similar to that observed on the maps of PGA (Figure 2), EPA (Figure 4) and EPV (Figure 6) for the same seismic event.

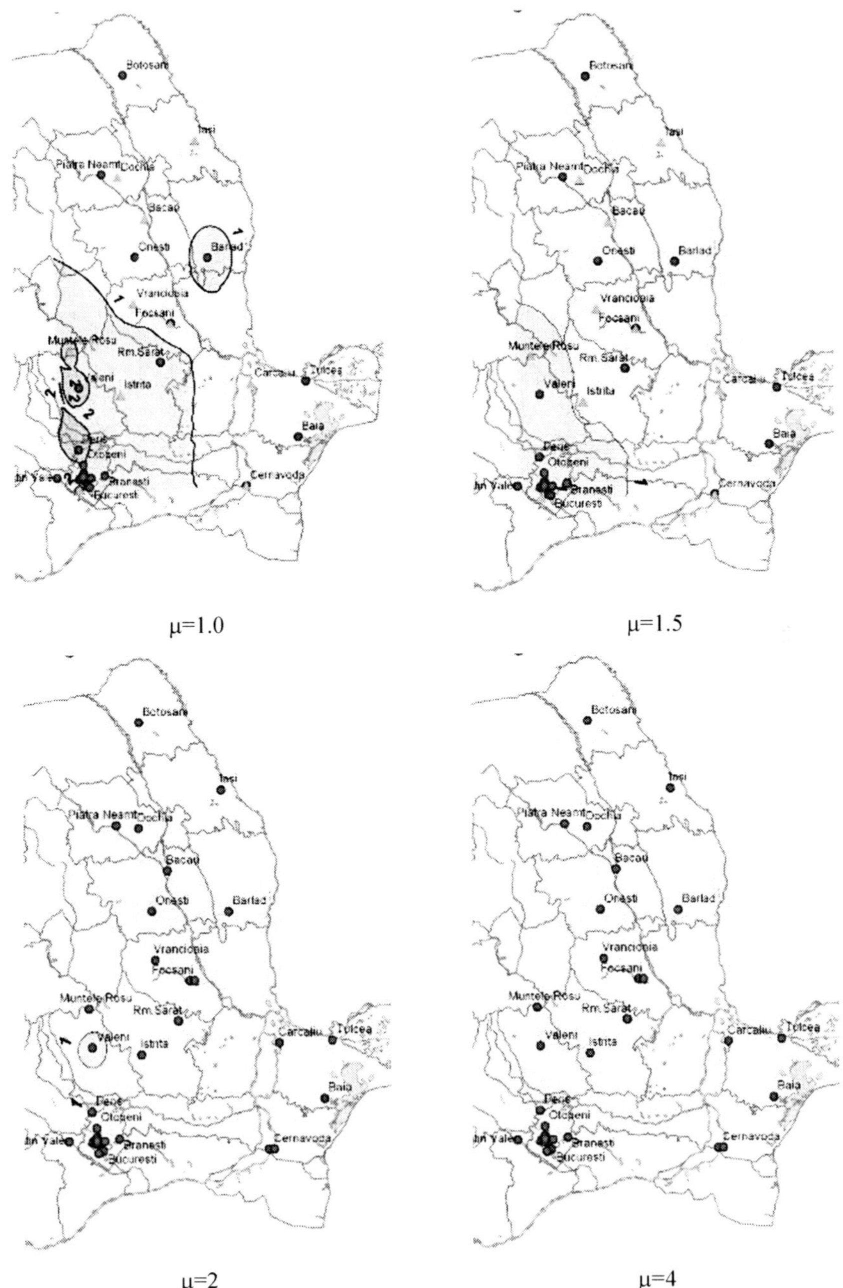

Figure 11: August 30, 1986. Distribution of elastic (μ = 1.0) and inelastic (μ = 1.5, 2.0, 4.0) spectral acceleration for structure period T = 1.5 s

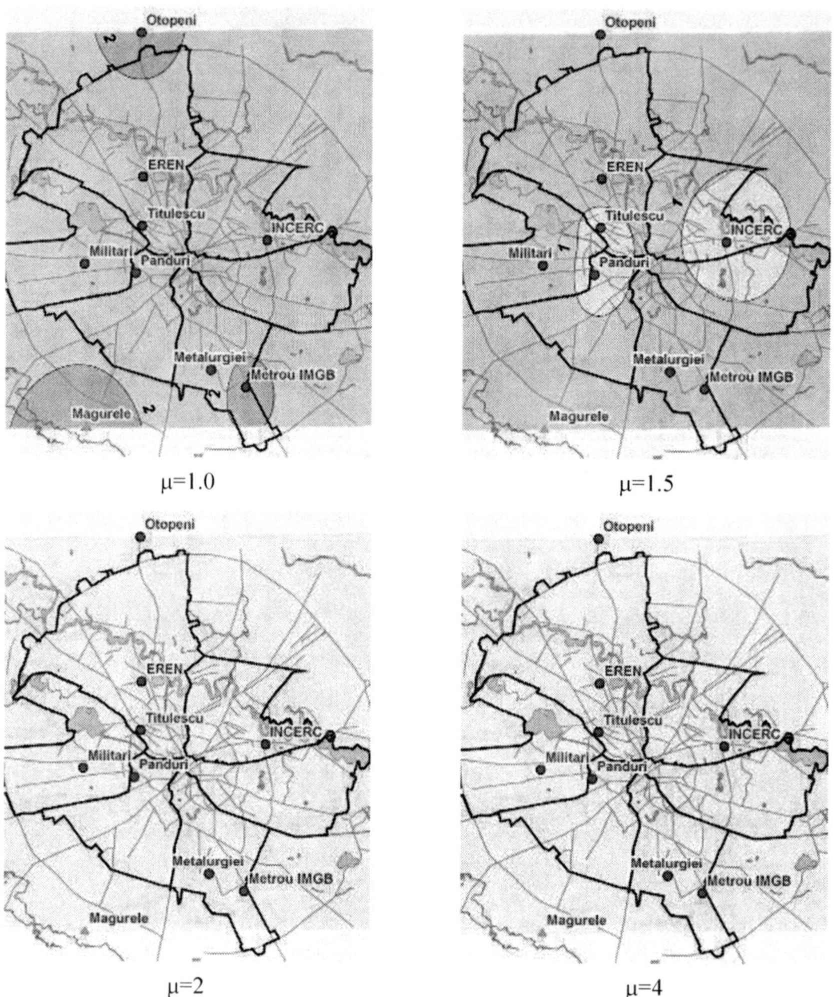

Figure 12: Bucharest, August 30, 1986. Distribution of elastic ($\mu = 1.0$) and inelastic ($\mu = 1.5$, 2.0, 4.0) spectral acceleration for structure period $T = 1.5$ s

5.2. SPECTRAL DISPLACEMENTS

Maps of elastic and inelastic spectral displacement ordinates are shown in Figures 13–16 for the August 30, 1986 earthquake.

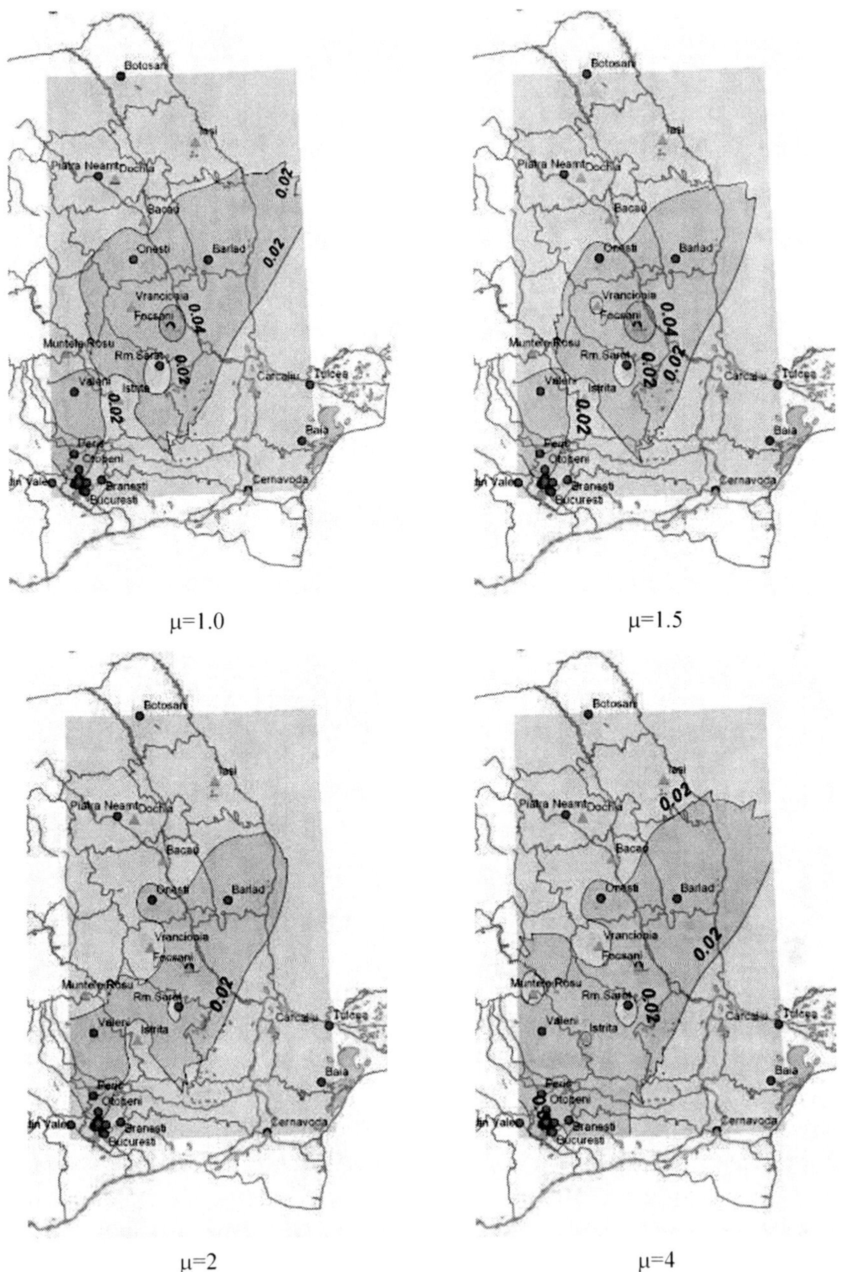

Figure 13: August 30, 1986. Distribution of elastic ($\mu = 1.0$) and inelastic ($\mu = 1.5$, 2.0, 4.0) spectral displacement for structure period $T = 0.5$ s

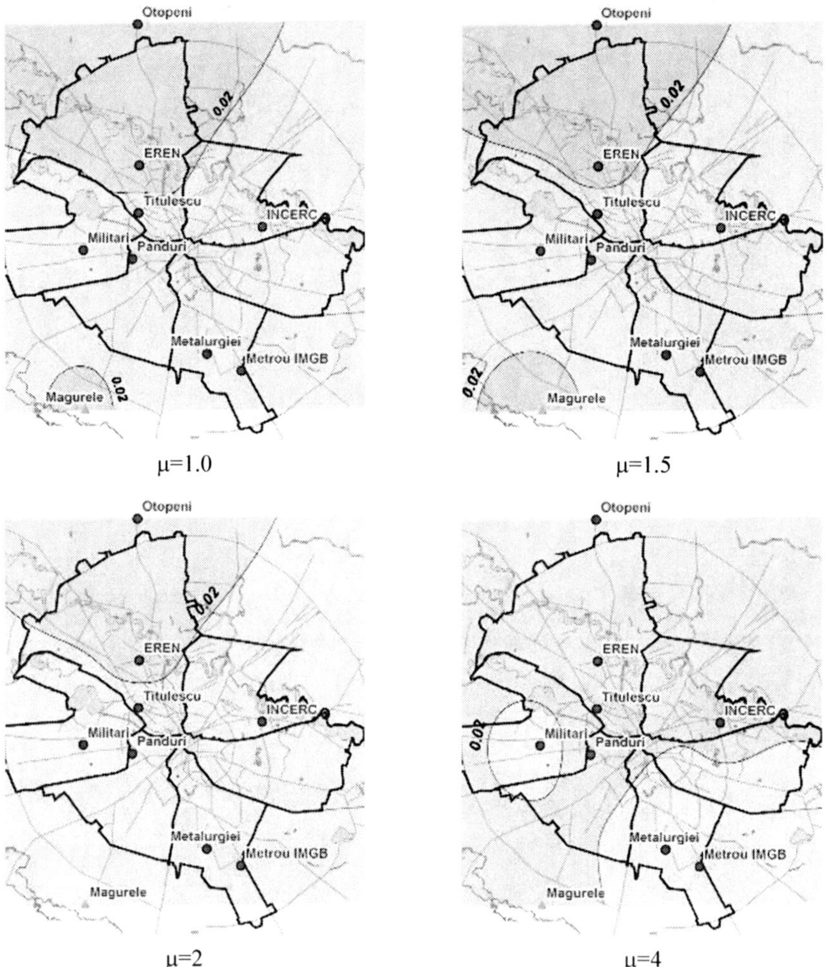

Figure 14: Bucharest, August 30, 1986. Distribution of elastic ($\mu = 1.0$) and inelastic ($\mu = 1.5$, 2.0, 4.0) spectral displacement for structure period $T = 0.5$ s

The variation of spectral ordinates with ductility is non-monotonous; therefore, a general conclusion cannot be drawn concerning the influence of this factor. Consequently, the beneficial effect of inelastic behavior described for spectral accelerations does not apply for spectral displacement maps. In what concerns the influence of the structure period, there is a marked increase of displacement ordinates with this parameter.

The above remarks, which are consistent with previous studies on inelastic displacement spectra, can be made with respect to both Romania and Bucharest contour maps.

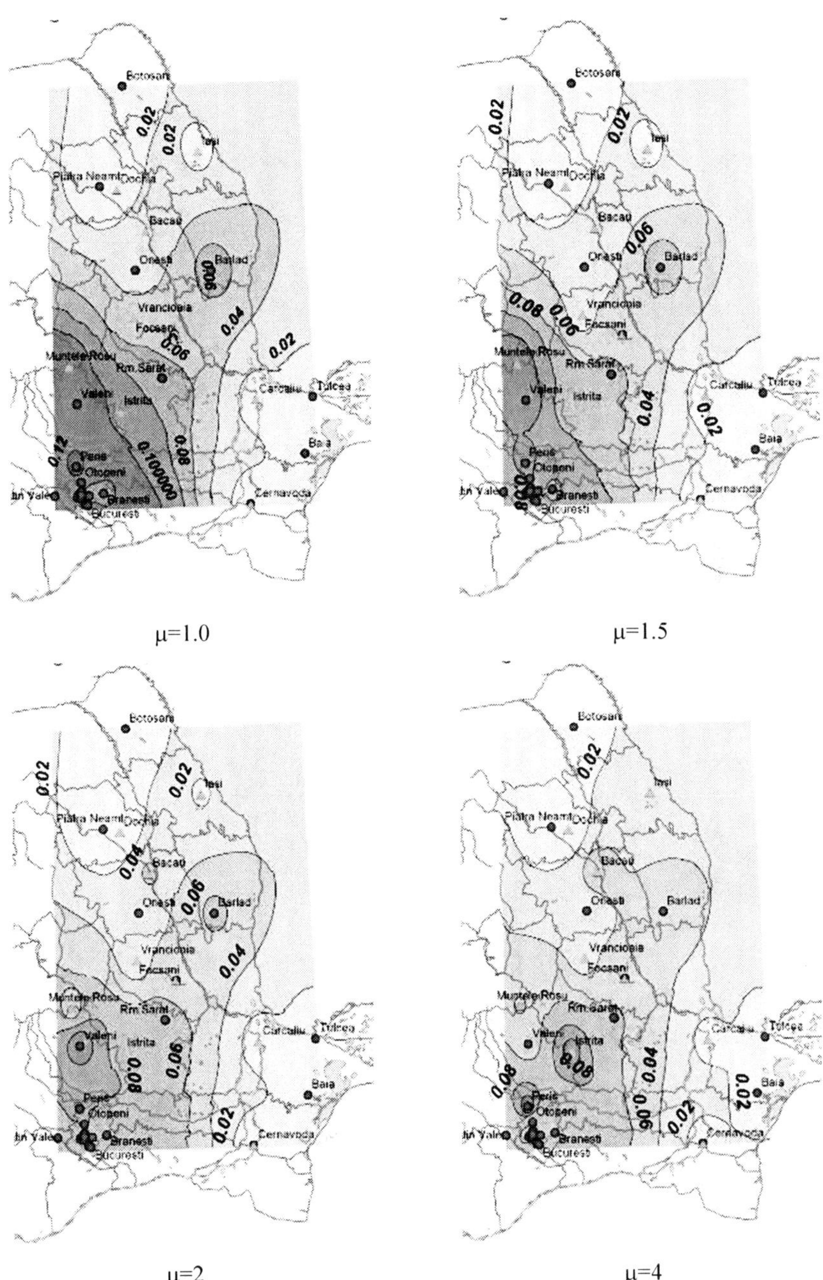

Figure 15: August 30, 1986. Distribution of elastic (μ = 1.0) and inelastic (μ = 1.5, 2.0, 4.0) spectral displacement for structure period $T = 1.5$ s

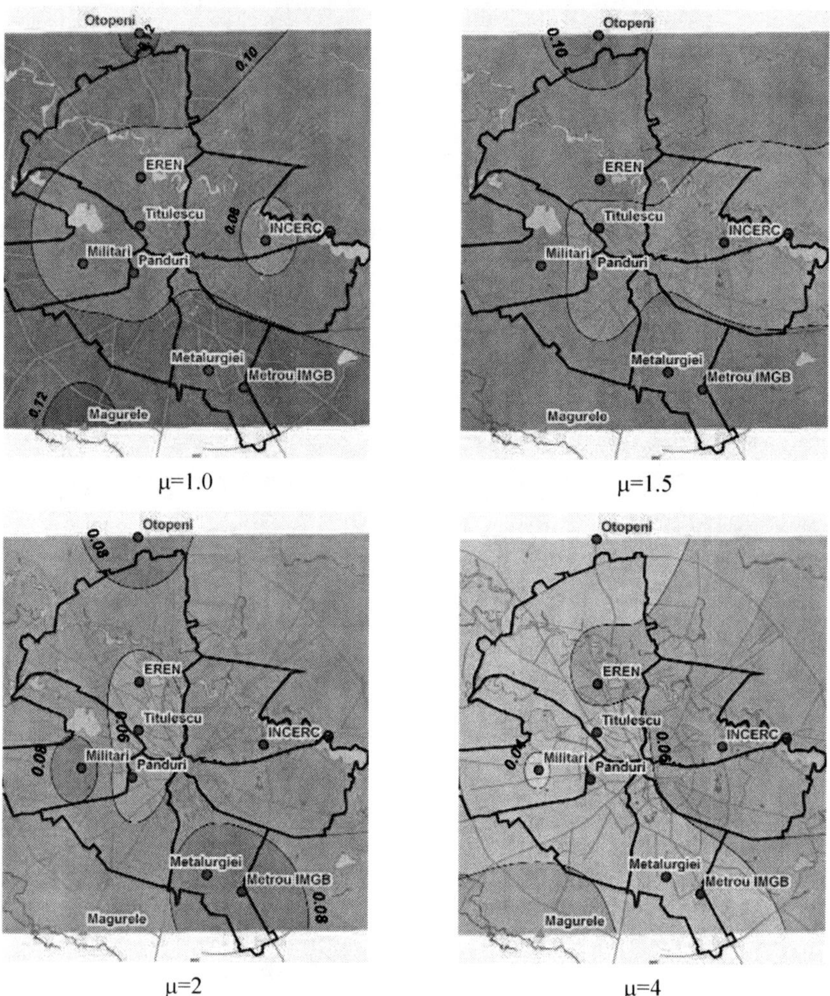

Figure 16: Bucharest, August 30, 1986. Distribution of elastic (μ = 1.0) and inelastic (μ = 1.5, 2.0, 4.0) spectral displacement for structure period T = 1.5 s

It is worth noting that the observation concerning the northeast-southwest orientation of map contours can also be made for spectral displacements.

6. Conclusions

1. The spatial distribution of all mapped parameters differs significantly between the seismic events considered in the study. The explanation for this phenomenon can be found in the mobility of the frequency contents and of

the intensity of ground motion with earthquake magnitude and risk characteristics.

2. The dynamic characteristics of the structure have a strong influence on the spatial distribution of strength and displacement demands. The influence observed on the map is consistent with the results obtained from the overall study of spectral response.

3. There is a significant attenuation of strength demands whose spatial distribution becomes more uniform with an increase in ductility.

The last two conclusions highlight the importance of structural characteristics on the spatial distribution of strength and displacement demands imposed by earthquakes. In certain cases, the influence of structural characteristics can prevail on those of factors related to ground motion, especially for buildings that exhibit relatively large post-elastic incursions during a seismic event.

Acknowledgements

This research was partly sponsored by NATO's Scientific Affairs Division in the framework of the Science for Peace Programme, project SfP-980468.

References

Craifaleanu, I. G. (1996). Contributions to the Study of Inelastic Seismic Response of Reinforced Concrete Structures, *Ph.D. thesis*, Technical University of Civil Engineering, Bucharest (in Romanian).

Craifaleanu, I. G. (1999). Studies on Inelastic Response Spectra for Vrancea Earthquakes, *Buletinul AICPS*, No. 3, 62–68 (in Romanian).

Craifaleanu, I. G. (2001). A Perspective on the Use of Inelastic Response Spectra in Building Design, *Buletinul AICPS*, No. 3, 8–19 (in Romanian).

Craifaleanu, I. G. (2005). Nonlinear Single-Degree-of-Freedom Models in Earthquake Engineering, *Matrix Rom Publishers*, Bucharest (in Romanian).

Craifaleanu, I. G., Lungu, D., Borcia, I. S. (2006). Shakemaps of Nonlinear Spectral Ordinates for Vrancea Earthquakes, *Proceedings of the First European Conference on Earthquake Engineering and Seismology*, Geneva, Switzerland, 3–8 September, Paper No.1257 (on CD-ROM).

Lungu, D., Cornea, T., Craifaleanu, I., Aldea, A. (1995). Uniform hazard response spectra in soft soil condition and Eurocode 8, *7th International Conference of Application of Statistics and Probability in Civil Engineering, Reliability and Risk Analysis*, ICASP-7, Paris, July 10–13, Proceedings, Vol. 1, pp. 619–626, A. A. Balkema, Rotterdam.

Lungu, D., Cornea, T., Craifaleanu, I., Demetriu, S. (1996), Probabilistic seismic hazard analysis for inelastic structures on soft soils. *Proceedings of the 11th World Conference on Earthquake Engineering, Acapulco, Mexico*, June 23–28, Paper No. 1768 (on CD-ROM).

Lungu, D., Cornea, T., Aldea, A., Nedelcu, C., Demetriu, S. (1999), Uncertainties in mapping frequency content of soil response to various magnitude earthquakes. *EUROMECH 372 Colloquium of the European Mechanics Society, Reliability in Nonlinear Structural Mechanics*, 21–24 Oct., Université Blaise Pascal, Institut Français de Mécanique Avancée, Clermont-Ferrand, France, pp. 39–48.

Lungu, D., Aldea, A., Demetriu, S. (2005), City of Bucharest: Seismic hazard and microzonation of site effects. *Proceedings of the International Conference dedicated to the 250th Anniversary of the 1755 Lisbon Earthquake*, CD-ROM.

A COMPARATIVE ANALYSIS OF ACCELEROGRAPHIC RECORDS OBTAINED IN THE REPUBLIC OF MOLDOVA AND ROMANIA DURING THE 1986 AND 1990 VRANCEA EARTHQUAKES

I.S. BORCIA[*]

National Institute for Building Research INCERC, Bucharest, Romania

Abstract. Interpreting instrument data obtained in the Republic of Moldova (Chisinau and Cahul) and Romania (Vaslui and Birlad) during the 1986 and 1990 Vrancea earthquakes using methodological elements developed in Romania (definition of intensities based on alternative instrumental criteria) is the main goal of this paper. The response spectra (for 12 azimuthally equidistant horizontal directions) of strong-motion records, the global parameters that characterize an individual component of a record (effective peak values and corner [control] periods) and instrumental intensity (global and averaged on a frequency interval based on the destructiveness spectrum and on the response spectrum) are the main numerical results obtained for the seismic records in the two countries. The results are evaluated.

Keywords: Accelerographic records, seismic intensity

1. Introduction

Numerous accelerographic records were obtained in Romania during the strong Vrancea earthquakes on 30 August 1986 (M_{GR} = 7.0), 30 May 1990 (M_{GR} = 6.7) and 31 May 1990 (M_{GR} = 6.1) (Table 1). M_{GR} denotes Gutenberg-Richter magnitudes. The Vrancea seismogenic zone is by far the most important source zone in Romania. According to (Bălan et al., 1982), it releases on average more than 95% of all seismic energy released per century in Romania. The wealth of

---

[*]National Institute for Building Research INCERC, Bucharest, Romania, e-mail: isborcia@ incerc2004.ro

A. Zaicenco et al. (eds.), *Harmonization of Seismic Hazard in Vrancea Zone,*   263

© Springer Science + Business Media B.V. 2008

instrumental data available made it possible to obtain a comprehensive picture of the features of Vrancea earthquakes.

In order to get a deeper insight into the features of ground motion, response spectra for 12 azimuthally different, equidistant directions and instrumental intensity spectra were determined for various events and sites according to techniques presented in Stancu and Borcia (1999) and Sandi and Floricel (1998) (Table 2 and Figure 1). The sequences of spectra for various recording stations and events showed a tendency to stability of dominant frequencies (Sandi and Borcia, 2006) (e.g. Cernavodă Town Hall and Vaslui sites were characterized by the presence of a strong contrast of S wave velocities at shallow depths) as well as a tendency to the strong variability of dominant frequencies (e.g. Bucharest and Focşani sites were characterized by deep, relatively soft ground).

TABLE 1: Characteristics of Vrancea earthquakes

No	Earthquake	LatN	LongE	Code EQ	h (km)	Date	M_w
1	Vrancea M $(G-R)$ = 7.0	45,53	26,47	861	133	1986.08.30	7,3
2	Vrancea M $(G-R)$ = 6.7	45,82	26,90	901	91	1990.05.30	7,0
3	Vrancea M $(G-R)$ = 6.1	45,83	26,89	902	79	1990.05.31	6,4

TABLE 2: Recording stations

Nr. crt.	Recording station (belonging to)	Lat. North	Long. East	Code station	1986. 08.30	1990. 05.30	1990. 05.31
1.	Chisinau – Iss1 (IGGASM)	47.058	28.872	CHI1	[a]		
2.	Chisinau - Iss2 (IGGASM)	46.989	28.860	CHI2		[a]	[a]
3.	Chisinau – Iss3 (IGGASM)	47.000	28.856	CHI3		[a]	[a]
4.	Chisinau Ul. Dimo (IGGASM)	47.061	28.874	CHI4		[a]	[a]
5.	Cahul (IGGASM)	45.905	28.200	CAH1		[a]	[a]
6.	Vaslui (INCERC)	46.637	27.733	VLS1	[a]	[a]	[a]
7.	Birlad (INCERC)	46.228	27.666	BIR1	[a]	[a]	[a]

[a]Existing record; INCERC: National Institute for Building Research, Bucharest, Romania;

IGGASM: Institute of Geology and Seismology, Academy of Sciences of Moldova, Chisinau, Republic of Moldova.

2.   Methodological Aspects

The investigation of the features of ground motion and of the reasons for these features was performed using the following main approaches:

1. Determining response spectra for strong motion records for 12 horizontal, azimuthally equidistant directions as presented in Borcia (2006)
2. Calculating corner periods of response spectra
3. Determining intensity spectra as defined in Sandi and Floricel (1998) and used in Borcia (2006)

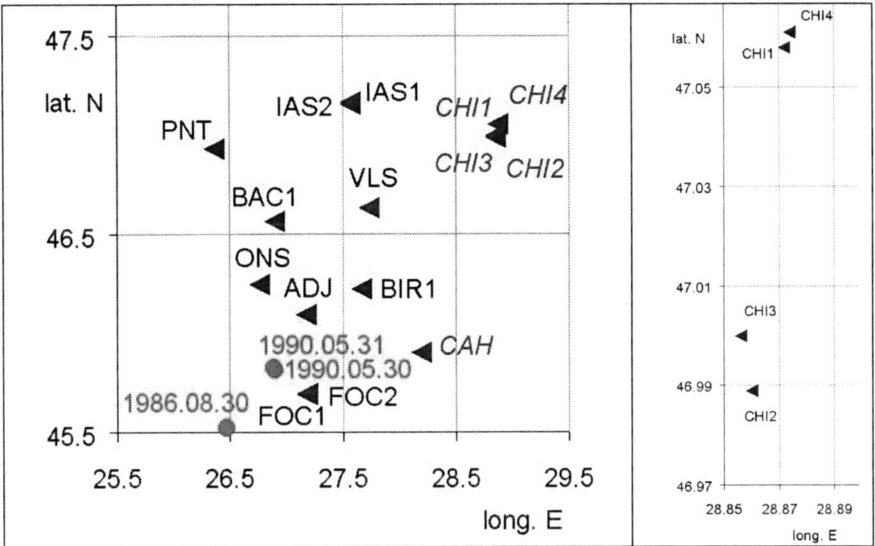

Figure 1: Map of instrumental epicenters and of recording stations (belonging to IGGASM [Md] and INCERC [Ro]) in Moldova and Northeastern Romania

The following definitions were used for global intensities, for frequency-dependent intensities and for intensities averaged on spectral bands.

1. For global intensities:

1.1 The global intensity based on response spectra, I_S:

the parameters

$$EPAM = \max_T s_{aa}(T, 0.05) / 2.5 \tag{1}$$

and

$$EPVM \,(\text{m/s}) = \max_T s_{av}(T, 0.05) / 2.5 \tag{2}$$

are introduced, where $s_{aa}(T, n)$ is the response spectrum of absolute accelerations related to periods (m/s^2) and $s_{av}(T, n)$ is the response spectrum of absolute velocities related to periods (m/s);
the global intensity I_S is given by the expression

$$I_S = \log_4 (EPAM \times EPVM) + 8.0 \tag{3}$$

1.2 The Arias intensity I_A is given by the expression

$$I_A = \log_4 \int [w_g(t)]^2 dt + 6.75 \tag{4}$$

2. For intensities depending on frequencies φ (Hz):
2.1. The intensity based on response spectra $i_s(\varphi)$:

$$i_s(\varphi) = \log_4 [s_{aa}(\varphi, 0.05) \times s_{va}(\varphi, 0.05)] + 7.70 \tag{5}$$

2.2. The intensity based on the destructiveness spectrum $i_d(\varphi)$:
where the (absolute) accelerogram $w_a(t, \varphi, 0.05)$ for a pendulum of (undamped) natural frequency φ and a 5% fraction of critical damping is used,

$$i_d(\varphi) = \log_4 (\int w_a^2(t, \varphi, 0.05)\, dt) + 5.75 \tag{6}$$

3. For intensities averaged on spectral bands the rule of averaging on a frequency band (φ', φ''):
3.1. For the spectrum-based intensity $i_s(\varphi)$:

$$i_s^*(\varphi', \varphi'') = \log_4 \{1 / \ln (\varphi''/\varphi') \int [s_{aa}(\varphi, 0.05) \times s_{va}(\varphi, 0.05)\, d\varphi / \varphi]\} + 7.70 \tag{7}$$

3.2. For the intensity based on the destructiveness spectrum, $i_d(\varphi)$:

$$i_d^*(\varphi', \varphi'') = \log_4 \{1 / \ln (\varphi''/\varphi') \int [(\int w_a^2(t, \varphi, 0.05)\, dt)\, d\varphi / \varphi]\} + 5.75 \tag{8}$$

The definitions adopted for effective peak values (P100-1/2006, 2006) were:

$$EPA = \frac{(s_{aa\ averaged\ on\ 0.4s})\max}{2.5} \tag{9}$$

$$EPV = \frac{(s_{rv\ averaged\ on\ 0.4s})\max}{2.5} \tag{10}$$

$$EPD = \frac{(s_{rd\ averaged\ on\ 0.4s})\max}{2.5} \tag{11}$$

and for the corner (control) periods

$$T_C = 2\pi \frac{EPV}{EPA} \tag{12}$$

$$T_D = 2\pi \frac{EPD}{EPV} \tag{13}$$

where s_{aa}, s_{rv}, s_{rd}, represent the response spectra for absolute accelerations, relative velocities and relative displacements depending on frequency φ or period T and for a 5% fraction of critical damping.

3. Determining Response Spectra and Corner Periods

Response spectra for the absolute acceleration S_{aa} (T, n) were determined for 12 horizontal, azimuthally equidistant directions as adopted in Stancu and Borcia (1999) (Figures 2–4). This was done for a 5% critical damping. The availability of response spectra along 12 equidistant directions made it possible to emphasize the differences in different directions of ground motion.

TABLE 3: Global characteristics of horizontal components of records obtained in Moldova

No	Record	Code axis	pga	pgv	epa	epv	Tc	I_S	I_S 1	I_A	I_D1
			m/s^2	m/s	m/s^2	m/s	s				
1	861CHI1	l: 11	1.874	0.0830	1.604	0.0993	0.39	7.59	7.23	7.40	7.38
2	861CHI1	t: 101	2.118	0.2094	1.578	0.2311	0.92	7.69	7.73	7.43	7.44
3	901CAH	l: 74	1.264	0.0614	0.844	0.0683	0.51	6.66	6.45	6.94	6.87
4	901CAH	t: 344	1.354	0.0886	1.366	0.0654	0.30	6.78	6.86	7.25	7.21
5	901CHI2	l: 132	1.882	0.0528	0.907	0.0409	0.28	6.80	6.54	7.12	6.98
6	901CHI2	t: 42	1.726	0.0579	0.947	0.0431	0.29	6.84	6.65	7.17	7.05
7	901CHI3	l: 100	1.213	0.0466	0.679	0.0325	0.30	6.12	6.20	6.80	6.66
8	901CHI3	t: 10	1.437	0.0389	0.651	0.0243	0.23	6.67	6.25	7.01	6.88
9	901CHI4	l: 0	0.750	0.0549	0.528	0.0412	0.49	5.77	5.85	6.00	5.92
10	901CHI4	t: 90	0.810	0.0611	0.722	0.0382	0.33	6.19	6.11	6.23	6.18
11	902CAH	l: 74	0.944	0.0364	0.635	0.0349	0.35	5.82	5.81	6.14	6.09
12	902CAH	t: 344	0.560	0.0294	0.453	0.0268	0.37	5.65	5.42	5.77	5.71
13	902CHI2	l: 132	0.756	0.0598	0.537	0.0478	0.56	6.11	5.92	6.01	5.92
14	902CHI2	t: 42	0.876	0.0255	0.443	0.0214	0.30	5.70	5.55	5.80	5.67
15	902CHI3	l: 100	0.534	0.0295	0.327	0.0353	0.68	5.69	5.43	5.73	5.62
16	902CHI3	t: 10	0.612	0.0293	0.255	0.0319	0.79	5.54	5.13	5.54	5.41
17	902CHI4	l: 0	0.396	0.0298	0.388	0.0331	0.54	5.18	5.30	5.20	5.18
18	902CHI4	t: 90	0.569	0.0563	0.382	0.0549	0.90	5.51	5.68	5.39	5.39

Note: indices used
l (longitudinal): first horizontal direction of record;
t (transversal): second horizontal (orthogonal with the first one) direction of record;
lt averaging upon two orthogonal horizontal directions; (no index): global intensities;
1 (one): averaging of frequency dependent intensities over the frequency interval (0.25, 16.0 Hz).

Attention was paid not only to the features of individual motions or spectra but also to sequences of spectra which made it possible to determine the tendencies to stability or to variability (from one event to another) of the features of ground motion.

The values of corner periods Tc in Table 3 are between 0.23 and 0.92 s making the value 0.5 s for the end of the flat portion of the spectral curve corresponding to the code in force in the Republic of Moldova (Alkaz, 2006) questionable. This fact is illustrated also in Figure 5 (left). Table 3 shows a very good correlation between values of Arias intensity I_A and values of intensity based on the destructiveness spectrum averaged over the frequency interval (0.25, 16.0 Hz), I_D1.

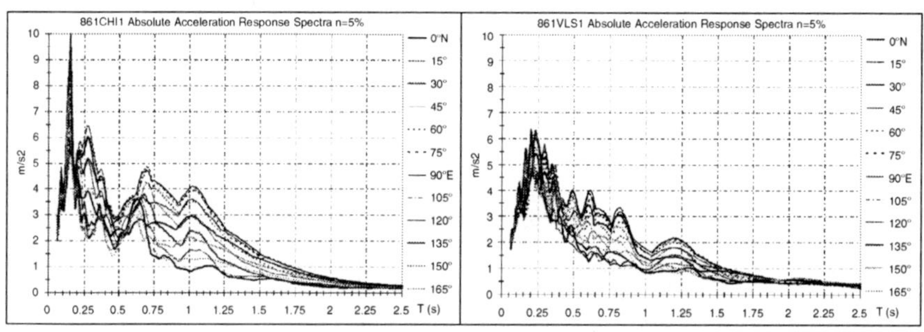

Figure 2: Response spectra for absolute accelerations along 12 equidistant azimuthal directions for records obtained in Chisinau (Md) and Vaslui (Ro) during the 1986 Vrancea earthquake

In spite of the fact that the ground conditions are the same for all directions of oscillation, there are important differences between spectral ordinates corresponding to different directions for the same event and place (the extreme ratios of ordinates reach or even exceed, the threshold 3.0 for some oscillation periods as illustrated in Figure 2 for the CHI1 station, and in Figure 3 for the CHI4 station).

4. Intensity Spectra Derived on the Basis of Accelerographic Records

The intensity spectra presented were derived on the basis of accelerographic records (Figures 5–7). The intensity spectra are organized as follows:

- The abscissa corresponds to lg *T*.
- The ordinate corresponds to (instrumental) intensity values.

An examination of Figure 7 shows that there are differences around 0.6° of intensity for intensities averaged over period intervals (0.0625–0.125 s) and

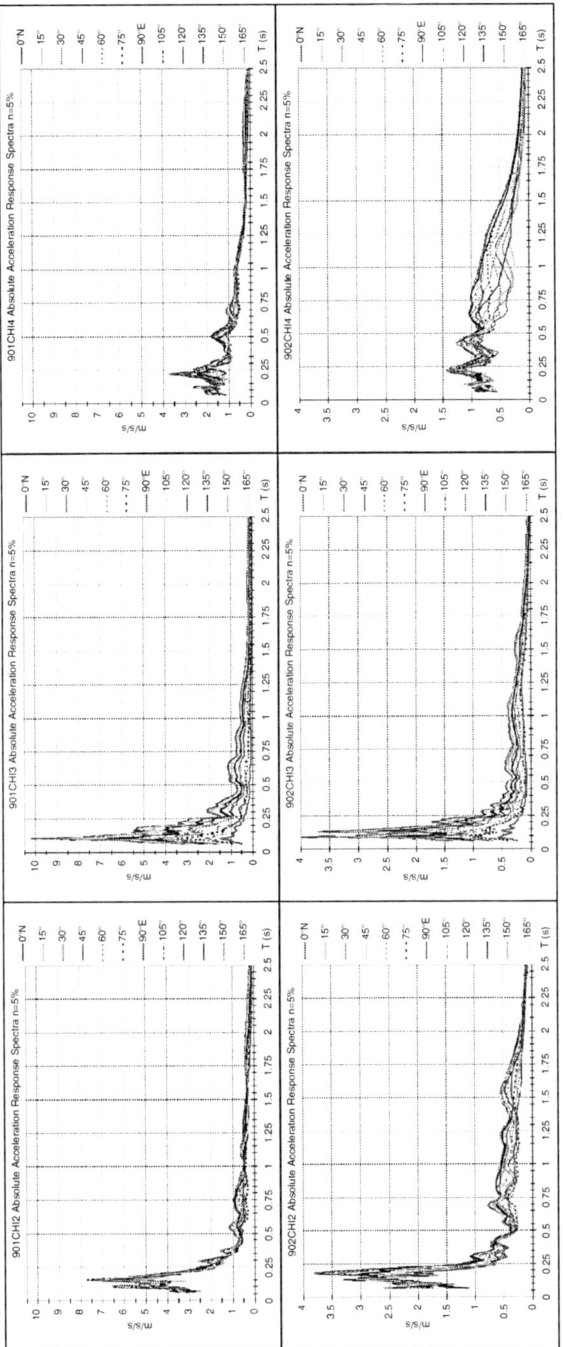

Figure 3: Response spectra for absolute accelerations along 12 equidistant azimuthal directions for records obtained in Chisinau (Md) during the 1900 Vrancea earthquakes

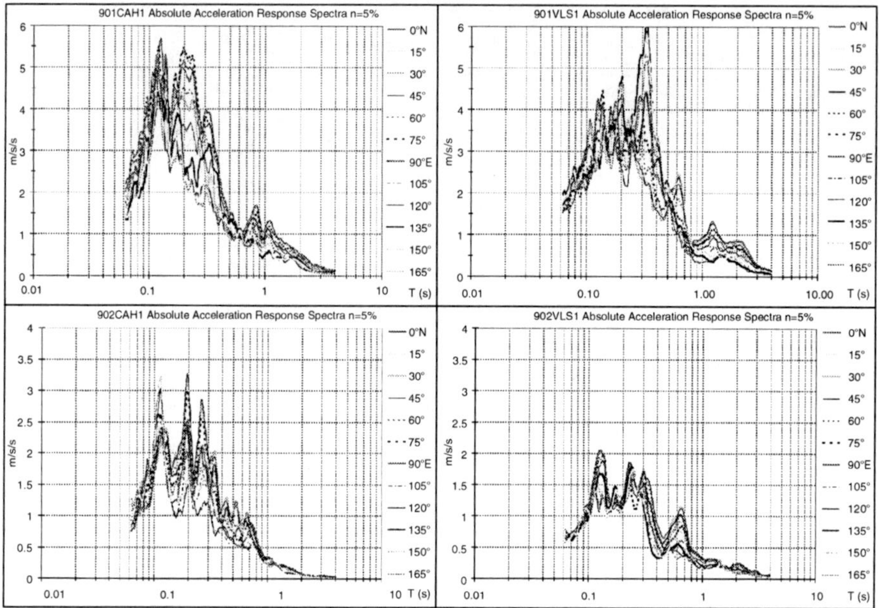

Figure 4: Response spectra for absolute accelerations along 12 equidistant azimuthal directions for records obtained in Cahul (Md) and Vaslui (Ro) during the 1990 Vrancea earthquakes

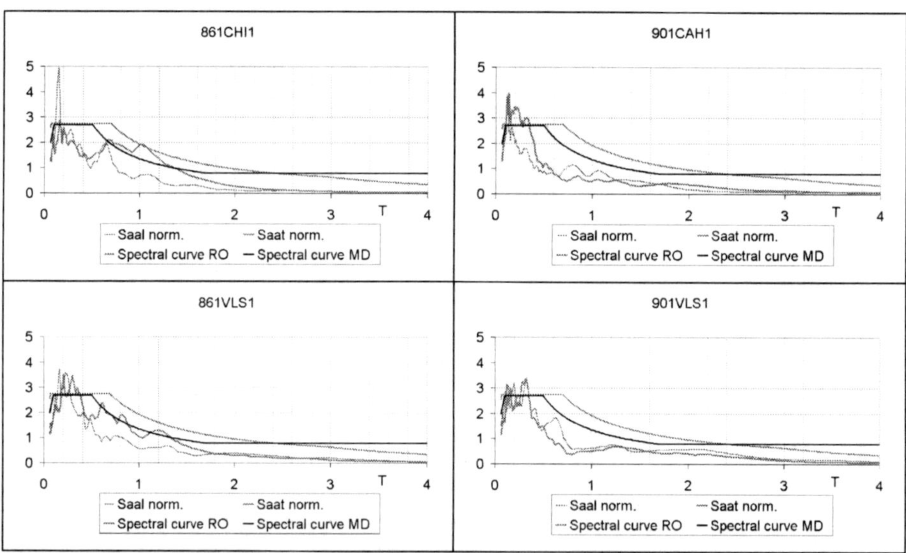

Figure 5: Normalized acceleration response spectra and spectral curves corresponding to the codes in force in the two countries (P100-1/2006, 2006; Alkaz, 2006) for Chisinau (CHI1) and Vaslui (VLS1) for the 1986 Vrancea earthquake (*left*) and for Cahul (CAH1) and Vaslui (VLS1) for 30 May 1990 Vrancea earthquake (*right*)

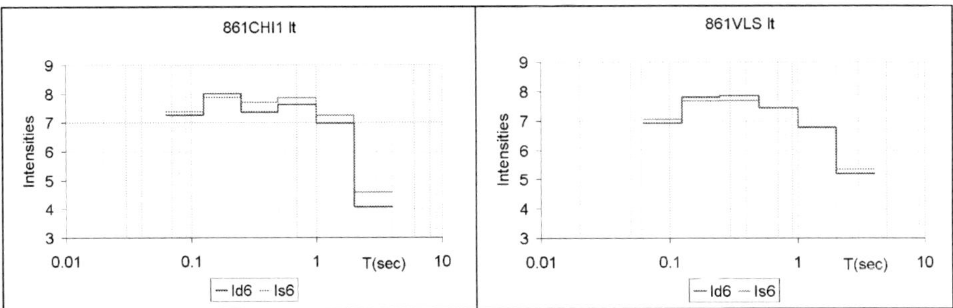

Figure 6: Average intensity spectra $i_s^{\sim}(\varphi', \varphi'')$ and $i_d^{\sim}(\varphi', \varphi'')$ for 6 dB intervals for Chisinau (recording station CHI1) and Vaslui (VLS1) for the 1986 Vrancea earthquake

(0.125–0.250 s) and differences around $-0.9°$ of intensity for intensities averaged over period intervals (1.0–2.0 s) between the recording stations CAH1 (Md) and VLS1 (Ro) for the 1990 earthquakes.

5. Final Considerations

An examination of the strong-motion records available and of the spectra determined indicate that it is particularly important to consider all available data because considering one station or even one event in isolation could lead to unrealistic conclusions. This is why the availability of strong-motion data for several events originating in the same source zone is so important.

An aspect of primary interest for this paper is the fact that for sites CHI1 and CHI4 there was a strong tendency to variability of the spectral contents of ground motion (as illustrated by the response spectra of Figures 2–4). Explaining the reasons for this is of obvious interest because it is directly connected with the ability to anticipate the spectral contents of future strong ground motions.

An examination of Figures 6–8 makes it possible to make comments provided the definitions of the average intensities $i_s^{\sim}(\varphi', \varphi'')$ and $i_d^{\sim}(\varphi', \varphi'')$ is accepted. Depending on period or frequency interval, there appears to be a significant variation of the spectral band-related average intensities. This provides a picture of the spectral intervals for which the intensities are higher and, consequently, the severity of seismic action appears to be higher. There are differences in outcome if alternative definitions of $i_s^{\sim}(\varphi', \varphi'')$ and $i_d^{\sim}(\varphi', \varphi'')$ are used, but the differences are moderate and definitely less than the possibilities for discrimination provided by the use of macroseismic intensities derived from visual post-earthquake surveys. It is to mention that the instrumental intensities are continuous quantities, differing of standard macroseismic intensity (discrete quantity).

I.S. BORCIA

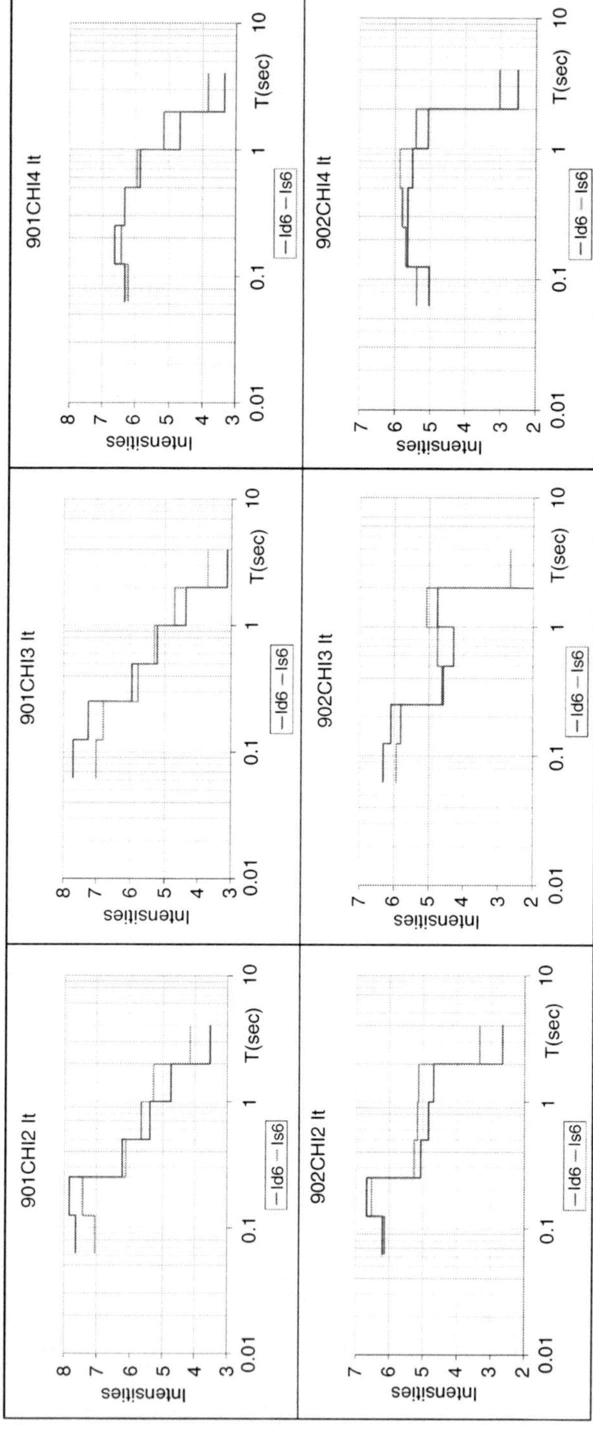

Figure 7: Average intensity spectra $i_s^{\sim}(\varphi', \varphi'')$ and $i_d^{\sim}(\varphi', \varphi'')$ for 6 dB intervals for Chisinau (recording stations CHI2, CHI3, CHI4) for the 1990 Vrancea earthquakes

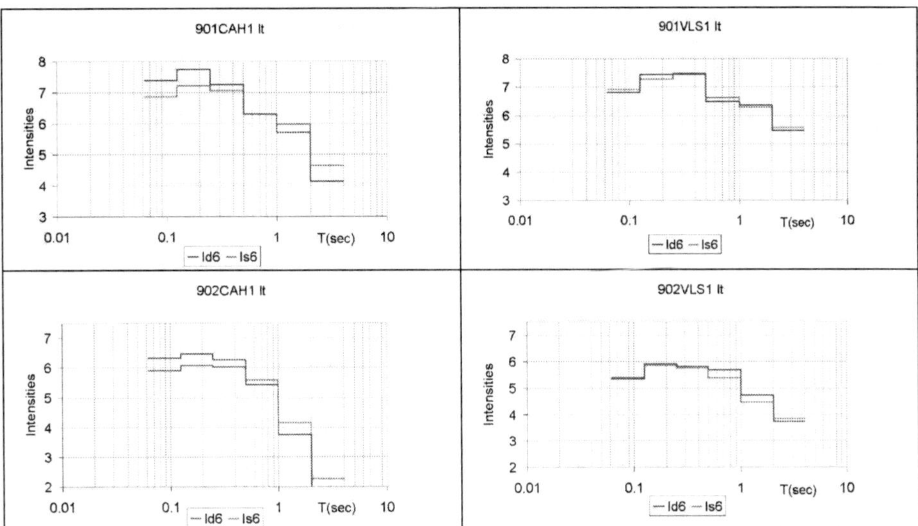

Figure 8: Average intensity spectra $i_s^\sim(\varphi', \varphi'')$ and $i_d^\sim(\varphi', \varphi'')$ for 6 dB intervals for Cahul (CAH1) and Vaslui (VLS1) for the 1990 Vrancea earthquakes

Acknowledgements

This work was prepared to a large extent as part of NATO Project SfP 980468 "Harmonization of Seismic Hazard and Risk Reduction in Countries Influenced by Vrancea Earthquakes." The author is grateful to Dr. V. Alkaz, E. Isiciko and I. Ilies for providing the strong ground motion data recorded in Moldova as part of NATO Project SfP 981619 "Quantification of Earthquake Action on Structures."

References

V. Alkaz, *Scientific – methodological bases of seismic hazard and risk evaluation of the territory of the Republic of Moldova*, Ph.D. thesis, Chisinau (extended abstract in Romanian), **2006**.

Ş.T. Bălan, V. Cristescu, I. Cornea (editors), *The Romania earthquake of 4 March 1977* (in Romanian), Editura Academiei, Bucharest, **1982**.

I.S. Borcia, *Data processing of strong motion records obtained during Romanian earthquakes* (in Romanian), Doctoral Thesis, UTCB, **2006**.

H. Sandi, I.S. Borcia, *Damage spectra and intensity spectra for recent Vrancea earthquakes*, Paper no. 574. Proceedings of the 1st European Conference on Earthquake Engineering and Seismology. Geneva, **2006**.

H. Sandi, I. Floricel, *Some alternative instrumental measures of ground motion severity.* Proceedings of the 11th ECEE, AFPS, Paris, **1998**.

M. Stancu, I.S. Borcia, *Studies concerning the directionality of seismic action for Vrancea earthquakes.* Proceedings of the International Workshop for Vrancea Earthquakes, Bucharest, 1997. Kluwer, Dordrecht, **1999**.

P100-1/2006: *Seismic Design Code. Part I. Design Rules for Buildings. UTCB – MTCT* (in Romanian), **2006**.

AN ASSESSMENT OF DAMAGE POTENTIAL AND BUILDING PERFORMANCE DEMANDS FOR ROMANIAN VRANCEA EARTHQUAKES

I. CRAIFALEANU[*,1], D. LUNGU[2]

[1]*Technical University of Civil Engineering, 124, Lacul Tei Blvd., 020396 Bucharest, Romania*
National Institute for Building Research, INCERC, Sos. Pantelimon 266, sector 2, 021652 Bucharest, Romania
[2]*Technical University of Civil Engineering, 124, Lacul Tei Blvd., 020396 Bucharest, Romania*
National Institute for Historical Monuments, 16, Enachita Vacarescu St., 040157 Bucharest, Romania

Abstract. The paper presents an evaluation of the damage potential of Romanian Vrancea earthquakes based on the use of damage spectra. Spectra of a derived damage index, $DM.\mu_u$, were computed for Vrancea events with moment magnitudes larger than 6.0. In addition, strength demand spectra were determined, in order to assess the strength capacity required to limit the damage index to a specified (target) value. The resulting strength demands were then mapped for several values of $DM.\mu_u$ and of the structural period. Some applications of the approach, from the structural engineering point of view, were pointed out.

Keywords: Damage index, damage potential, Vrancea earthquakes, damage spectra, damage maps, strength demand spectra, maps of strength demands

1. Introduction

The use of damage indices is presently considered a promising approach for assessing the building damage potential of ground motions. Several such

---

[*]Iolanda-Gabriela Craifaleanu, Reinforced Concrete Dept., Technical University of Civil Engineering Bucharest, 124 Lacul Tei Blvd., RO-020396 & Earthquake Engineering Dept., National Institute for Building Research, INCERC, 266 Pantelimon St., Sector 2, RO-021652, Bucharest, Romania, e-mail: i.craifaleanu@gmail.com

A. Zaicenco et al. (eds.), *Harmonization of Seismic Hazard in Vrancea Zone,*
© Springer Science + Business Media B.V. 2008

indices have been developed during the past two decades. One of the most widely used indices, mainly due to its extensive calibration against experimentally observed seismic structural damage, is the Park-Ang damage index (Park and Ang, 1985). Mapping of damage spectra ordinates (Bozorgnia and Bertero, 2001) has been used more recently and has proved to provide useful information in assessing the spatial distribution of damage for a given earthquake and for buildings with specified strength and stiffness characteristics.

The damage index proposed by Park and Ang is given by the following relationship:

$$DM = \left(u_{max}/u_u\right) + \beta\, E_H \big/\left(F_y u_u\right) \tag{1}$$

where

u_{max} = maximum deformation demand during the ground motion

u_u = ultimate deformation capacity of the system under monotonically increasing lateral deformation

E_H = hysteretic energy

F_y = yield strength

β = constant depending on structural characteristics

As it results from Equation (1), DM takes into account both damage due to the maximum deformation attained during the ground motion and damage due to repeated cycles of inelastic deformations. According to experimental results and field observations in earthquakes, β can be taken as equal to 0.15 (Cosenza et al., 1993; Fajfar, 1992). A value of DM less than 0.4 can be considered as the limit of repairable damage; from 0.4 to 1.0 as non-repairable damage and larger than 1.0 as failure (Teran-Gilmore, 1996).

The Park-Ang damage index has some drawbacks that have been pointed out in the literature (Bozorgnia and Bertero, 2001). For instance, for elastic response, when E_H is 0 and the damage should be zero, the value of DM is greater than zero Moreover, the index does not provide correct results for a system subjected to monotonic deformation. However, DM is still largely used for different applications due to its simplicity and its experimental validation.

Based on the following relationships:

$$\mu = u_{max}/u_y \tag{2}$$

$$\mu_u = u_u/u_y \tag{3}$$

$$\mu_E = 1 + E_H\big/\left(F_y u_y\right) \tag{4}$$

where μ is the displacement ductility, μ_u is the ductility under monotonically increasing lateral deformation and μ_E is the equivalent ductility (Mahin and

Lin, 1983), it results from Equation (1) that DM can be expressed as a function of the three ductilities above:

$$DM = \frac{\mu}{\mu_u} + \beta \frac{\mu_E - 1}{\mu_u} \tag{5}$$

or

$$DM \cdot \mu_u = \mu + \beta(\mu_E - 1) \tag{6}$$

The use of the product $DM.\mu_u$ instead of DM is more convenient for the present study, as it separates on the right-hand side of Equation (6) the values corresponding to alternate loading. Therefore, throughout the paper, damage is expressed by using $DM.\mu_u$. The interpretation of results is made by considering relevant values of μ_u.

2. Damage Spectra

Damage spectra were calculated for several ground motions recorded during the strong earthquakes of March 4, 1977 (moment magnitude $M_w = 7.5$, focal depth $h = 109$ km) and August 30, 1986 ($M_w = 7.1$, $h = 133$ km). Data provided by 35 seismic stations were used. Of these stations, 34 were located in Romania (24 in the INCERC network, 10 in the NIEP network (Seismic Database, 2001)) and one was located in the Republic of Moldova (station Chisinau, IGG network). Only the horizontal components of ground motions were taken into account in the study. The total number of seismic records analyzed was 72, including the two horizontal components of the INCERC Bucharest record of March 4, 1977.

The spectra were determined by considering the bilinear, elastic-perfectly plastic hysteretic behavior of the SDOF systems and a damping ratio of 5%. The above hypotheses were used throughout the entire study presented in this paper.

As a parameter of spectral curves, the yield strength coefficient C_y, was chosen. The coefficient C_y can be expressed as

$$C_y = \frac{F_y}{G} \tag{7}$$

where G is the weight of the SDOF system and F_y has the same signification as in Equation (1). The coefficient C_y is a simple measure of the yield strength of the system and can be quite easily related to the code-specified base shear coefficient C_s. By denoting the overstrength factor as R_{ov} the following expression can be written (Craifaleanu, 2005):

$$C_y = C_s R_{ov} \tag{8}$$

A typical spectrum of the product $DM.\mu_u$ is shown in Figure 1 (Craifaleanu, 2001) for the reference north-south component of the INCERC, March 4, 1977 ground motion record and for different values of C_y.

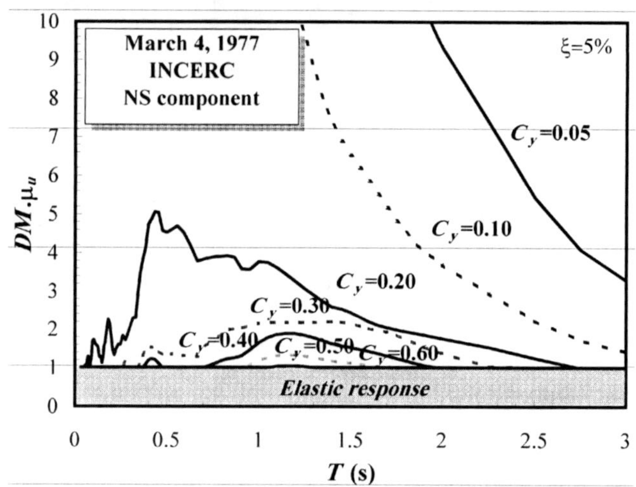

Figure 1: Spectra of the product $DM.\mu_u$, computed for different values of the yield strength coefficient, C_y. March 4, 1977, INCERC Bucharest, NS component

As it can be observed from Figure 1, at short periods, for C_y values lower than 0.2 (which corresponds to the PGA, in g's, for this record), the values of $DM.\mu_u$ are very large. They tend to infinity as SDOF system period and C_y tend to zero. With the increase of yield strength, the values of $DM.\mu_u$ gradually decrease, reaching values below unity for elastic response.

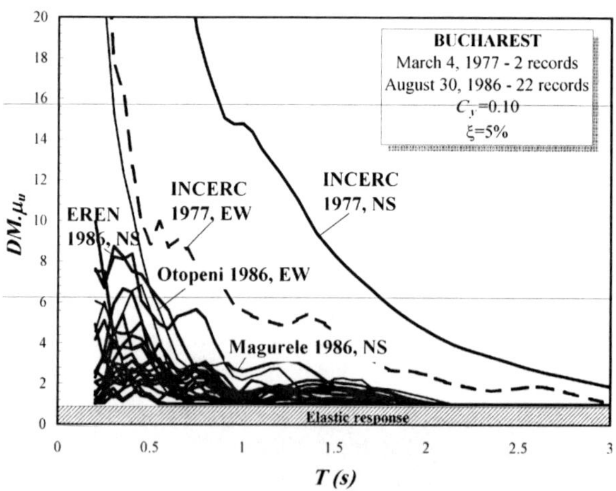

Figure 2: Comparison of $DM.\mu_u$ values for the two horizontal components of the INCERC Bucharest record of March 4, 1977 and for 22 Bucharest records of the August 30, 1986 earthquake. $C_y = 0.10$

A comparison between the $DM.\mu_u$ spectra of ground motions recorded in Bucharest during the previously mentioned seismic events (Figure 2) reveals large differences between their damage potential. As expected, the largest values of $DM.\mu_u$ were obtained for the records of 1977.

An interesting observation to be made on the diagram in Figure 2 concerns the larger ordinates corresponding to the seismic stations located in the sparsely built, peripheral areas of the city (EREN, Otopeni, Magurele), as compared with the stations in the central area.

3. Strength Demand Spectra for Target Values of Damage Index

An alternate approach in the study of damage indices is to determine the yield strength that would ensure that for buildings with certain characteristics, damage corresponding to the considered ground motion is limited to a desired (target) level, DM.

For the case of $DM.\mu_u$ spectra, this requires that the values of μ_u are specified.

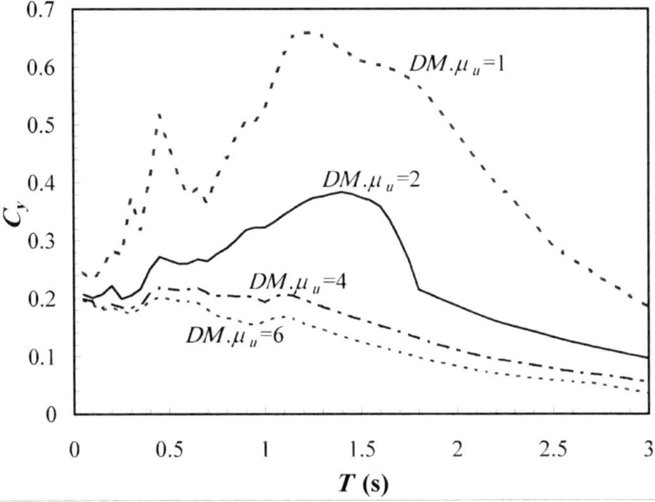

Figure 3: Spectra of C_y, determined for different values of $DM.\mu_u$. March 4, 1977, INCERC, NS component

For the purpose of the present study, strength demand spectra for a set of relevant values of $DM.\mu_u$ were determined for all 72 ground motions analyzed.

Figure 3 shows the strength demand spectrum of the NS component of the INCERC March 4, 1977 record, computed for four values of $DM.\mu_u$, among which the value 1, corresponding to the upper limit of elastic behavior. It is worth noticing on this diagram the evolution of the shape of the curves with the increase of $DM.\mu_u$, and the progressive attenuation of the "fingerprint" of the predominant period of the ground motion.

4. Distribution of Yield Strength Demands for Target Values of $DM.\mu_u$

The spectral ordinates of C_y, obtained for the records of the August 30, 1986 earthquake, were mapped in order to assess the spatial distribution of yield strength demands. Maps were generated for two values of the structure period, $T = 0.5$ s and $T = 1.0$ s, and for three values of the product $DM.\mu_u$, i.e. 2, 4 and 6 (Figures 4 and 5).

The extent of the interpolation surface was limited to the extreme northern, southern, western and eastern seismic stations considered. The epicenter of the earthquake is shown on the maps with a star-shaped maker.

One of the most distinct features of the maps is the orientation of contours along the northeast-southwest direction. This feature, which was previously observed as well on maps generated for other parameters, such as the peak ground acceleration or the spectral acceleration (Craifaleanu et al., 2006; Lungu and Craifaleanu, 2007), corresponds to the results of previous studies concerning the predominant direction of propagation of seismic waves for the August 30, 1986 seismic event.

The C_y ordinates calculated for $T = 0.5$ s are larger than the values corresponding to $T = 1.0$ s, as a consequence of the spectral contents of the analyzed records. For instance, for $T = 0.5$ s and $DM.\mu_u = 2$, the C_y values vary from 0.02 to 0.35, while for $T = 1.0$ s and the same value of $DM.\mu_u$, the range of variation is between 0.01 and 0.15.

The most evident consequences of increasing the values of $DM.\mu_u$ are the narrowing of the range of variation of the C_y ordinates and the general decrease of their values, i.e. the flattening of the interpolation surfaces.

The detailed spatial distribution of yield strength demands for the city of Bucharest is shown in Figures 6 and 7 for the same values of $DM.\mu_u$ and of the structure period mentioned above.

The tendencies observed on the maps drawn for the whole territory of Romania are also present here.

The largest C_y values occur in the three stations already mentioned in the previous section of the paper: Otopeni, Magurele and EREN.

In order exemplify the interpretation of the above maps, a value of $\mu_u = 6$ is assumed. For the three values of $DM.\mu_u$, it results in $DM = 0.33$, 0.67 and 1.00. It can then be observed that, in the central area of the city, a damage level corresponding to $DM = 0.33$ (in the range of repairable damage) could be

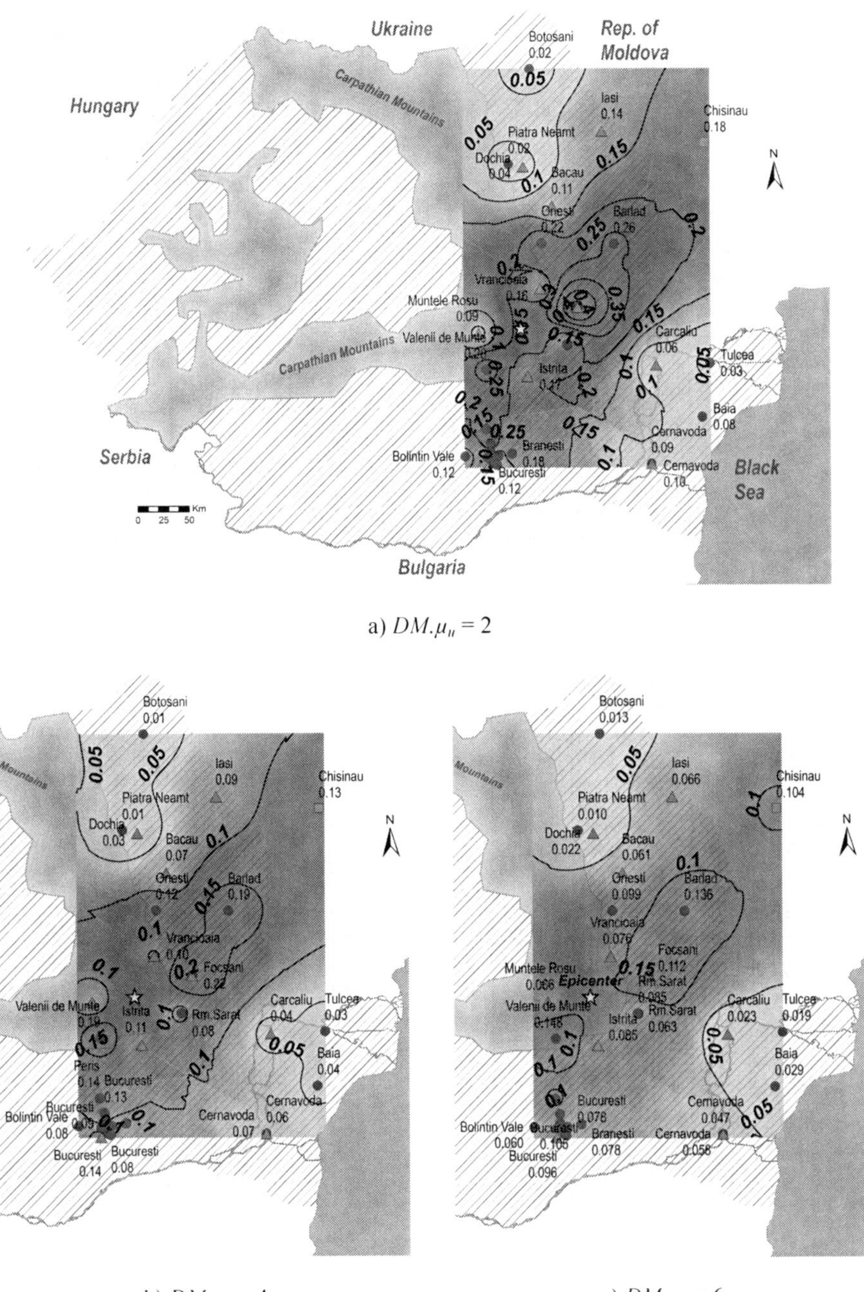

a) $DM.\mu_u = 2$

b) $DM.\mu_u = 4$ c) $DM.\mu_u = 6$

Figure 4: Spatial distribution of yield strength demands (C_y) for the Vrancea earthquake of August 30, 1986. Structure period $T = 0.5$ s

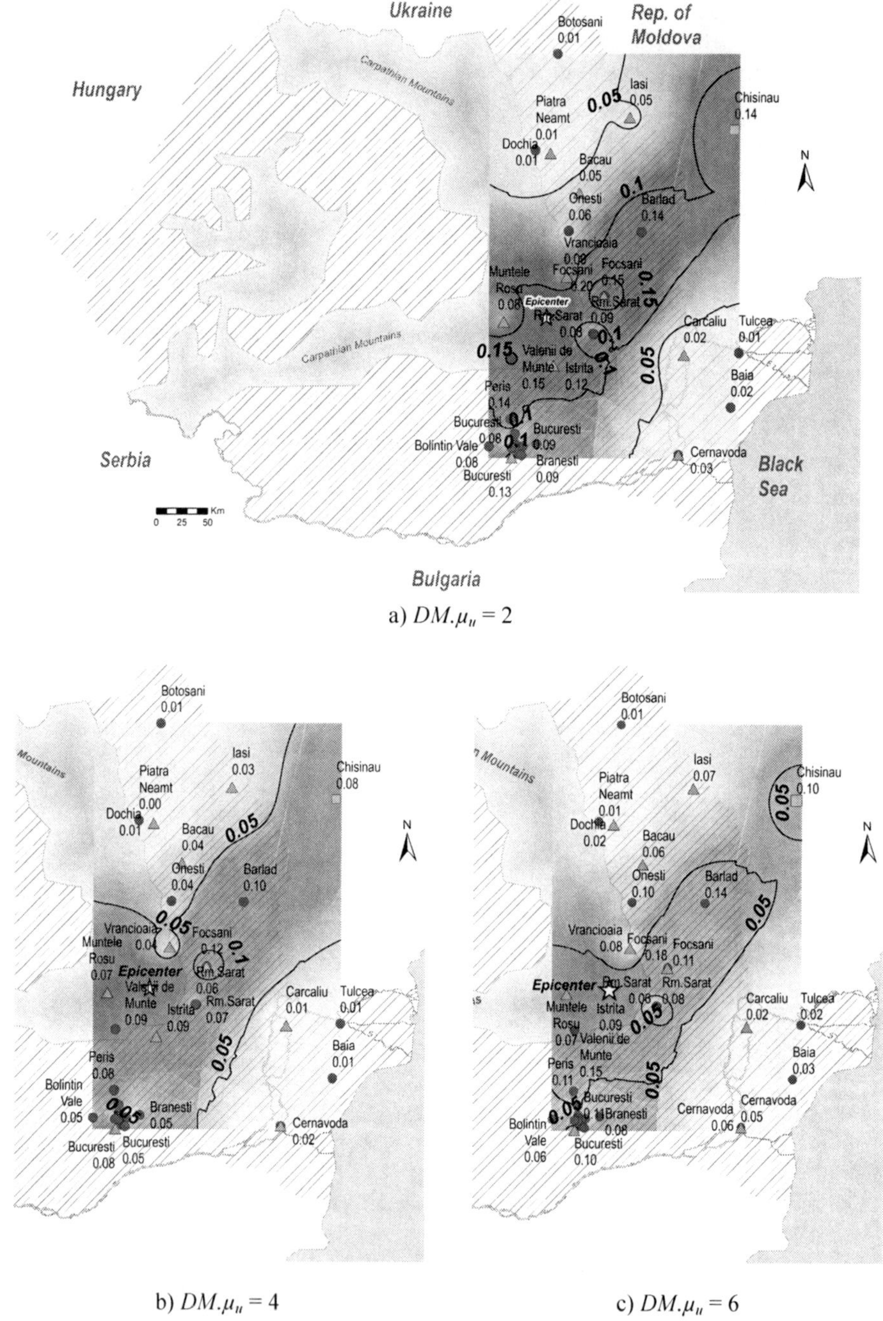

a) $DM.\mu_u = 2$

b) $DM.\mu_u = 4$ c) $DM.\mu_u = 6$

Figure 5: Spatial distribution of yield strength demands (C_y) for the Vrancea earthquake of August 30, 1986. Structure period $T = 1.0$ s obtained by providing the building an overall yield strength equal to 12–17% of the building weight, for $T = 0.5$ s (Figure 6a), and equal to 7–9% of the building weight, for $T = 1.0$ s (Figure 7a). For $DM = 0.67$, the same values were 8–10% (Figure 6b) and b 5–6%, respectively (Figure 7b), while for $DM = 1.00$, values of 7–8% (Figure 6a) and 4–5% (Figure 7c) were found.

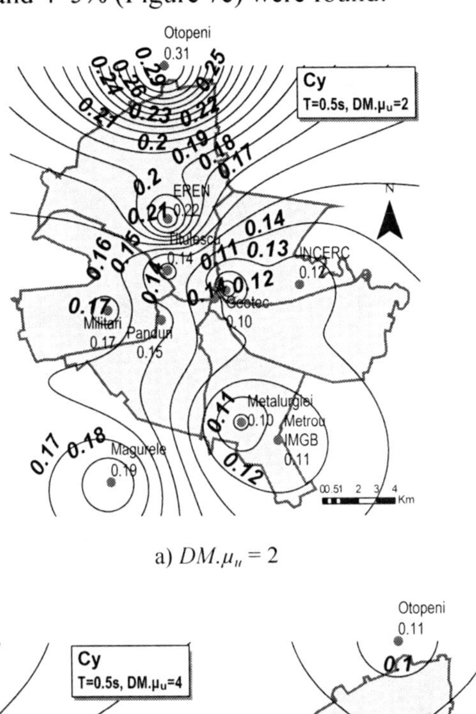

a) $DM.\mu_u = 2$

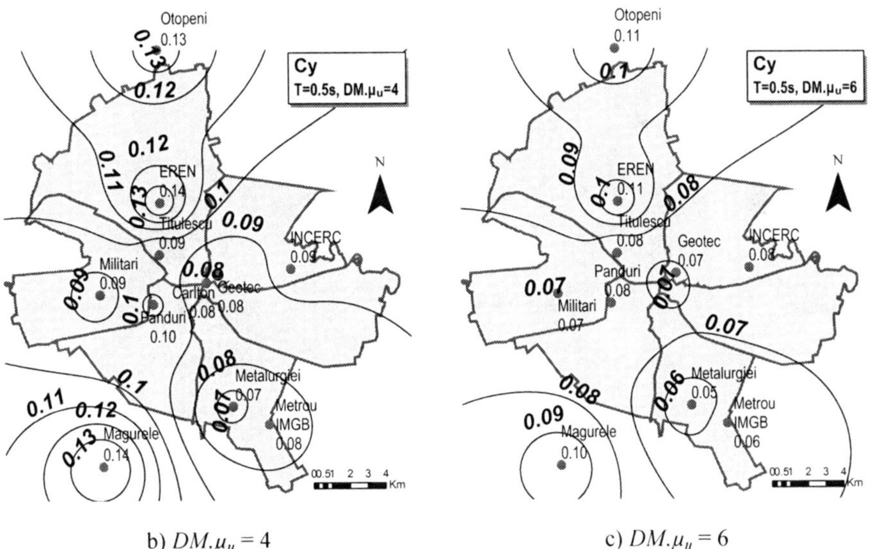

b) $DM.\mu_u = 4$

c) $DM.\mu_u = 6$

Figure 6: Spatial distribution of yield strength demands (C_y) in Bucharest, for the Vrancea earthquake of August 30, 1986. Structure period $T = 0.5$ s

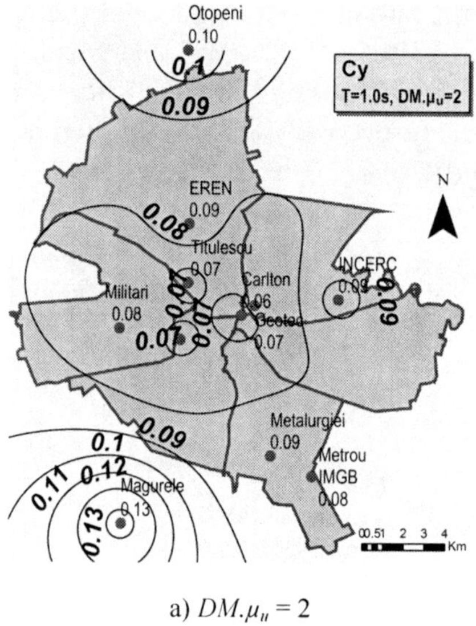

a) $DM.\mu_u = 2$

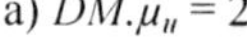

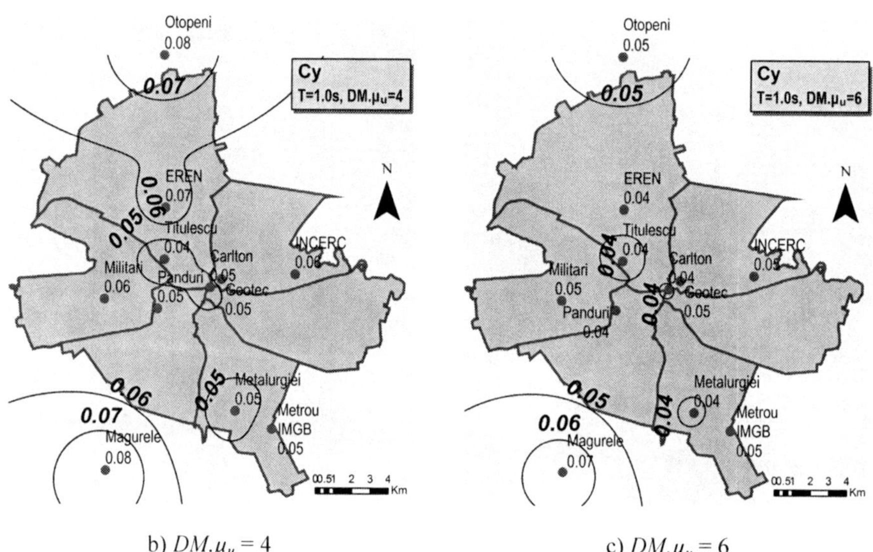

b) $DM.\mu_u = 4$ c) $DM.\mu_u = 6$

Figure 7: Spatial distribution of yield strength demands (*Cy*) in Bucharest, for the Vrancea earthquake of August 30, 1986. Structure period *T* = 1.0 s

The above results should be interpreted with reference to the typology, characteristics and spatial distribution of the building stock in Bucharest. Also, the limitations of the damage index considered should be taken into account.

5. Concluding Remarks

The paper presented some possible applications of damage indices in estimating the damage potential and the building performance demands of seismic ground motions. Spectra and maps of yield strength demands were generated for records of strong Romanian Vrancea earthquakes, taking into account different values of a target damage index. The study focused particularly on the earthquake of August 30, 1986, the strongest seismic event for which distributed records were available.

Despite the relatively small number of stations which provided data, interesting conclusions could be drawn. The mapping of strength demand spectral ordinates showed a spatial distribution pattern which is consistent with that obtained by the authors in previous mapping studies of ground motion parameters and of linear/nonlinear spectra. Also, it was shown that, for higher target (acceptable) damage levels, strength demands decrease and, furthermore, their spatial variation attenuates. Hence, in this case, the influence of structural behavior tends to prevail over that of ground motion characteristics.

Strength demand spectra, determined for specified (target) values of the damage index, appear to represent a more complex and promising way for assessing seismic demands, as compared with constant ductility spectra. The obtained values need, however, further validation based on experimental and analytical data.

Acknowledgements

This research was sponsored by NATO's Scientific Affairs Division in the framework of the Science for Peace Programme, project SfP-980468.

References

Bozorgnia, Y., Bertero, V. V. (2001), Improved shaking and damage parameters for post-earthquake applications, Proceedings of the SMIP01 Seminar on Utilization of Strong-Motion Data, Los Angeles, California, September 12, p. 1 22.

Cosenza, E., Manfredi, G., Ramasco, R. (1993), The use of damage functionals in earthquake engineering: a comparison between different methods, Earthquake Engineering and Structural Dynamics, Vol. 22, pp. 855–868.

Craifaleanu, I. G. (2001), Characterisation of seismic response using damage indices, Buletinul AICPS, No. 2, pp. 71–79 (in Romanian).

Craifaleanu, I. G. (2005), *Nonlinear Single-Degree-of Freedom Models in Earthquake Engineering*, Matrix Rom Publishers, Bucharest (in Romanian).

Craifaleanu, I. G., Lungu, D., Borcia, I. S. (2006), Shakemaps of nonlinear spectral ordinates for Vrancea earthquakes, Proceedings of the First European Conference on Earthquake Engineering and Seismology, Geneva, Switzerland, 3–8 September, Paper No. 1257 (on CD-ROM).

Lungu, D., Craifaleanu, I. G. (2007), Spatial distribution of strength and displacement demands for Romanian earthquakes, Proceedings of the International Symposium on Seismic Risk Reduction, The JICA Technical Cooperation Project in Romania, Bucharest, April 26–27, pp. 407–412, Paper ID: 79.

Fajfar, P. (1992), Equivalent ductility factors, taking into account low-cycle fatigue, *Earthquake Engineering and Structural Dynamics,* Vol. 21 (10), pp. 837–848.

Mahin, S. A., Lin, J. (1983), *Construction of Inelastic Response Spectra for Single-Degree-of-Freedom Systems: Computer Program and Applications.* Report No. UCB/EERC-83/17, Earthquake Engineering Research Center, University of California, Berkeley, California.

Park, Y.-J., Ang, A. H.-S. (1985), Mechanistic seismic damage model for reinforced concrete, *Journal of Structural Engineering,* Vol. 111 (4), pp. 722–739.

Teran-Gilmore, A. (1996), *Performance-Based Earthquake-Resistant Design of Framed Buildings Using Energy Concepts.* Ph.D. thesis, University of California, Berkeley.

Seismic Database for Romanian Earthquakes (2001), National Institute for Building Research, INCERC, and National Institute for Earthquake Physics, NIEP; research founded by the Romanian Ministry of Education and Research.

ESTIMATING AN UPPER LIMIT ON PREHISTORIC PEAK GROUND ACCELERATION USING THE PARAMETERS OF INTACT SPELEOTHEMS IN CAVES IN SOUTHWESTERN BULGARIA

K. GRIBOVSZKI[*,1], I. PASKALEVA[2], K. KOSTOV[3], P. VARGA[1], G. NIKOLOV[3]

[1]*Hungarian Academy of Sciences, Geodetic and Geophysical Research Institute, Seismological Observatory, e-mails: kgribovs@ggki.hu, varga@seismology.hu*
[2]*Bulgarian Academy of Sciences, Central Laboratory Seismic Mechanics and Earthquake Engineering, e-mail: paskalev_2002@yahoo.com*
[3]*Bulgarian Academy of Sciences, Geological Institute, e-mails: kskostov@geology.bas.bg, gabor@geology.bas.bg*

Abstract. Karst regions are very vulnerable to natural and human-activity hazards. The examination of broken and slim intact speleothems in Snezanka and Eminova caves in southwestern Bulgaria in the Rhodope Massif allows the estimation of an upper limit for horizontal peak ground acceleration generated by Paleolithic earthquakes. The density, Young's modulus and the tensile failure stress of samples from broken speleothems were measured in the laboratory, and the fundamental frequency and damping of speleothems were measured *in situ* by observation. The value of the upper limit horizontal ground acceleration resulting in failure and the natural frequency of speleothems were assessed by theoretical calculations using mechanical parameters – the density, Young's modulus and the tensile failure stress – of the samples originating from broken speleothems in Snezanka and Eminova caves. Pure elastic behavior in analytical modeling and calculations was used. According to our modeling results, the speleothem investigated have not been excited by a horizontal acceleration higher than 0.9 g in Snezanka Cave and 0.6 g in Eminova Cave in the last few thousand years. This acceleration level is much higher than the PGA value determined by the historical earthquake data of this territory.

[*]Hungarian Academy of Sciences, Geodetic and Geophysical Research Institute, Seismological Observatory, e-mail: kgribovs@ggki.hu

A. Zaicenco et al. (eds.), *Harmonization of Seismic Hazard in Vrancea Zone,*
© Springer Science + Business Media B.V. 2008

Keywords: Speleothem, stalagmite, prehistoric, frequency, seismic hazard, PGA

1. Introduction

Though Bulgaria suffers strong earthquakes, there are also areas where moderate seismic activity is experienced from the same source zone as the large earthquakes with a recurrence period of 10,000 years (Scholz, 1990). This means that no assumptions can be made regarding the maximum magnitude of an earthquake to be expected in a given area, even using observations over a 1,000–2,000-year period. Therefore, we cannot make well-grounded inferences for seismic hazard assessments by using only the data in earthquake catalogues.

To obtain more reliable and realistic data regarding the frequency and magnitude of earthquakes, we have to investigate earthquakes that occurred before historic times. Broken stalagmites could be an indication that a larger earthquake occurred in a given area during Paleolithic times. The connection between the growth, breaking and leaning of such dripstones has not yet been investigated in Bulgaria. Age is very important for determining the stalagmite's seismic vulnerability. Before any physical measurements are made, the geological environment has to be examined thoroughly in order to exclude other causes of breakage such as landslides or cave floods.

Research on the relationship between earthquakes and the growth, tilting and breaking of speleothems is promising and has been initiated in recent times. A short summary of recent research on this topic can be found in Szeidovitz et al. (2008a, b). The authors of this paper were the first to carry out non-intrusive *in situ* measurements in Bulgarian caves in order to determine fundamental frequencies and horizontal ground acceleration that would result in the failure of slim stalagmites to remain intact. This research was carried out in the framework of bilateral cooperation between the Central Bulgarian Academy of Sciences and the Hungarian Academy of Sciences.

Karstified rocks cover about 22.7% of Bulgaria (Popov, 1970), and about 5,500 caves and potholes have been discovered and surveyed. This paper presents our investigations in Snezanka and Eminova caves situated in the southwest (Figure 1a–c) including part of the ridge of Rhodope Mountains. This is known as one of the richest cave regions in the country.

Figure 1: (a) Location of Snezanka and Eminova caves. (b) Map of the potential seismogenic zones (MPSZ) for western Bulgaria (Bonchev et al., 1982). The values in the adjacent zones are in magnitude ranges 4.1–4.5, 4.6–5.0, 5.1–5.5, 5.6–6.0, 6.1–6.5, 6.6–7.0, 7.1–7.5, 7.6–8. The sites of the target caves are shown. (c) Earthquakes with magnitudes 2.5–5.4 in Bulgaria between 1981 and 2000 (Botev et al., 2002).

2. General Characteristics of Historical Seismicity in Bulgaria

Bulgaria is located in the central part of the Balkan Peninsula and is part of the Alpo-Himalayan seismic belt which is characterized by a high level of seismic activity. This area of the peninsula is part of the intensively deforming continental marginal zone of Eurasia. Bulgaria has been divided into three main seismic zones according to seismological and tectonic criteria: Northeast, Srednogorie and Rila-Rhodopes (Grigorova et al., 1979; Stanishkova and Slejko, 1994). Available seismological information indicates that strong (M ≥ 7), shallow crustal earthquakes occur in all three zones.

- The Northeast includes the northern Black Sea coast and is characterized by a significant number strong (M $\geq$ 7) and moderate (5.0 $\leq$ M $\leq$ 7.0) earthquakes.
- The Srednogorie region includes two major seismogenic zones: Sofia (covering the Sofia Kettle and its vicinity) and Maritza (covering the Upper Thracian Lowland). The strongest earthquake (M = 7.8) registered in Bulgaria occurred in the Kresna seismic zone in 1904. The Sofia Zone is characterized by moderate earthquakes (6.0 $\leq$ M < 7.0).
- The Rila-Rhodope region covers southwestern Bulgaria and is characterized by the highest seismic activity.

In past centuries, the territory of Bulgaria has suffered from strong earthquakes. It is important to mention the quakes of 1818 (VIII-IX MSK) and 1858 (M $\approx$ 6.3, I_0 = IX MSK) that occurred in the vicinity of Sofia. The 1818 earthquake caused significant damage in Sofia; the thermal springs in the western part of the city are a result of this quake.

Accurately assessing strong earthquake ground motions in the Kresna region is important since it is the most seismically active region in Europe (Bonchev, 1987) and is very close to Snezanka Cave (Figure 1b). (The fault symbols used in Figure 1c are the same as those in Figure 2) The Rhodope domain in Figure 2 has fewer seismic events, but the pattern of stress axes is similar to that of the Rila domain. Only the west northwest-striking Dospat fault (Figure 2, see Kotzev et al., 2006) appears to be an active fault with a probable but undetermined strike–slip component.

2.1. SEISMICITY OF THE KRESNA REGION

The region of Kresna (Figure 1b Zone 5) has been of interest since the serious earthquake of 1904 (Watzov, 1905). Large scale generalizations (Kirov and Palieva, 1961; Matova and Rigikova, 1980) have contributed to very detailed investigations to better understand the seismological network (Sokerova et al., 1995). The seismic source in the zone is characterized by very high geodynamic activity as indicated by recent geodetic and morphological analyses (Totomanov, 1985; Milev et al., 1997). The zone is the most seismically active part of Bulgaria; the seismic energy realized since the 1904 earthquake is the maximum in the country. Recently, information on the seismic hazard in the zone has been published (Stanishkova and Slejko, 1994; van Eck and Stoyanov, 1996).

The seismogenic zone Kresna-Krupnik is characterized by strong earthquakes, i.e., 1961 (Ms = 7.3), 1866 (Ms = 7.1) and 1904 (M = 7.8) and foreshock (Ms = 7.2). The order of energy realized in the 1904 earthquake made it one of the strongest events on continental Europe in the last century. Maximum focal depths reached 50 kilometers (km), but most were distributed at depths of 25–35 km. Recent investigations indicate a tendency toward increasing seismic activity. Data collected as part of the ASPELEA – EC (ERBIC97CT0200) project are the most complete from the catalogue and macroseismic point of view (Ranguelov et al., 1998). The re-establishment of the horizontal and vertical accelerations of the 1904 earthquake (Ranguelov and Paskaleva, 1998) show that the destructive potential of the sources in Kresna-Krupnik are very serious.

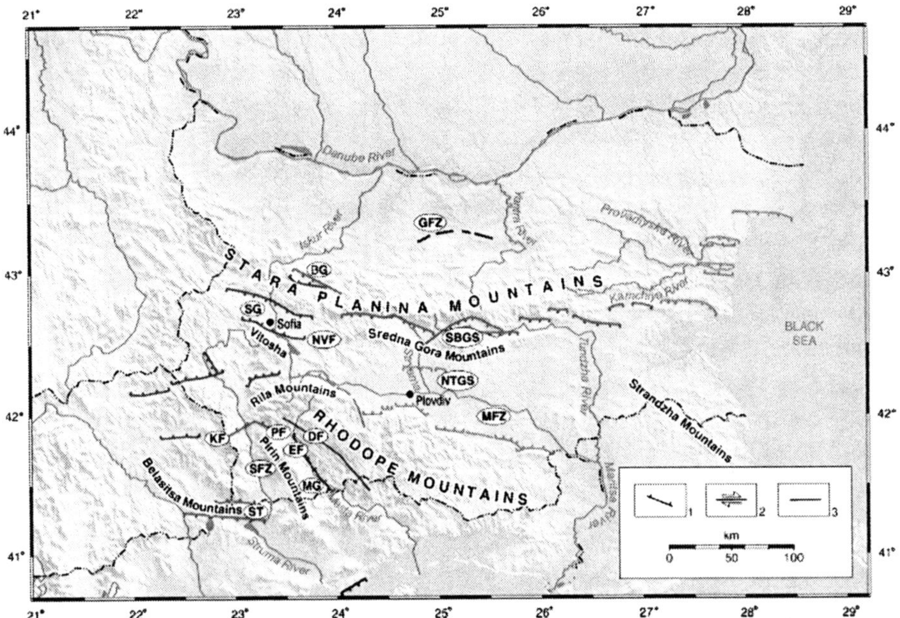

Figure 2: Major extensional structures in Bulgaria. Faults: 1: normal; 2: strike–slip with direction of relative motion; 3: fault with unclear type of displacement. BG, Botevgrad graben; DF, Dospat fault; EF, East Pirun fault; GFZ, Gorna Oriakhovitsa fault zone; KF, Krupnik fault; MFZ, Maritsa fault zone; MG, Mesta graben; NTGS, North Thracian graben system; NVF, North Vitosha fault; PF, Predela fault; SBGS, Sub-Balkan graben system; SFZ, Struma fault zone; SG, Sofia graben; ST, Strumeshnitsa graben. Colors indicate the geological evidence for recent activity (see Kotzev et al., 2006)

3. The Caves

3.1. SNEZANKA CAVE

Snezanka is a show cave located on the left slope of the Novomahlenska River, about 5 km southwest of the town of Peshtera in Plovdiv District in the western

Rhodope Mountains. The coordinates of the entrance are N 42° 00' 40.6" and E 24° 16' 30.7". The cave is highly karstified Proterozoic marble from the Dobrostan marble formation period (Kozhuharov, 1984) and consists of one gallery formed along two basic systems of fractures directed southeast/northwest and south/north. From a morphological point of view, the cave can be divided into six chambers; the stalagmites to be studied are situated in the biggest one – the Great Hall – which is 60 x 40 x 12 meters (m).

The length of Snezanka Cave is 230 m, the denivelation is 18 m and the total surface area of the cave is 3,150 m^2. (See Figure 3) Deposits are presented by detritus (breakage with a maximum block size of 5.5 x 2.8 m), thin clay and a variety of different speleothems. The morphological landmarks of Snezhanka are the highly decorated stalactite ceiling (more then 800,000 stalactites of various shapes) and the availability of thin cylindrical stalagmites suitable for our research.

The cave was discovered on January 2, 1961 by cavers G. Kotzev, G. Zlatarev and B. Evtimov from the town of Peshtera. After archaeological excavations and bore-holing in the same year by the team of N. Djambazov, artifacts from the early Iron Age and Roman-Thracian era including pottery, the remains of a fireplace and bronze needles were found in the Great Hall (Djambazov, 1977).

Snezanka cave has been a show cave for tourists since 1968 and is managed by the Kupena Tourist Society in Peshtera. It was declared a protected natural monument by Decree 504/11. 07. 1979. Since 1983, the cave has been included in the buffer zone of the Kupena Natural Reserve.

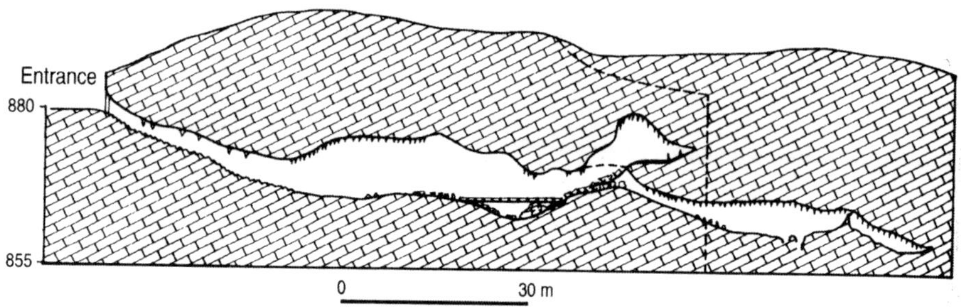

Figure 3: Profile of Snezanka Cave (Beron et al., 2006)

3.2. EMINOVA CAVE

Eminova Cave is located in the southern part of the Western Rhodope Mountains about 5 km northeast of the village of Borino. The two entrances are situated on the left bank of the deep Kastrakli Gorge and are difficult to access. It is easier to enter the cave from the upper, small, vertical entrance that is on the edge of a 20 m cliff. After difficult penetration through a 25 m narrow passage, we reached the main gallery of the cave – formed by a *diaclase* – where the stalagmites to be studied are located.

The cave is Proterozoic marbles of the Dobrostan marble formation period (Kozhuharov, 1984). The total length is 635 m with a denivelation of 35 m. The cave consists of one large northwest/southeast gallery with a height of up to 12 m and a width of 8–10 m in some places (Figure 4). Typical formations are the rich speleothems of different types: stalactites, stalagmites, draperies, flowstones and sinter lakes (gurs).

The cave was discovered by Sofia cavers S. Petkov and Ts. Ostromski in 1992. After digging the 25 m narrow passage near the entrance, Eminova was explored and surveyed in the same year by Ts. Ostromski, S. Petkov, M. Zlatkova, Z. Iliev, G. Raichev and others. The cave is rarely visited and is well protected from negative human impact.

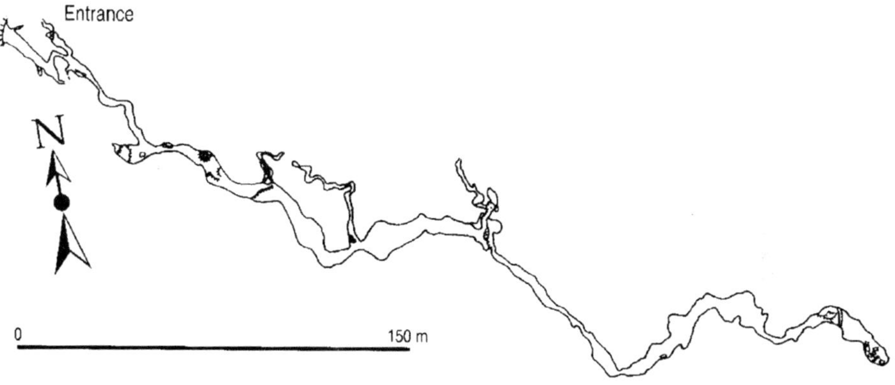

Figure 4: Map of Eminova Cave (Beron et al., 2006)

It is known that with the deepening of caves, the attenuation of seismic waves rises (Becker et al., 2006); therefore, it is important to mention that these two caves are situated at shallow depths not farther than 20–30 m from the surface.

4. Non-intrusive *in situ* Measurements of Speleothems in the Caves

Literature on dripstones from visits to Snezanka and Eminova indicated there were stalagmites in both that would be suitable (Figures 5a, 5b, 7a, b, 9) for investigating the occurrence of paleo-earthquakes. The stalagmites were examined non-intrusively to determined their dimensions and natural frequency.

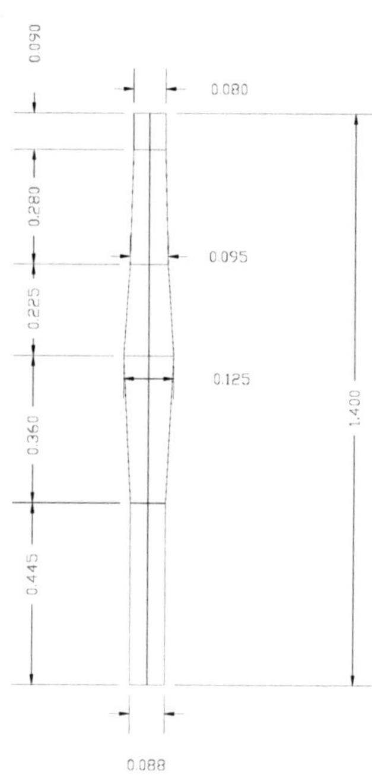

Figure 5a: Stalagmite No. 1 (*right hand side*) in Snezanka Cave, height = 1.40 m. It has a complicated flaring middle. Measured perimeters $P(z_i)$ (and diameters D_i) at different hights Z_i:

Figure 5b: Schematic view of stalagmite No. 1 in *Snezanka* cave used for numerical analysis (dimensions in m)

1. $Z_1 = 17$ cm $P(z_1) = 27$ cm; $D_1 = 8.6$ cm

2. $Z_2 = 60$ cm $P(z_2) = 39$ cm; $D_2 = 12.4$ cm

3. $Z_3 = 106$ cm $P(z_3) = 28$ cm; $D_3 = 8.9$ cm

Minimum perimeter $P_{min} = 25$ cm; ($D_{min} = 8.0$ cm)

$11.2 \leq H/D \leq 17.5$

4.1. SNEZANKA CAVE

Two different stalagmites – No. 1 (Figure 5a) and No. 2 (Figure 7a) – were investigated. Both are 140 centimeters (cm) tall, but No. 1 has a complicated shape with a flaring middle and has a very low natural frequency. The height (H) vs diameter (D) ratios are below (H/D<) 20 in both cases. The diameter of stalagmite No. 1 ranges between 8.0 and 12.4 cm where the widest part is its middle part. The Hvs D ratio is between $11.2 \leq H/D \leq 17.5$. The results of measuring the dimensions are in Table 1 and Figure 5a, b.

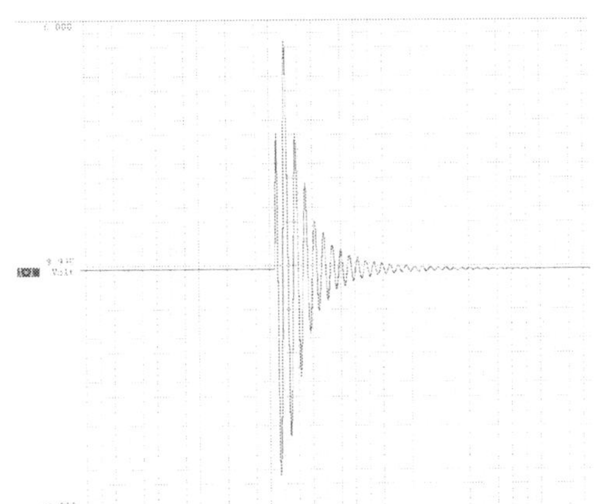

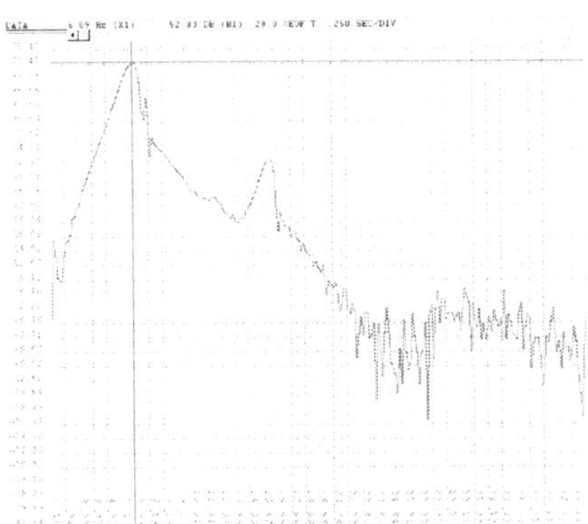

Figure 6a: In situ measurements of stalagmite No. 1 in Snezanka Cave – free oscillation

Figure 6b: In situ measurements of stalagmite No. 1 in Snezanka Cave– power spectra density

The diameter of stalagmite No. 2 ranges between 8.3 and 13.7 cm. The H vs D ratio is between $10.2 \leq H/D \leq 16.9$. Its shape is more or less cylindrical. The results of measuring the dimensions are in Table 1 and Figure 7a, b.

4.2. EMINOVA CAVE

The diameter of the stalagmite investigated in Eminova Cave ranges between 3.0 and 5.1 cm, and its height is 1.17 m. The H vs D ratio is between $22.9 \leq H/D \leq 39.0$. The results of measuring the dimensions are in Table 1 and Figure 9.

It is important to note that in papers on similar stalagmite investigations, exceptionally large H vs D ratios ($25 \leq H/D \leq 72$) for stalagmites in Baradla Cave (Szeidovitz et al., 2008a) were found; however the maximum H/D ratio found by Cadorin et al. (2001) for the speleothems in Hotton Cave was only 20.

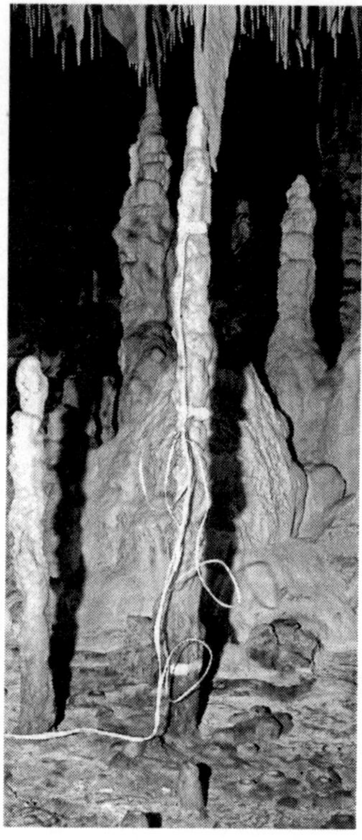

Figure 7a: General view of stalagmite No. 2 in Snezanka Cave, *Figure 7b*: Close view of stalagmite
height = 1.40 m. Measured perimeters $P(z_i)$ (and diameters D_i) No. 2 in Snezanka Cave
at different hights Z_i:

1. $Z_1 = 17$ cm $P(z_1) = 43$ cm; $D_1 = 13.7$ cm

2. $Z_2 = 73$ cm $P(z_2) = 26$ cm; $D_2 = 8.3$ cm

3. $Z_3 = 113$ cm $P(z_3) = 39$ cm; $D_3 = 12$ cm

Minimum perimeter $P_{min} = 26$ cm; ($D_{min} = 8.3$ cm)

$10.2 \leq H/D \leq 16.9$

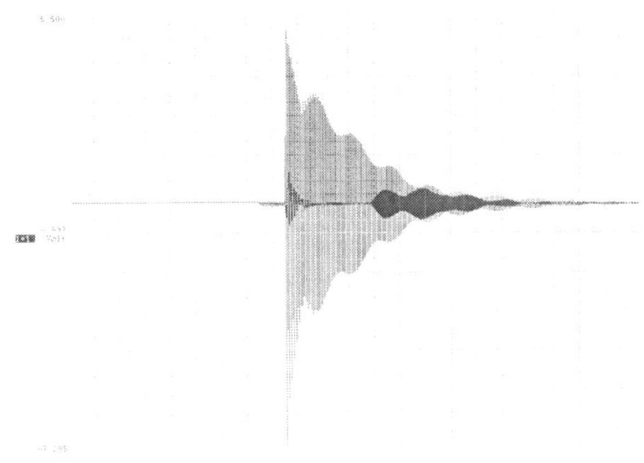

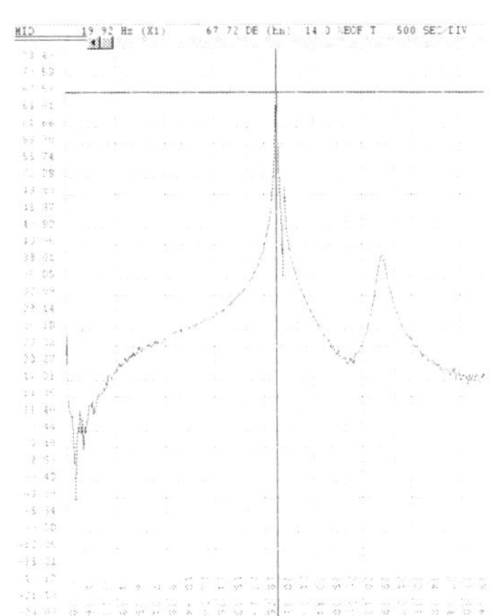

Figure 8a: *In situ* measurements of stalagmite No. 2 in Snezanka Cave – free oscillation

Figure 8b: *In situ* measurements of stalagmite No. 2 in Snezanka cave – power spectra density

Figure 9: Stalagmite in Eminova Cave, height 1.17 m

Its shape is almost cylindrical.

Measured perimeters $P(z_i)$ (and diameters D_i) at different heights Z_i:

1. $Z_1 = 15$ cm $P(z_1) = 15$ cm; $D_1 = 4.8$ cm

2. $Z_2 = 46$ cm $P(z_2) = 12$ cm; $D_2 = 3.8$ cm

3. $Z_3 = 65$ cm $P(z_3) = 9.5$ cm; $D_3 = 3.0$ cm

4. $Z_4 = 90$ cm $P(z_4) = 16$ cm; $D_4 = 5.1$ cm

Minimum perimeter $P_{min} = 9.5$ cm; ($D_{min} = 3.0$ cm)

$22.9 \leq H/D \leq 39.0$

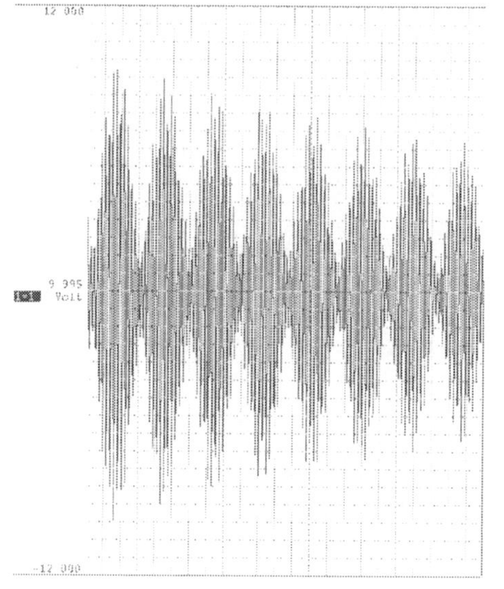
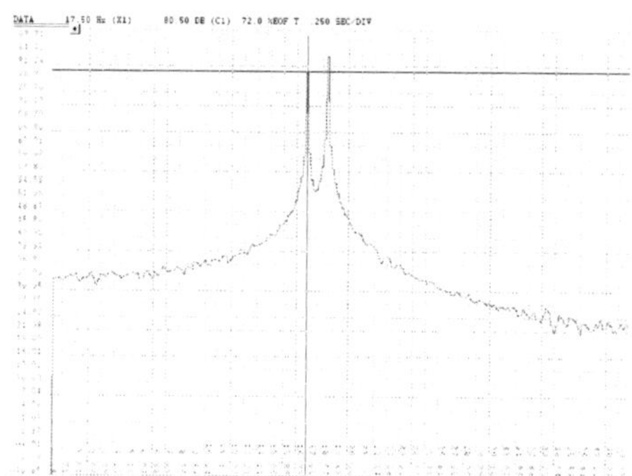

Figure 10a: *In situ* measurements in Eminova Cave – free oscillation

Figure 10b: *In situ* measurements in Eminova Cave – power spectra density

TABLE 1: The parameters of the stalagmites investigated in Snezanka and Eminova caves

	H (cm)	$D_{Average}$ (D_{Min}–D_{Max}) (cm)	Measured natural freq. f_0 (Hz)	Theoretical natural freq. f_0 (Hz)	Elast. modul E (GPa)	Density (g/cm^3)	Tensile failure stress (MPa)	a_g (m/s^2)
1	2	3	4	5	6	7	8	9
Snezanka No. 1	140	8.6–12.4 complicated form, middle part flaring	6.1	5.1	6.240	2.32 ± 0.1 5	2.61 ± 0.28	9.2 / 10.0–14.0
Snezanka No. 2	140	11.3 8.3–13.7	19.9	13.2	6.240	2.32 ± 0.15	2.61 ± 0.28	10.0–19.0
Eminova	117	4.2 3.0–5.1	17.5	5.7	3.735	2.14 ± 0.25	3.83 ± 0.81	6.0–10.5

Note: The results in the grey cells came from a three-dimensional analytical model. The theoretical natural frequencies, f_0 in white cells were calculated by using Equation 1 (given below), the theoretical horizontal ground accelerations resulting in failure, a_g by Equation (2) (given below).

In order to measure natural frequency, a small amplitude forced vibration was obtained by a gentle hit using one's hand or a rubber hammer. The horizontal frequency measuring equipment was mounted on the stalagmites, and by hitting the stalagmite, a horizontal vibration was caused (Figures 7b and 9). The horizontal acceleration of the speleothem was registered by an SM6 geophone (its natural frequency is 4.5 Hz) and a SIG SMACH SM-2 digitiser. The geophones were mounted with adhesive tape at different heights of the speleothems. The sampling rate of the analog–digital converter was set to 256 Hz, whereas the cut-off frequency of the anti-aliasing filter was 50 Hz. The power spectral density of the vibration was determined by a fast Fourier transform. The traces measured and their spectra for certain speleothems are displayed in Figures 6a, b, 8a, b and 10a, b. The natural frequency measured of the speleothems investigated in Snezanka and Eminova caves was in the range of 6.9–19.9 Hz (Table 1).

The natural frequency measured in Snezanka stalagmite No.1 was exceptionally low, which can be explained by its unusual shape. The low natural frequency was confirmed by analytical calculations. To determine the natural frequency of the stalagmite in Eminova Cave, we used only one geophone, mounted at the bottom because due to the low weight of this stalagmite, the geophones could distort the eigen-frequency.

Although dripstones are seemingly simple structures, their tensile strength, ages and natural frequencies change during their formation. In some dripstones, their natural frequencies may rise as high as the frequency domain of earthquakes. If the natural frequency is below 20 Hz and is the approximate upper limit of the frequency range of nearby earthquakes, then resonance can occur (which is the case of Snezanka No. 1 and the Eminova stalagmites), which means that in reality, the dripstones would break at a lower value of horizontal acceleration than the computed one (where static, horizontal ground acceleration is assumed) (see Table 1).

5. Mechanical Properties of the Speleothems Investigated

Mechanical laboratory measurements were performed on samples originating from speleothems in Snezanka and Eminova caves that were found lying broken on the ground in the same chambers where the stalagmites stood.

5.1. RESULTS FOR SNEZANKA CAVE

According to experimental results, the average density was 2.32 g/cm^3, and the standard deviation of eight values measured was 0.15 g/cm^3. Young's modulus was determined using data from a uniaxial compressive strength test. The value was 6,240 MPa; the standard deviation could not be assessed since only one sample was tested. The mean tensile failure stress of the four samples was 2.61 MPa with a standard deviation of 0.28 MPa. The Brazilian test was used.

5.2. RESULTS FOR EMINOVA CAVE

In the same chamber where the one stalagmite we investigated stood, only re-calcified broken stalagmites could be collected on the floor which means that the reliability of the mechanical laboratory measurement is questionable. According to the results, the average density was 2.14 g/cm^3 and the standard deviation of 12 values measured was 0.25 g/cm^3. The mean value from two specimens of Young's modulus was 3,735 MPa; the standard deviation was 35 MPa. The mean tensile failure stress from eight samples was 3.83 MPa with a standard deviation of 0.81 MPa. Results from the laboratory measurements are summarized in Table 1 (columns 6, 7 and 8).

5.3. THE OSCILLATION OF STALAGMITES: THEORETICAL CONSIDERATIONS

For stalagmite No. 2 in Snezanka cave and the stalagmite in Eminova cave, the following theoretical considerations and calculations were applied. In our modeling, the stalagmites were considered as vertical cylinders of height H and diameter D with a circular cross section. We supposed that the bottoms of the cylinders were firmly fixed to the ground and that the top could move freely. The material of the stalagmites was considered to be homogenous.

The natural frequency of a stalagmite (Szeidovitz et al., 2008b) is:

$$f_0 \approx \frac{1}{\pi} \sqrt{\frac{3.1ED^2}{16\rho H^4}} \tag{1}$$

where E is Young's moduls and ρ is the density of the speleothem.

The static, horizontal ground acceleration resulting in failure (Cadorin et al., 2001) is:

$$a_g = \frac{D\sigma_u}{4\rho H^2} \tag{2}$$

where σ_u is the tensile failure stress of the stalagmite. This equation does not take into consideration the phenomenon of resonance.

It can be seen that both the natural frequency and horizontal ground acceleration resulting in failure depend on the geomctrical properties of the stalagmite in the same way, i.e. they are proportional to D/H^2.

Equations (1) and (2) were used to calculate the theoretical natural frequency and the theoretical static horizontal ground acceleration resulting in failure for stalagmite No. 2 in Snezanka and for the stalagmite in Eminova. Due to its very irregular shape, a three-dimensional analytical model was used for calculating the natural frequency and the theoretical static horizontal ground acceleration resulting in failure for stalagmite No. 1 in Snezanka.

6. Discussion of the Results

The theoretical and *in situ* geometrical parameters (height and diameter) natural frequencies measured, the mechanical parameters of the broken stalagmites and the minimum horizontal acceleration needed to break them taking into account the average diameter and mean tensile failure stress are summarized in Table 1. The theoretical natural frequency and the minimum horizontal acceleration needed to break stalagmite No. 1 were calculated by using a three-dimensional analytical model.

The natural frequency of the speleothems investigated changed in the range of 6.9–19.9 Hz. The theoretical natural frequencies ranged from 5.1 to 13.2 Hz by applying Equation (1) for stalagmite No. 2 in Snezanka Cave and for the stalagmite in Eminova Cave and by applying the analytical model for stalagmite No. 1 in Snezanka Cave (Table 1, column 5).

The observed and theoretical natural frequencies (Table 1, columns 4, 5) differ at most by a factor of 3.1 (the theoretical values were in every case smaller than the measured ones). The difference probably comes from the approximations used because the shapes of the stalagmites differ from a cylinder in varying degrees, their material is not homogeneous, and the material parameters are based on measurements performed on specific broken speleothems.

The destructive acceleration was determined by applying Equation (2) at a mean tensile stress $\sigma_u = 2.61$ MPa in the stalagmite depending on its height and level; stalagmite diameter is given in Figure 11a. The vertical lines show expected destructive peak ground acceleration for Snezanka Cave at $H = 1.40$ m ($D_{min} = 0.08$ m) and for Eminova cave at $H = 1.17$ m ($D_{min} = 0.03$ m). Breaking a stalagmite in Snezanka can be expected for peak ground acceleration in the range 10.0–14.0 m/s^2 (for $D = 0.08–0.10$ m). Breaking of a stalagmite in Eminova can be expected for peak ground acceleration in the range of 6.0–10.5 m/s^2 (for $D = 0.03–0.05$ m). For Eminova, this conclusion is valid if we accept that the mean tensile stress $\sigma_u = 2.61$ MPa characterized the stalagmite. (It was mentioned previously that the tensile stress for Eminova cave was determined using unreliable, re-calcified broken specimens.)

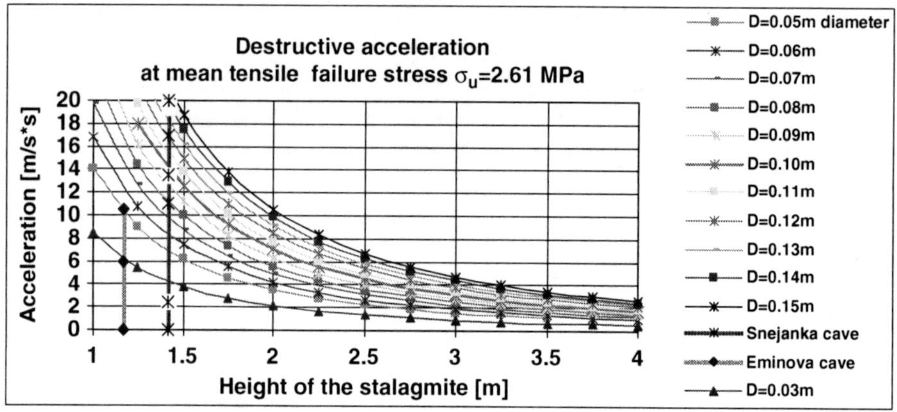

Figure 11a: Destructive acceleration <u>at mean</u> tensile stress σ_u = 2.61 MPa in the stalagmite depending on height, level and diameter. The vertical lines show expected destructive peak ground acceleration for Snezanka at H = 1.40 m and for Eminova at H = 1.17 m

The destructive acceleration determined by applying Equation (2) <u>at mean +</u> <u>1σ</u> (standard deviation) tensile stress σ_u = 2.89 MPa in the stalagmite depending on its height and diameter is given in Figure 11b. The vertical lines show expected destructive peak ground acceleration for Snezanka at H = 1.40 m (D_{min} = 0.08 m) and for Eminova at H = 1.17 m (D_{min} = 0.03 m). Breaks in the stalagmite in Snezanka can be expected at a peak ground acceleration in the range 11.5–14.3 m/s^2 (for D = 0.08–0.10 m). Breaks in the stalagmite in Eminova can be expected at a peak ground acceleration in the range 6.5–11.0 m/s^2 (for D = 0.03–0.05 m).

In Figure 12a, the destructive acceleration <u>at mean</u> tensile stress σ_u = 3.83 MPa in the stalagmite depending on its height, level and diameter is given. The vertical lines show expected destructive peak ground acceleration for Eminova at H = 1.17 m (D_{min} = 0.03 m). Breaks in the stalagmite in Eminova Cave can be expected for a peak ground acceleration of 10.0 m/s^2. Figure 12b provides the destructive acceleration <u>at mean + 1σ</u> tensile stress σ_u = 4.64 MPa in the stalagmite depending on its height, level and diameter. The vertical lines show expected destructive peak ground acceleration for Eminova at H = 1.17 m (D_{min} 0.03 m). Breaks in the stalagmite in Eminova Cave can be expected for a peak ground acceleration of about 11.0 m/s^2.

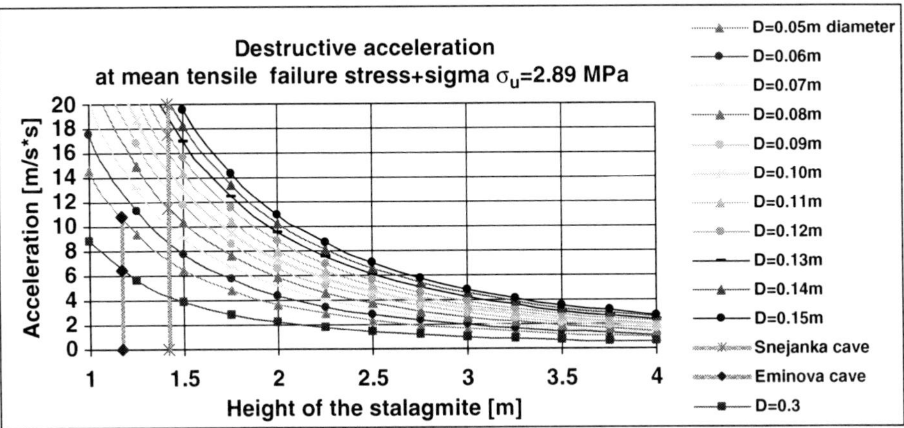

Figure 11b: Destructive acceleration <u>at mean + 1σ</u> tensile stress σ_u = 2.89 MPa in the stalagmite depending on its height and diameter. The vertical lines show expected destructive peak ground acceleration for Snezanka Cave at H = 1.40 m and for Eminova cave at H = 1.17 m

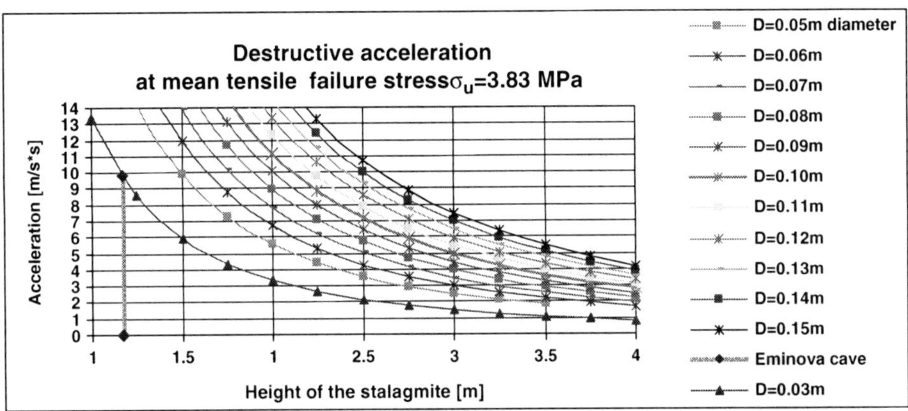

Figure 12a: Destructive acceleration <u>at mean</u> tensile stress σ_u = 3.83 MPa in the stalagmite depending on its height, level and diameter. The vertical lines show expected destructive peak ground acceleration for Eminova Cave at H = 1.17 m (D_{mi} = 0.03 m)

The horizontal acceleration values assessed in Figures 11, 12 are of a X–XI intensity on an MSK macroseismic scale as in Meskó (1990) which can occur in a strong earthquake.

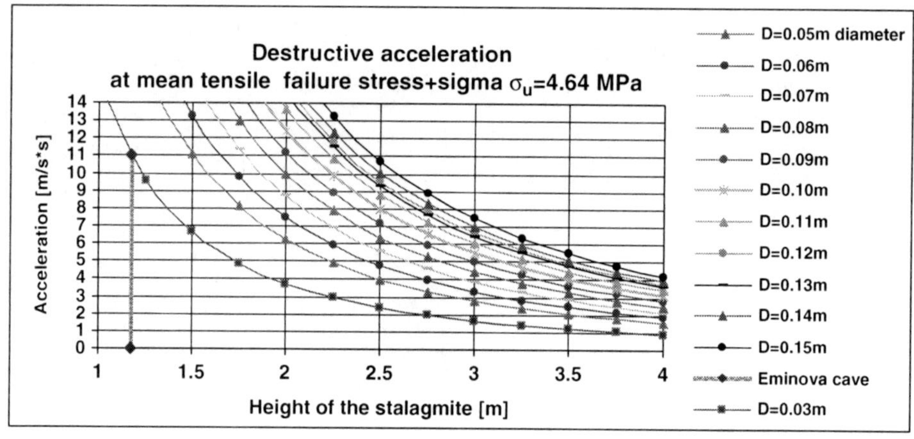

Figure 12b: Destructive acceleration <u>at mean + 1σ</u> tensile stress σ_u = 4.64 MPa in the stalagmite depending on its height, level and diameter. The vertical lines show expected destructive peak ground acceleration for Eminova Cave at H = 1.17 m (D $_{min}$ = 0.03 m)

6.1. COMPARISON WITH THE RESULTS OF LACAVE ET AL. (2003)

Our results and those of Szeidovitz et al. (2008b) are in close agreement with those of Lacave et al. (2000) and Lacave et al. (2003). Figure 13, is a combination of the results of this study, the recent one by Szeidovitz et al. (2008b) and Figure 1 in Lacave et al. (2004). The natural frequency of the speleothems is shown as a function of their length and diameter. Figure 13 can be used as a rough estimate of the natural frequency of a speleothem as a function of its type and length based on *in situ* measurements in French caves.

The frequencies measured in this study and in Szeidovitz et al. (2008b) for "small" speleothem from Varteshkata Cave with a diameter D ~ 6 cm fall in line with diameter D = 5 cm. The result for Eminova cave is between diameters 2–5 cm; the actual diameters for this stalactite are 3–5 cm.

Frequencies for "tall" speleothem from Varteshkata Cave and the speleothem from Elata Cave are situated above the line representing the results for the 10 cm diameter stalactite in Figure 1 in Lacave et al. (2004) which is normal since the diameters of the speleothems from Bulgarian caves are D > 10 cm. The result from Snezanka Cave (stalagmite No. 1.) is below the line which can be explained by its complicated, flaring middle. It can be concluded that the frequency values determined in this study are in logical agreement in spite of the different measuring techniques.

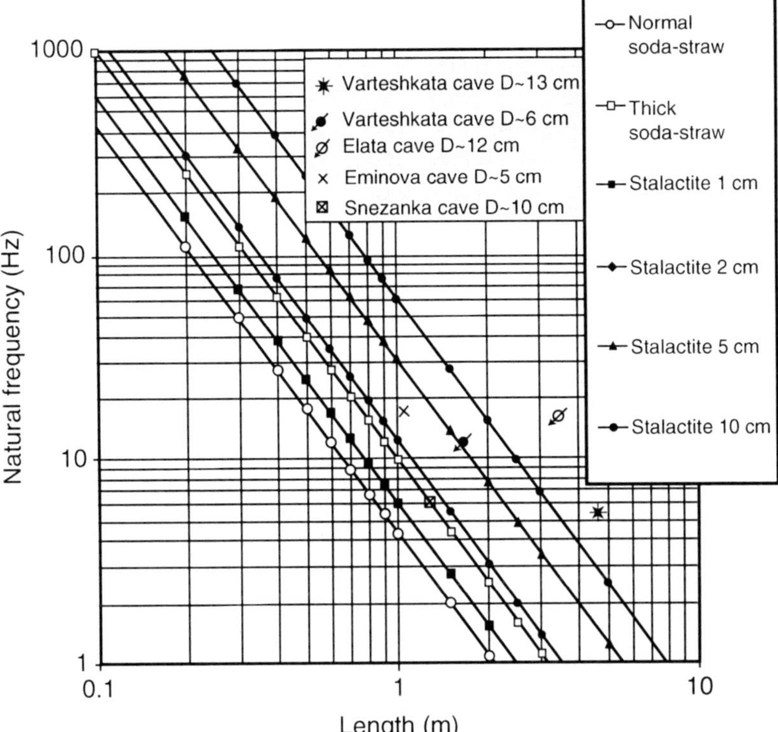

Figure 13: Rough estimate of the natural frequency of a speleothem as a function of its type, length and diameter based on *in situ* measurements in French caves in Lacave et al. (2000, 2003, 2004) and the results for Varteshkata, Elata, Snezhanka and Eminova caves

7. Conclusions

While data in earthquake catalogues for Bulgaria record strong, destructive earthquakes, the seismicity in some parts of the country is nevertheless moderate. For this reason, the investigation of low natural frequency, slim stalagmites (stalagmites with high height/diameter ratios) for evidence of Paleolithic earthquakes could be successful in improving seismic risk policies for karst regions.

Stalagmites with large height/diameter ratios (H/D > 20) and low natural frequencies ($f_0 < 20$) have been found in four Bulgarian caves: Varteshkata, Elata, Snezanka and Eminova. Investigations of stalagmites in Varteshkata and Elata caves are presented in Szeidovitz et al. (2008b). Investigations of stalagmites in Snezhanka and Eminova caves were made in the following manner.

(a) The natural frequencies of three speleothems in Snezanka (2) and Eminov (1) caves were measured *in situ*.

(b) The mechanical properties i.e. the density, Young's modulus and the tensile failure stress of speleothem specimens were determined by laboratory tests.

(c) Based on a simple mechanical model and a three-dimensional analytical model, the theoretical natural frequency (f_0) and the horizontal ground acceleration values resulting in failure (a_g) were calculated for the stalagmites.

The natural frequencies of the speleothems investigated fell in the range of 6.9–19.9 Hz. The horizontal frequencies for the stalagmites differed from the measured ones. In the worst case, the observed natural frequency was three times greater than the theoretical one.

The a_g values computed for the stalagmites studied fell in the range of 6.0–16.2 m/s^2; such high acceleration values can arise in case of very strong earthquakes. Because the natural frequency of the stalagmites generally is in the frequency range of nearby earthquakes, failure acceleration can be even smaller because of the resonance effect.

According to our measurements and theoretical calculations, the speleothem investigated have not been excited by a horizontal acceleration higher than 6.0 m/s^2 in Eminova Cave and 9.2 m/s^2 in Snezanka Cave in the last few thousand years. According to the map for the 1,000-year return period in the Bulgarian code (1987) Snezanka Cave is in a region with $a_g = 0.27$ g and Eminova Cave is in a region with $a_g = 0.15$ g. The maximal horizontal ground accelerations in both caves are much higher than those listed the Bulgarian code. Further investigation would be required to determine if taking the effect of resonance into account would decrease the horizontal ground acceleration resulting in failure in the stalagmites we studied.

Acknowledgements

The study was supported by: NATO project N980468 "Harmonization of the Seismic Hazard and Risk Reduction in Countries Influenced by Vrancea Earthquakes"; Project INTAS-Moldova 200505-104-7584 "Numerical Analysis of 3-D seismic wave propagation using Modal Summation, Finite Elements and Finite Differences Method"; NATO linking project ESP CLG 981966 – Seismic "Vulnerability Estimates of High Risk Structures"; NZ – MU 1504 "Traces of paleo-seismicity in karstic caves"; bilateral project between the Hungarian and Bulgarian Academy, No.: 42. – "Assessment of the PGA generated by paleo-earthquakes from failure tensile stress of speleothems. Study of seismicity of the remote past with the use of engineering seismology"; Hungarian Scientific Reserch Fund projects: K6180 and K60394.

The authors wish to thank L. Petrova and G. Kiev from the Kupena Tourist Society for providing the professional guide in Snezhanka Cave and to the people of the town of Peshtera for their kind assistance.

References

Becker A., Davenport C.A., Eichenberger U., Gilli E., Jeannin P.-Y., Lacave C. (2006) Speleoseismology: A Aritical Perspective. Journal of Seismolology 10, 371–388.

Beron P., Daliev T., Jalov A. (2006) Caves and Speleology in Bulgaria. Sofia, Pensoft, pp. 508.

Bonchev E. (1987) Main Ideas in the Tectonic Synthesis of the Balkans. 1. The Lithosphere Plates and the Collision Space Between Them, Geologica Balcanica 17(4), 9–20.

Bonchev E., Bune V., Christoskov L., Karagjuleva J., Kostadinov V., Reisner G., Rizhikova S., Shebalin N., Sholpo V., Sokerova D. (1982) A Method for Compilation of Seismic Zoning Prognostic Maps for the Territory of Bulgaria. Geologica Balcanica 12(2), 2–48.

Botev E., Samardjieva E., Milushev R., Dimitrov B., Babachkova B., Donkova K., Alexandrova I., Delibaltova B., Velichkova S., Genov K., Hristova C., Rizhikova S., Toteva T., Tzoncheva I. (2002) Preliminary Data on Seismic Events Recorded by NOTSSI (in Bulgarian with English summary). *Bulg. Geophys. J.* (1982–2001) ISSN: 1311-753X.).

Cadorin J.F., Jongmans D., Plumier A., Camelbeeck T., Delaby S., Quinif Y. (2001) Modelling of Speleothem Failure in the Hotton Cave (Belgium). Is the Failure Earthquake Induced? Netherlands Journal of Geosciences 80(3–4), 315–321.

Djambazov N. (1977) Archeological Studies of Caves in Bulgaria. "Speleologia" – Proceedings of National Speleological Conference 10.12.1976, Sofia, pp. 102–111 (in Bulgarian).

Grigorova E., Sokerova D., Christoskov L., Rijikova S. (1979) Catalogue of the Earthquakes in Bulgaria for the Period 1900–1977, Geoph. Inst., BAS, Sofia.

Kirov K., Palieva K. (1961) On the Seismisity of the Struma Valley. Journal of the Geophysical Institute 2, 57–93 (in Bulgarian with English summary).

Kotzev V., Nakov R., Georgiev Tz., Burchfiel B.C., King R.W. (2006) Crustal Motion and Strain Accumulation in Western Bulgaria. Tectonophysics 413, 127–145.

Kozhuharov D. (1984) Litostratigraphy of the Precambrian Metamorphic Rocks of the Rhodopean Supergroup in the Central Rhodopes Mts. Geologica Balcanika, 14(1), 43–92 (in Russian).

Lacave C., Levret A., Koller M. (2000) Measurements of Natural Frequencies and Damping of Speleothems. Proceedings of the 12th World Conference on Earthquake Engineering, Auckland, New-Zealand, paper 2118.

Lacave C., Koller L., Eichenberger M.W., Jeannin U. (2003) Prevention of Speleothem Rupture During Nearby Construction. Environmental Geology 43, 892–900.

Lacave C., Koller M.G., Egozcue J.J. (2004) What Can Be Concluded About Seismic History from Broken and Unbroken Speleothems? Journal of Earthquake Engineering 8(3), 431–455.

Matova M., Rigikova Sn. (1980) On the Tectonics and Seismisity at Southwest Bulgaria. BGJ VI(N3), 73–86 (in Bulgarian).

Meskó A. (1990) Introduction to Geophysics, University textbook, Tankönyvkiadó, pp. 510.

Milev G., Matova M., Shanov S., Minchev M., Vassileva K., Dobrev N. (1997) Geodynamic Investigations in the Crossing Zone of the Strouma and the Krupnik Faults. The Earth and the Universe, Ziti Editions, Aristotle Univ., Thessaloniki, pp. 483–492.

Popov V. (1970) The Distribution of Karst in Bulgaria and Some of Its Features. Review of the Geographical Institute, Bulgaria Academy of Science 13, 5–19 (in Bulgarian).

Ranguelov B., Paskaleva I. (1998) Strong Ground Motion Reconstruction for the Earthquakes (M = 7.8 and M = 7.2) in Kresna on 4-th April 1904. Book of papers, XXVI Gen. Ass. ESC, Tel Aviv. pp. 231–235.

Ranguelov B., Gospodinov D., Rizhikova S., Toteva T. (1998) Kresna Seismic Zone as a Test Site for Seismic Potential Research, Book of papers, XXVI Gen. Ass. ESC, Tel Aviv. pp. 41–44.

Scholz H. (1990) The Mechanics of the Earthquakes and Faulting, Cambridge University Press, Cambridge, p. 467.

Sokerova D., Dineva S., Velichkova S. (1995) On the Nature of the "Krupnik" Seismic Source. BGJ XXI(N2), 32–39 (in Bulgarian).

Stanishkova I., Slejko D. (1994) Seismic Risk of Bulgaria. Natural Hazards, Kluwer, Dordrecht, The Netherlands, Vol. N9, pp. 247–271.

Szeidovitz Gy., Surányi G., Gribovszki K., Bus Z., Leél-O″ssy Sz, Varga Zs. (2008a) Estimation of an Upper Limit on Prehistoric Peak Ground Acceleration Using the Parameters of Intact Speleothems in Hungarian Caves. Journal of Seismology 12(1), 21–33.

Szeidovitz Gy., Paskaleva I., Gribovszki K., Kostov K., Surányi G., Varga P., Nikolov G. (2008b) Estimation of an Upper Limit on Prehistoric Peak Ground Acceleration Using the Parameters of Intact Speleothems in Caves Situated at the Western Part of Balkan Mountain Range, North-West Bulgaria. *Acta Geofizika e Geod., Hungary* (in print).

Totomanov I. (1985) The Regional Peculiarity of the Fields and Correlation for the Maximal Magnitudes of the Earthquakes and Resent Vertical Crust Movements in Southwest Bulgaria. BGJ XI(N4), 113–123 (in Bulgarian).

van Eck T., Stoyanov T. (1996) Seismotectonics and Seismic Hazard Modeling for Southern Bulgaria. Tectonophysics 262, 77–100.

Watzov S. (1905) The Earthquakes in Bulgaria during 1904, Journal CMI, 180.

MEMS-BASED DATA LOGGER FOR SEISMIC ARRAYS AND STRUCTURAL HEALTH MONITORING

R. MOHNIUC[*,1], A. ZAICENCO[1], C. KUENDIG[2],
T. KURMANN[2]

[1]*Institute of Geology and Seismology, Academy of Sciences,
Republic of Moldova, e-mail:ruslan.mohniuc@gmail.com,
anton.az@gmail.com*
[2]*GeoSIG Ltd., Europastrasse 11, 8152 Glattbrugg, Switzerland,
e-mail:ckuendig@geosig.com, TKurmann@geosig.com*

Abstract. One of the hottest technology growth areas is microelectromechanical systems (MEMS). According to the market research firm Frost and Sullivan (Mountain View, CA), MEMS is one of a handful of new technologies that could revolutionize the 21st century. The advantage of MEMS for these applications is the relatively low cost and simplicity of the devices. The paper presents a MEMS-based data logger designed for seismic arrays and structural health monitoring.

Keywords: MEMS, accelerometer, seismic arrays, structural health monitoring

1.   Basic Accelerometer Theory

The equation of the motion for an SDOF system with base excitation is in the form:

$$m\ddot{\upsilon} + c\dot{\upsilon} + k\upsilon = -m\ddot{\upsilon}_g \tag{1}$$

If the excitation is the real part of $e^{i\omega t}$, then the system response to this excitation is a real part of the complex function:

$$\upsilon(t) = H(i\omega)e^{i\omega t} \tag{2}$$

---

[*]Institute of Geology and Seismology, Academy of Sciences, Republic of Moldova, e-mail: ruslan.mohniuc@gmail.com

A. Zaicenco et al. (eds.), *Harmonization of Seismic Hazard in Vrancea Zone,*
© Springer Science + Business Media B.V. 2008

and Equation (1) will become:

$$m\ddot{\upsilon} + c\dot{\upsilon} + k\upsilon = -m e^{i\omega t} \tag{3}$$

Noting that $\omega^2 = k/m$ and $\xi = c/2\sqrt{km}$, the above equation is transformed into:

$$-[H(i\omega)e^{i\omega t}]\omega^2 + 2\xi\omega_0[H(i\omega)e^{i\omega t}]i\omega + \omega_0^2[H(i\omega)e^{i\omega t}] = -e^{i\omega t} \tag{4}$$

$$-H(i\omega)\omega^2 + 2\xi\omega_0 H(i\omega)i\omega + \omega_0^2 H(i\omega) = -1 \tag{5}$$

$$H(i\omega) = \frac{1}{\omega^2 - 2\xi\omega_0 i\omega - \omega_0^2} \tag{6}$$

$$H(i\omega) = \frac{1}{\omega_0^2(\omega^2/\omega_0^2 - 2\xi i\omega/\omega_0 - 1)} \tag{7}$$

which is complex, containing both magnitude and phase information. Recalling that $|a + ib| = \sqrt{a^2 + b^2}$, we get the magnitude:

$$|H(\omega)| = \frac{1}{\omega_0^2\sqrt{(\beta^2 - 1)^2 + 4\xi\beta^2}} = \frac{1}{\omega_0^2}D \tag{8}$$

where $\beta = \omega/\omega_0$, and D is the dynamic magnification factor (Figure 1).

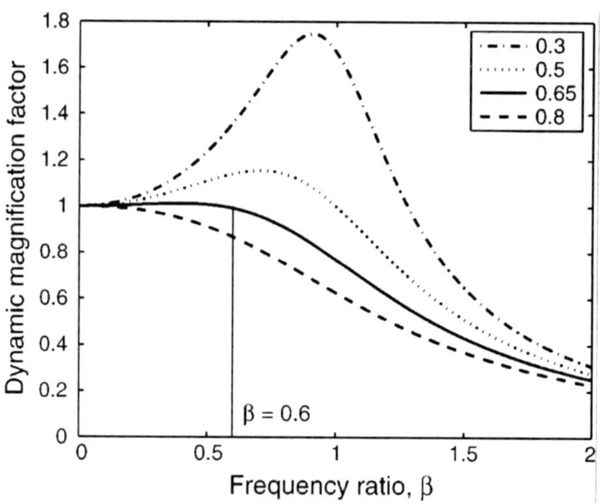

Figure 1: Dynamic magnification factor, $D(\beta)$

Note that for $\xi \approx 0.65$ and $\beta < 0.6$ the response of the system is directly proportional to the base excitation. This property of the SDOF system is used to build acceleration sensors, which are usually designed to have high natural frequency ω_0.

2. Sensing Element in MEMS Accelerometer

Current technologies used to manufacture MEMS sensors can be divided into the following categories (as function of performance): (a) piezo-film (the cheapest technology), (b) surface MM capacitive, (c) bulk capacitive, (d) piezo-electric, electro-mechanical servo vibrational (the most expensive), and (e) piezoresistive.

Correspondingly, the costs will vary in the limits $\approx$1–1,000 USD. In near future type (b) technology will gain popularity providing superior performance at prices below 100 USD per component.

The technology for creating micro-machined accelerometers allows one to suspend silicon structures which are attached to the substrate in a few points called anchors (Figure 2) and are free to move in the direction of the sensed acceleration, (STMicroelectronics LIS3L02AL sensor). The distance between the suspended beam and substrate is of order 1 μm, the proof mass is $\approx$0.1 μg, and the gap between capacitor plates is $\approx$1 μm. To be compatible with the traditional packaging techniques a cap is placed on top of the sensing element to avoid blocking the moving parts during the moulding phase of plastic encapsulation. When an acceleration is applied to the sensor the proof mass displaces from its nominal position, causing an imbalance in the capacitive half-bridge. This imbalance is measured using charge integration in response to a voltage pulse applied to the sense capacitor. The minimum detectable beam deflection is $\approx$0.2 $\overset{\circ}{A}$ (1/10-th of an atomic diameter), the smallest detectable capacitance change is $\approx 10^{-18}$ F, and total capacitance change for full-scale is 100fF for about five orders of magnitude potential dynamic range or $\approx$100 dB.

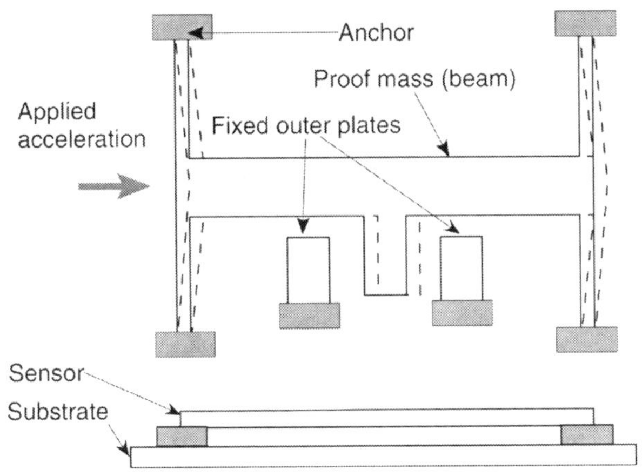

Figure 2: MEMS sensor operation

Mechanical characteristics of the LIS3L02AL three-axial MEMS sensor are provided in Table 1. Dynamic range is calculated for RMS noise over 35 Hz bandwidth.

3. ADC and Resolution

To convert the input analog voltage from the MEMS sensor to the digit numbers, a 16-bit analog-to-digital converter (ADC) is used in our accelerograph design. The parameters of the sensor and ADC given in

TABLE 1: Mechanical characteristics of the MEMS sensor

Range (g)	Dynamic range (dB)	Shock limit (g, ms)	Cross axis (%FS)	Resonance frequency (Hz)
±2	77	3,000, 0.5	2	1,500

Table 2 suggest that the ADC can resolve:

$$N = \frac{V_{max} - V_{min}}{V_B / 2^R} = 2.88 \cdot 10^4 \tag{9}$$

and that the resolution of the converter plus MEMS sensor is:

$$C_R = \frac{2 \times A_r}{N} = 1.39 \cdot 10^{-4}\, g \tag{10}$$

The value of RMS noise of the sensor at 100 Hz bandwidth is 0.5 mg, which can be resolved with the 16-bit ADC.

An example of noise measurements at a site with level of noise close to 1 mg rms is shown in Figure 3. Raw data and their Fourier spectra for horizontal axis are shown at the left, while low-passed time-series (elliptic filter with corner frequency 100 Hz) are given in the right. It can be seen that applying the low-pass filter reduces noise to the level below 1 mg.

The MEMS sensor has a flat frequency response with bandwidth $0 \div 0.6$. 1,500 Hz = 900 Hz. The sampling rate of the ADC, f_s, allows resolution of up to 500 Hz, which is acceptable for the applications the accelerograph is designed for.

It is also important to mention that MEMS accelerometer, like most sensors, is sensitive to the temperature changes. The zero-g level drift vs temperature is shown in Figure 4 and suggests a need to calibrate the sensor before taking measurements in a given environment, or to incorporate an additional temperature sensor to compensate for zero-g drift. Sensitivity change vs temperature is about 0.01% per °C (Δ from +25°C), and can be neglected for the usual operating range (+5…35°C).

TABLE 2: Characteristics of the MEMS sensor and ADC

Symbol	Parameter	Value	Unit
MEMS sensor			
V_{dd}	Supply voltage	3.3 (typ.)	V
A_r	Acceleration range	±2	g
V_{max}	Voltage at 2 g for V_{dd}	2.97	V
V_{min}	Voltage at −2 g for V_{dd}	0.33	V
V_{off}	Zero-g level (at T = 25°)	$V_{dd}/2$ (typ.)	V
ADC			
V_B	Reference voltage	3	V
R	Resolution	16	bit
f_s	Sampling rate	1,000	Hz

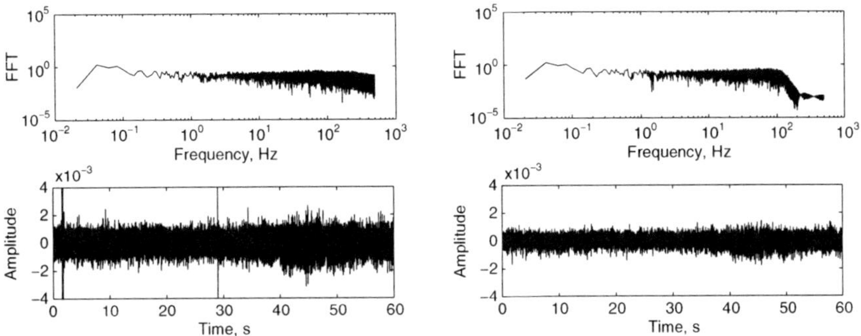

Figure 3: Accelerograph noise measurements: raw data (*left*) and low-passed with fourth order elliptic filter 0–100 Hz data (*right*)

4. Applications of MEMS-Based Accelerographs

4.1. NESTED SEISMIC ARRAYS

Spatial variability of PSV response spectra given in terms of site-to-site PSV response-spectrum variance versus station separation for soil sites, (Evans et al., 2003), a need to supply PGA and PGV in near-real-time to construct ShakeMap (Wald et al., 1999); and loss-estimation models such as HAZUS (Nishenko et al., 1998), generate the need for dense arrays of strong ground-motion instruments with station separation distances of about 1 km.

Strong-motion instruments are generally divided into Class A, B, and C devices (Table 3); (Evans et al., 2003). In Table 3 resolution is defined for the RMS noise versus full-scale range.

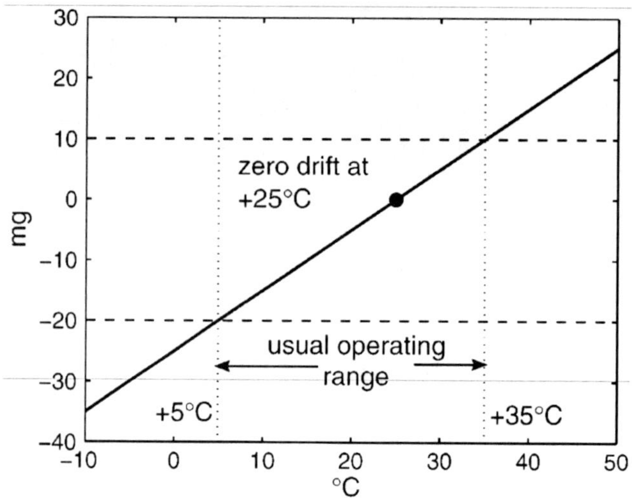

Figure 4: Zero-g level change vs temperature: Δ from +25°C (mean curve)

TABLE 3: A classification of strong-motion instruments

Class	Resolution		Retail cost (USD)
	Bits	dB	
A	20 bits	120	≈7,000, quantity 100
B	16–19 bits	96–114	≈2,500, quantity 1,000
C	12–15 bits	72–90	<1,000, quantity 10,000

Class A are the state-of-the-art instruments with maximum amplitude fidelity, are high-priced and form the "backbone" of each seismic network, Figure 5.

The quality of the Class B instruments improves each year, and the gap between Class A units is being bridged very fast. These accelerometers filling the amplitude- and spatial-resolution between Class A and C stations in the nested arrays.

Class C sensors are gap "fillers" between the rest of the nodes in the array, Figure 5. They are very big in numbers but provide less accurate data. Nevertheless, they are critical to some applications. For example, (Evans and Rogers, 1995), looked at a range of MEMS accelerometers, demonstrating that even the noisiest MEMS accelerometer was adequate for extending ground-motion models and meeting emergency response needs.

There is a wide range of architectures, performance specifications, and prices among MEMS accelerometers. Very low-cost MEMS sensors are good candidates for Class C networks providing coverage where there is no other opportunity for it.

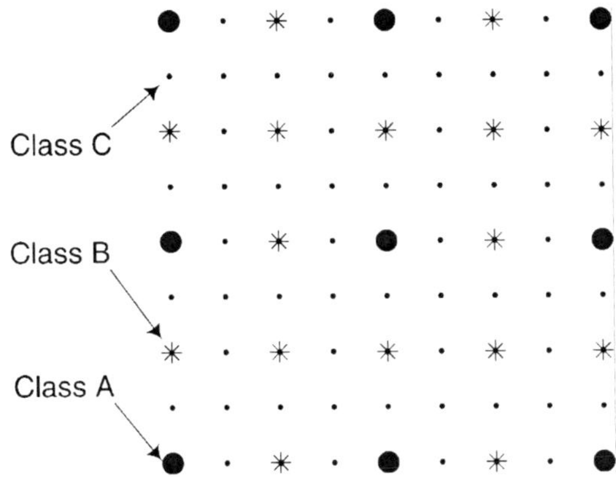

Figure 5: Schematic nested array with three grades of strong-motion accelerographs, (Evans et al., 2003)

5. 16-bit MEMS Data Logger

An acceleration data logger incorporating MEMS sensor has recently been built by the authors, Figure 6. Technical specifications are provided in Table 4. It might be viewed as a perfect candidate for Class C networks discussed in the previous section. Being a low-power device it operates from AAA-type batteries and is equipped with a GPS sensor.

A 16-bit ADC and MEMS sensor from STMicroelectronics© allow to resolve accelerations at sub milli-g level with a bandwidth of 100 Hz.

In early 2008 this data logger was tested against a Guralp CMG-5TD accelerometer on a shake table and demonstrated good performance. Another advantage of this device is low weight – about 250 g.

Figure 6: 16-bit data logger with MEMS sensor

TABLE 4: Technical specifications of data logger

	Description
Sensor	Three-axis linear accelerometer LIS3L02
Acceleration range	±2.0 g
Acceleration noise density	0.05 mg · sqrt (Hz)
ADC	16 bit (full range is not used by sensor)
Sampling rate	up to 1 kHz per channel
Triggering	Amplitude/IR/RF/PC-link
Memory	32 Mb
Recording time	90 min per 32 Mb
Power supply	via USB connector or three batteries type AAA
Time base	1 ms (GPS synchronization)
Communication protocol	Binary, MODBUS-RTU
Communication	USB
Size	200 × 95 × 50 mm
GPS antenna	For internal GPS receiver

6. Conclusions

MEMS accelerometers have recently gained popularity in various motion-sensing application. Surface micro-machined capacitive technology will, probably, dominate the market during the next few years providing the best price/performance ratio.

Our newly designed accelerograph based on MEMS sensors allows use of low-voltage circuits powered by 3 AAA-type batteries to measure sub-mg levels of acceleration with bandwidth 0–100 Hz. A 16-bit ADC is used and ensures more than adequate amplitude resolution for this sensor. The accelerograph can resolve frequencies up to 500 Hz, which is sufficient for many engineering applications.

The compact accelerograph is designed mainly for use in structural health monitoring and nested seismic arrays. Moreover, due to its flexible architecture and compact size other applications are also possible.

Acknowledgements

This research was partly sponsored by NATO's Scientific Affairs Division in the framework of the Science for Peace Programme, project SfP-980468. Authors are grateful to Dr. John R. Evans (USGS) for providing valuable comments on the paper.

References

Evans, J. R. and J. A. Rogers. (1995). Relative performance of several inexpensive accelerometers. U.S. Geological Survey Open File Report, 38 pp.

Evans, J. R., R. H. Hamstra, Jr., P. Spudich, C. Kundig, P. Camina, and J. A. Rogers. (2003). Additional information for "TREMOR: A Wireless, MEMS Accelerograph for Dense Arrays". U.S. Geological Survey Open File Report 2003-159, 13 pp.

LIS3L02AL MEMS® Inertial Sensor: 3-axis − +/−2g ultracompact linear accelerometer. ©2006 STMicroelectronics, datasheet (rev. May 2006).

Nishenko, S., C. Drury, and J. Milheizler. (1998). Recent FEMA activities in earthquake risk analysis and mitigation, in: Wind and Seismic Effects, Proc. of the 30th Joint Meeting of the U.S.–Japan Cooperative Program in Natural Resources, ed. N. J. Raufaste, 30th joint meeting of the U.S.–Japan Cooperative Program in Natural Resources, panel on Wind and seismic effects, Gaithers-burg, MD, May 12–15, 1998. *Nat. Inst. of Standards and Tech. Spec. Pub.,* **931**, 300–305.

Wald, D. J., V. Quitoriano, T. H. Heaton, H. Kanamori, C. W. Scrivner, and C. B. Worden. (1999). TriNet "ShakeMaps": Rapid generation of instrumental ground motion and intensity maps for earthquakes in Southern California, *Earthq. Spectra,* **15**, 537–556.

NUMERICAL SOLUTION OF AN ELASTIC WAVE EQUATION USING THE SPECTRAL METHOD

A. ZAICENCO[*], V. ALKAZ

Institute of Geology and Seismology, Academy of Sciences of Moldova

Abstract. One of the primary goals in seismology is to predict realistic ground motion during an earthquake. The problem of seismic wave propagation based on partial differential equations is the part of continuum physics that deals with the medium at large scales, and the desired precision of the solutions allows approximating the matter as a continuum. Discrete methods for the solution of a wave equation have a practical interest to earthquake engineering community dealing with seismic risk problems. This paper presents solutions based on a spectral method involving Chebyshev grids.

Keywords: Elastic waves, numerical solution, Chebyshev grid

1. Introduction

A second-order partial differential equation (PDE) of a form:

$$a \cdot u_{xx} + 2b \cdot u_{xy} + c \cdot u_{yy} + d \cdot u_x + f \cdot u = g \tag{1}$$

can be represented by a polynomial:

$$P(\alpha, \beta) = a\alpha^2 + 2b\alpha\beta + c\beta^2 + d\alpha + e\beta + j \tag{2}$$

Solutions of Equation (1) are largely determined by the properties of the polynomial $P(\alpha, \beta)$, and namely its principal part that includes high-order derivatives (a, $2b$, c). Hyperbolic PDEs, characterized by positive discriminant $D = b^2 - a \cdot c$, are most computationally challenging. The wave equation is of practical importance in seismology and geophysics:

$$\frac{\partial^2 p}{\partial t^2} = c\nabla^2 p \tag{3}$$

---

[*]Institute of Geology and Seismology, Academy of Sciences of Moldova, e-mail: anton.az@gmail.com

A. Zaicenco et al. (eds.), *Harmonization of Seismic Hazard in Vrancea Zone,*
© Springer Science + Business Media B.V. 2008

where c is the propagation speed of the wave. Equation (3) describes the acoustic wave in homogeneous media; the unknown is the acoustic pressure p.

3-D seismic waves are described by the equation in an isotropic homogeneous elastic medium:

$$\rho\ddot{\mathbf{u}} = (\lambda + 2\mu)\nabla(\nabla \cdot \mathbf{u}) - \mu\nabla \times (\nabla \times \mathbf{u}) \tag{4}$$

where λ and μ are Lame constants that define the elastic moduli:

$$C_{ijkl} = \lambda\delta_{ij}\delta_{kl} + \mu(\delta_{ik}\delta_{jl} + \delta_{il}\delta_{jk}) \tag{5}$$

Equation (4) approximates the wave propagation in many non-linear, non-isotropic, and non-elastic materials if the heterogeneities of the material and the amplitude of the wave are small compared to the wavelength.

For complicated boundary conditions (BCs) and media properties, analytical solutions are usually not possible. Therefore, discrete methods are employed in many practical applications.

The greatest challenges for applied numerical methods are accuracy and optimal use of computational resources. In this regard spectral differentiation based on Chebyshev polynomials becomes very attractive.

2. Finite Difference, Finite Element and Spectral Method

The major classes of the discrete methods are finite difference (FD), finite element (FE) and spectral. The FD method works directly with the PDEs. The derivatives are approximated by finite differences, and a truncated Taylor series expansion of the function is employed. Derivation is performed via Toeplitz matrices, which are large and sparse. For example, a 1-D Laplacian operator is a tridiagonal matrix for a three-point centered difference. The convergence rate is related to the order of the Taylor series truncation.

In the finite element method (Zienkiewicz and Cheung, 1967), a general form of the solution is assumed. It is usually a polynomial:

$$\hat{u}(x) = a + a_1 x + a_2 x + \ldots + a_n x^n \tag{6}$$

The weak form of the PDE is the integral over the computational domain of the residual multiplied by the weighting function w_i:

$$\int_{x_0}^{x_L} (u(x) - \hat{u}(x))\omega_i(x)\,dx = 0 \tag{7}$$

The Galerkin formulation of the FE method states that the weighting functions are partial derivatives of the assumed solution. The solution domain is discretized into elements (weak form), and a global matrix $\mathbf{K}_{gl}$ is assembled from the element matrices. Again, $\mathbf{K}_{gl}$ is large and sparse.

One of the major disadvantages of the FD scheme is poor handling of the complex geometries, since regular meshes are employed. On the contrary, the FE method can easily operate on an unstructured grid and discretizes the constitutive relations well (Mattiussi, 2000). On the computational side, for field problems the FD method offers an advantage – it is possible to compute -curl-, -del- and -div- for each direction separately in Equation (4), while the FE method requires assembly of the $\mathbf{K}_{gl}$ matrix all at once.

The spectral method, as the extension of the FD approach, is derived by fitting a high-order polynomial through more grid points. The drawback is that the FD stencil gets larger as the order of the polynomial approximation increases (Peiró and Sherwin, 2005). Applying a higher order polynomial expansion within every element is known as *p-type* FEM in structural mechanics or the *spectral element method* in fluid mechanics.

If the solution to a PDE on a simple domain is required with high accuracy, and if data defining the problem are smooth, then spectral methods are usually the best tool (Trefethen, 2000). For many seismological problems, a computational box with a known velocity structure of the matter inside is commonly employed.

2.1. SPECTRAL METHOD ON A CHEBYSHEV GRID

In spectral collocation methods the following principles are formulated:
- A single function p is chosen in such a way that $p(x_j) = u_j$ for all j.
- The derivative is $w_j = p'(x_j)$.

For non-periodic domains, algebraic polynomials on the irregular grids are selected ("Chebyshev" methods [Canuto et al., 1987]). Derivation is performed by solving the same system of linear equations as in the FD method. Yet now, the Toeplitz matrix is dense. This disadvantage in terms of memory requirement is compensated for by a very fast convergence or *spectral accuracy*, Figure 1. As Chebyshev polynomials are specified for the interval $[-1,+1]$, an arbitrary function $y(x)$, $x \in [-a, +b]$ needs first to be "shifted" and "scaled" to match this domain.

The disadvantage of using the "standard" Chebyshev grid that is clustered at the boundaries and gives poor resolution in the center of the domain can be overcome by a transformation that maps collocation points to a new set of interpolation points, (Kosloff and Tal-Ezer, 1993). This method allows using the equal-spaced grid that avoids the problem of poor resolution in the center of the domain and numerical stability in time integration. If standard, the

Chebyshev method requires at least π points per wave before it starts to resolve an oscillatory function like $\sin(m\pi x)$. The modified Chebyshev method begins to resolve a function when we have at least two points per wave in the center of the domain (Don and Solomonoff, 1994).

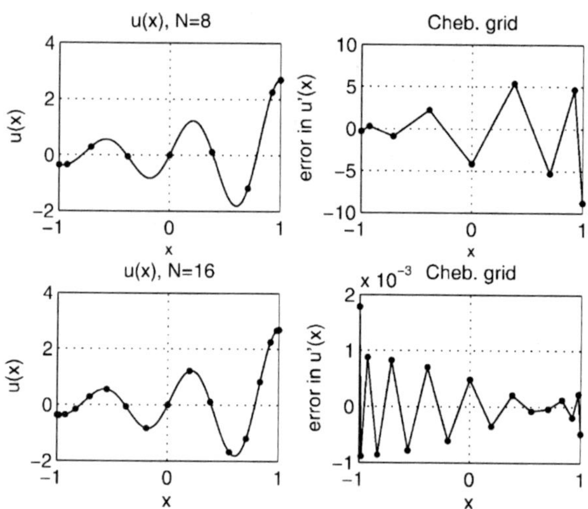

Figure 1: Chebyshev differentiation of $u(x) = e^x \cdot \sin(8x)$

3.   Absorbing Boundary Conditions for Wave Equation

In order to delimit the computational domain to a manageable size, artificial BC's should be imposed, (Fevens and Jiang, 1996). Such BCs must meet certain requirements. They must (a) be coupled with an interior solution, (b) be well-posed with respect to the interior solution (no exponentially growing solutions), (c) be stable (bounded), (d) annihilate incident waves to produce no reflections, and (e) contain a minimal number of points.

Radiation BCs initially formulated by (Sommerfeld, 1949) were used to construct absorbing layers. Such first-order BCs are a function of the effective phase velocity c of the incident wave. If the acoustic equation (3) gives a single value of c, the seismic wave equation (4) indicates different phase velocities for compressional and shear waves. Since a single differential equation can absorb only waves of a certain group velocity, there might be two strategies: (a) to target only one group of waves, or (b) to use a higher-order absorbing BC's. The second approach has been formulated by (Jiang and Wong, 1990) for incidental waves with several group velocities. In a 1-D case for the left boundary:

$$\left[\prod_{j=0}^{p}\left(\frac{\partial}{\partial x}-b_{j}\frac{\partial}{\partial t}\right)\right]u\Big|_{x=0}=0 \tag{8}$$

where b_j are the approximations of group velocities.

Application of Sommerfeld's absorbing BCs for the initial value problem given by a 2-D acoustic equation (Equation (3)) is demonstrated in Figure 2. The plot of the computed response in the middle of the domain is given in Figure 3 which suggests that a small reflection from the boundaries caused by incident waves with different angles to the normal of the boundary can be tolerated.

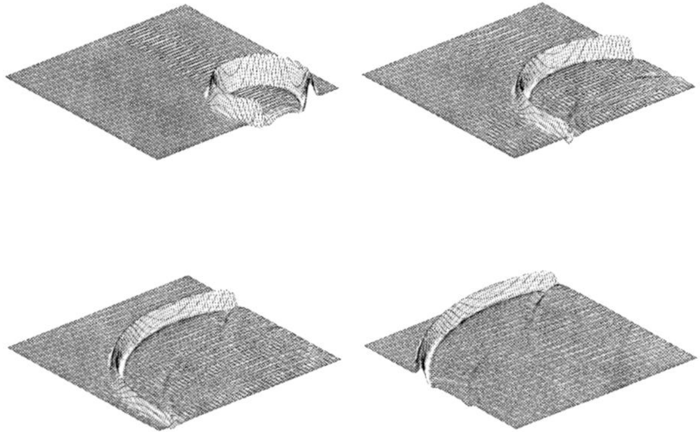

Figure 2: Solution of the acoustic wave equation in 2-D on modified Chebyshev grid with absorbing first-order BCs

This type of BCs is applied in the following to the 3-D seismic wave equation (4).

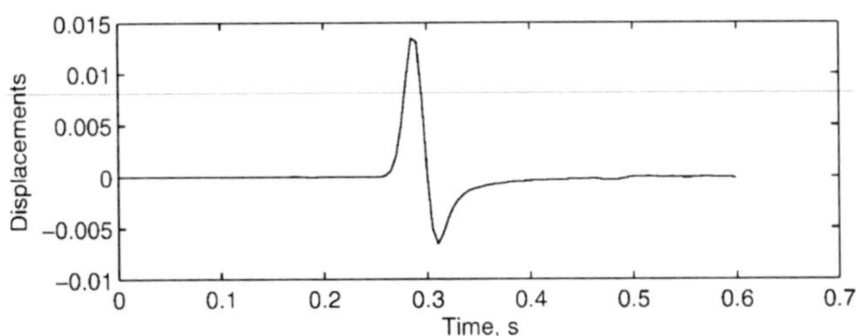

Figure 3: Computed response in the center of domain: solution of the acoustic wave equation in 2D on modified Chebyshev grid with absorbing first-order BCs

4. 3-D Waves in an Isotropic, Homogeneous, Elastic Medium

The numerical solution of a wave equation (4) is computationally more demanding than that of an acoustic equation (3), as the curl, del and div of the displacement vector field must be computed.

Space derivatives are obtained using the Chebyshev differentiation method described previously. To integrate the PDE (4) in the time domain, we reduce it to an ODE via state vector $\mathbf{y} = [\mathbf{u}\ \dot{\mathbf{u}}]^{\mathrm{T}}$:

$$\dot{\mathbf{y}} = \left\{ \begin{array}{c} \dot{\mathbf{u}} \\ \ddot{\mathbf{u}} \end{array} \right\} = \left[\begin{array}{cc} 0 & \mathbf{I} \\ \mathbf{M}^{-1}(\mathbf{L}\nabla\nabla\cdot\mathbf{I} - \mathbf{N}\nabla\times\nabla\times\mathbf{I}) & 0 \end{array} \right] \left\{ \begin{array}{c} \dot{\mathbf{u}} \\ \ddot{\mathbf{u}} \end{array} \right\} \tag{9}$$

where $\mathbf{M}$ is the matrix of media density and $\mathbf{L}$ and $\mathbf{N}$ are matrices of compressional and shear-wave component velocities. This ODE is solved with the Cash-Carp Runge-Kutta method keeping in view CFL stability conditions.

At each time integration step, the curl, del and div of the displacement vector field are evaluated using the Chebyshev differentiation. Computing $\mathbf{M}^{-1}$ is inexpensive, as only the main diagonal of $\mathbf{M}$ has non-zero elements.

Initial conditions in the numerical example are specified as a displacement pulse centered at the point $[x_0, y_0, z_0]$:

$$\mathbf{R} = \sqrt{(\mathbf{x} - x_0)^2 + (\mathbf{y} - y_0)^2 + (\mathbf{z} - z_0)^2}$$

$$\mathbf{u}(x, y, z, 0) = e^{-\mathbf{R}\cdot\alpha} \tag{10}$$

$$\dot{\mathbf{u}}(x, y, z, 0) = \mathbf{0}$$

where $\alpha = 50$ is the weighting coefficient for the pulse. Non-reflecting BCs are given by Equation (8) which is solved on the boundary faces of the computational domain only. Computed in this way, the solution of the wave equation is shown in Figure 4. To simplify visualization, the displacement vector field is converted to a scalar field.

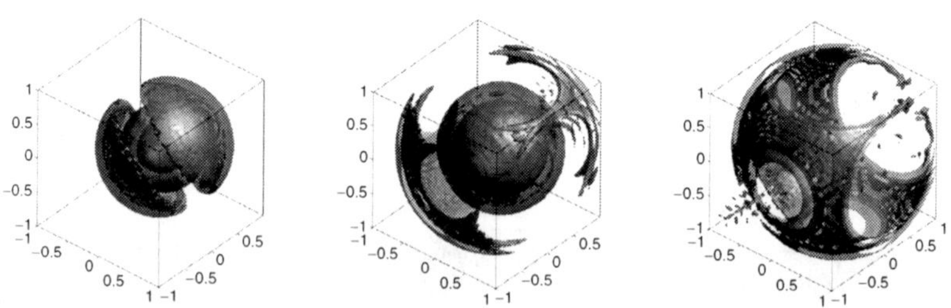

Figure 4: Isotropic homogeneous elastic waves in 3-D on a modified Chebyshev grid (45 × 45 × 45) with absorbing first-order BCs: spherical waves

5.   Aspects of Implementation on SMP Computers Using Octave

Since the algorithm of the numerical solution of Equation (4) is easily parallelizable, it is important to discuss the aspects of its implementation on symmetric multiprocessing (SMP) and multi-core computers, that have recently become quite popular.

The Octave high-level language (Eaton, 2002) has been chosen as the platform for code development. Dynamically linked functions provide access to matrix operations while multiple processes can be handled by a Parallel Virtual Machine (PVM, 2004) via message passing.

The challenging task is to create effective multi-threaded computational process in the the section of the code that numerically integrates Equation (4) reduced to the ODE via state vector, Equation (9).

Integration is performed with the **ode4** Runge-Kutta solver:

y = ode4(odefun, tspan, y0, varargin)

where **odefun** takes the following global variables: **K** (Chebyshev differentiation matrix), **M**, **L**, **N**, Δt, and data structure **G** containing indices of nodes at the faces of the computational domain. Global variables are accessed using **get_global_value**, and the data structure is accessed using the standard template library **map** and **Cell** variable:

```
octave_value Sv = get_global_value(S, true);
Octave_map OX = Sv.map_value();
Cell BoB = OX. contents("BottomFace");
RowVector BoB_vect ( BoB(0).vector_value());
```

After reading all input data, **odefun** starts j PVM processes to compute curl, del and div in the rhs of Equation (4) for each dimension separately. An example of passing matrix **K** to the process on the same machine will yield the following:

```
pvm_spawn("dxx",(char**)0,1,"localhost.localdomain",1,&slave_tid[0]);
pvm_initsend(PvmDataDefault);
SK = int(pow(K.rows(),2) );
pvm_pkdouble(static_cast<double*>(&K(0) ),SK,1);
```

PVM function **pvm_pkdouble** packs the double precision array, but Octave stores **K** in **Matrix** format. The conversion of formats is performed with the type-casting operator **static_cast** on the master and slave sides of the program, which creates a computational overhead. If the computational domain size is $n \times n$, the overall speedup factor will depend on this overhead ($3 \cdot j \cdot n^3$ flops) and the gain obtained from parallelizing curl, del and div operations, which are based on partial derivatives, or on the matrix-vector product $\mathbf{K} \cdot \mathbf{u}_{x,y,z}$ ($2 \cdot n^2$

flops). Derivation is performed on the slave process inside the for loop, which makes use of the higher speed of the oct-files:

```
pvm_upkdouble(static_cast<double*>(&K(0) ),SizeK,1);
int zk=K.rows();
int jj=0;
for(int i=0; i<zk*zk; i++)
{
   accel.insert((K*displ.extract(zk*jj,zk*jj+zk−1)),0,i);
   jj++
}
```

Taking one partial derivative for the entire computational domain will result in $2 \cdot n^4$ flops. Equation (4) contains 18 partial derivatives illustrated below, totaling $36 \cdot n^4$ flops. Speedup factor for two computational threads ($j = 2$) – one for $\nabla(\nabla \cdot \mathbf{u})$, and another for $\nabla \times (\nabla \times \mathbf{u})$ – is $18/12 = 1.5$.

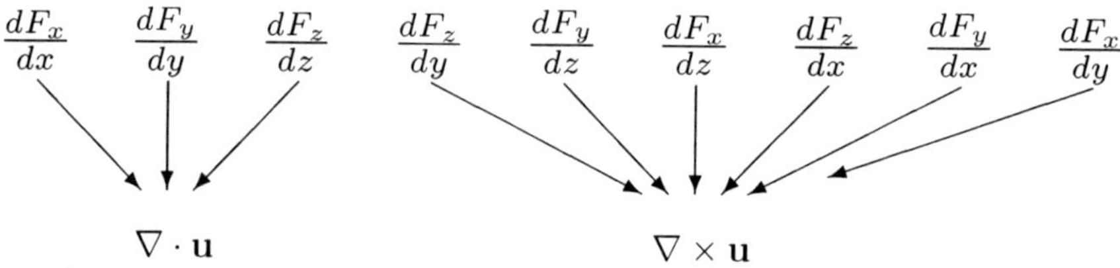

Nine partial derivatives for the entire domain or sub-domains can be computed in parallel ensuring the high scalability of the algorithm. The speedup factors for different cases are provided in Table 1. It can be seen from this figure that a decrease in the speedup for the domain size $n < 200$ and number of processes $i > 3$ is essential due to the overhead.

TABLE 1: Speedup factors

Nr. of parallel processes (i)	Nr. of sub-domains	Speedup factor (s)
$(1 \div 9) \cdot N$	$N \geq 1$	$\dfrac{9 \cdot 2n^4}{(18n^4)/i + i \cdot 3n^3}$

Acknowledgements

This research is sponsored by NATO's Scientific Affairs Division in the framework of the Science for Peace Programme, project SfP-980468, and INTAS project 05-104-7584.

References

Canuto, C., Hussaini, M.Y., Quarteroni, A., Zang, T.A. (1987). *Spectral Methods in Fluid Dynamics*, Springer, New York.

Don, W.S., Solomonoff, A. (1994). Accuracy Enhancement for Higher Derivatives Using Chebyshev Collocation and a Mapping Technique, Institute for Mathematics and Its Applications, University of Minnesota, Minneapolis, MN.

Eaton, J.W. (2002). GNU Octave Manual, Network Theory Limited, ISBN 0-9541617-2-6.

Fevens, T. and Jiang, H. (1996). Absorbing boundary conditions for the Schrodinger equation, *Technical Report 95–376*, Department of Computing and Information Science, Queen's University at Kingston, Kingston, Ontario, Canada.

Jiang, H. and Wong, Y.S. (1990). Absorbing Boundary Conditions for Second-Order Hyperbolic Equations, *Journal of Computational Physics*, **88**(1), 205–231.

Kosloff, D. and Tal-Ezer, H. (1993). Modified Chebyshev Pseudospectral Method with $O(N^{-1})$ Time Step Restriction, *Journal of Computational Physics*, **104**(2), 457–469.

Mattiussi, C. (2000). The Finite Volume, Finite Element, and Finite Difference Methods as Numerical Methods for Physical Field Problems, *Advances in Electronics and Electron Physics* (P. Hawkes, Ed.), **113**, 1–146.

Parallel Virtual Machine v.3.4.5: http://www.csm.ornl.gov/pvm/

Peiró, J. and Sherwin, S. (2005). Finite Difference, Finite Element and Finite Volume Methods for Partial Differential Equations, In: *Handbook of Materials Modeling*, Vol. 1: Methods and Models, S.Yip (Ed.), Springer, Berlin, pp. 1–32.

Sommerfeld, A. (1949). *Partial Differential Equations in Physics*, Academic Press, New York.

Trefethen, L.N. (2000). *Spectral Methods in Matlab*. SIAM, Philadelphia.

Zienkiewicz, O.C. and Cheung, Y.K. (1967). *The Finite Element Method in Structural and Continuum Mechanics*, McGraw-Hill, London.

A COMPARISON BETWEEN THE PROVISIONS OF PRESENT AND PAST ROMANIAN SEISMIC DESIGN CODES BASED ON REQUIRED STRUCTURAL OVERSTRENGTH

I. CRAIFALEANU*
*Technical University of Civil Engineering Bucharest, 124 Lacul
Tei Blvd., Sector 2, RO-020396 Bucharest, Romania
National Institute for Building Research, INCERC, 266
Pantelimon St., Sector 2, RO-021652, Bucharest, Romania*

Abstract. Due to several factors, the lateral resistance of a structure is usually greater than the lateral force used in seismic design. As a result, a structure has an overstrength that can be mobilized in variable amounts when the structure is subjected to lateral seismic loads. The available structural overstrength is difficult to evaluate analytically; a more promising quantity for research is required overstrength, i.e. the overstrength necessary to a structure in order to withstand specified seismic forces. Based on required overstrength, the severity of provisions concerning seismic force evaluation was assessed comparatively for the two versions of the Romanian seismic design code.

Keywords: Overstrength, seismic design code, response spectra, design spectra

1. Introduction

As part of the process of harmonization with European standards, a substantial effort has been made in Romania in recent years to implement regulations concerning the seismic design of buildings. Most of the provisions of Eurocode 8 Part 1 (Eurocode 8, 2004), were adopted (with a number of required adjustments) in the new Romanian seismic design code, P100-1/2006 (P100,

---

*Reinforced Concrete Dept., Technical University of Civil Engineering Bucharest, 124 Lacul Tei Blvd., RO-020396/Earthquake Engineering Dept., National Institute for Building Research, INCERC, 266 Pantelimon St., Sector 2, RO-021652, Bucharest, Romania, e-mail: i.craifaleanu@ gmail.com

A. Zaicenco et al. (eds.), *Harmonization of Seismic Hazard in Vrancea Zone,*
© Springer Science + Business Media B.V. 2008

2006). This new code introduces important changes in comparison with the previous one, P100-92 (P100, 1992), one of the most significant being the evaluation of seismic forces.

A comparative analysis of behavior factors and seismic forces specified by the two versions of the code is made in this paper with reference to the provisions of Eurocode 8. Then, required overstrength is evaluated for both versions of the Romanian seismic design code. Based on the results, severity assessments are made.

2. Design Spectra and Behavior Factors in Eurocode 8 and in the Romanian Codes for Seismic Design

2.1. EUROCODE 8

The elastic response spectrum for the horizontal components of seismic action specified by Eurocode 8 has the shape in Figure 1 and is defined by the following expressions:

$$0 \leq T \leq T_B : \qquad S_e(T) = a_g \cdot S \cdot \left[1 + \frac{T}{T_B}(\eta \cdot 2.5 - 1) \right] \qquad (1)$$

$$T_B \leq T \leq T_C : \qquad S_e(T) = a_g \cdot S \cdot \eta \cdot 2.5 \qquad (2)$$

$$T_C \leq T \leq T_D : \qquad S_e(T) = a_g \cdot S \cdot \eta \cdot 2.5 \left(\frac{T_C}{T} \right) \qquad (3)$$

$$T_D \leq T \leq 4s : \qquad S_e(T) = a_g \cdot S \cdot \eta \cdot 2,5 \left(\frac{T_C T_D}{T^2} \right), \qquad (4)$$

where:

$S_e(T)$	= elastic response spectrum
T	= vibration period of a linear SDOF system
a_g	= design ground acceleration on type A ground
T_B, T_C, T_D	= control periods
S	= soil factor
η	= damping correction factor with reference value $\eta = 1$ for 5% viscous damping

The recommended values for S, T_B, T_C, T_D for 5% damping are tabulated in the code with the explicit stipulation that the definitive values are to be specified by the National Annexes of each country.

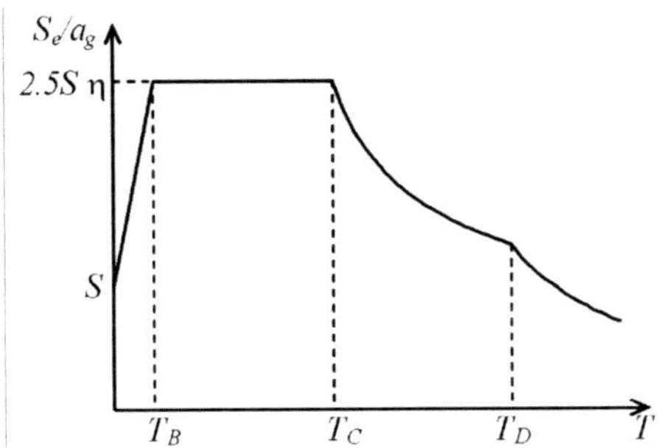

Figure 1: Eurocode 8 elastic response spectrum for the horizontal components of the seismic action

In order to reduce the elastic spectrum for design values, Eurocode 8 uses the behavior factor q. This factor approximates, "the ratio of the seismic forces that the structure would experience if its response was completely elastic with 5% viscous damping to the minimum seismic forces that may be used in design – with a conventional elastic analysis model – still ensuring a satisfactory response of the structure." The behavior factor also accounts for the effect of viscous damping ratios other than 5%. Its values depend on the material, structural type and ductility class.

For reinforced concrete structures, the behavior factor is determined for each horizontal direction of the structure with the expression:

$$q = q_0 \cdot k_w \geq 1.5, \tag{5}$$

in which
q_0 = basic value of the behavior factor and
k_w = factor reflecting the prevailing failure mode in structural systems with walls

For certain structural system types and ductility classes, q_0 also depends on the overstrength ratio α_u/α_1 where

- α_1 = the multiplier of the horizontal seismic design action at first attainment of member flexural resistance anywhere in the structure while all other design actions remain constant.
- α_u = the multiplier of the horizontal seismic design action with all other design actions constant at the formation of plastic hinges in a number of sections sufficient for the development of overall structural instability.

If the overstrength ratio is not obtained by calculations, Eurocode 8 provides approximate values ranging between 1.0 and 1.3 for different structural types. The upper limit of α_u/α_1 is fixed to 1.5, even if larger values result from a nonlinear static global analysis.

For steel structures, Eurocode 8 recommends general ranges of values for q factors depending on the ductility class of the structure. For the medium and high ductility classes, q values are tabulated for different structural types. Some of these values depend on the overstrength ratio α_u/α_1. If more sophisticated calculations are not performed, Eurocode 8 specifies a set of values for α_u/α_1, ranging between 1.0 and 1.3. An upper limit of 1.6 is specified for all cases.

2.2. OLD ROMANIAN SEISMIC DESIGN CODE (P100-92)

The old Romanian seismic design code took into account inelastic behavior by multiplying the elastic design spectra by a specific coefficient denoted by ψ. The coefficient, with values ranging from 0.15 to 0.65, was differentiated by structure and material type.

The elastic design spectra for the horizontal components of the seismic action normalized by peak ground acceleration (PGA) are plotted in Figure 2 for the three seismic zones specified by the code characterized by control (corner) periods T_C, of 0.7, 1.0 and 1.5 s.

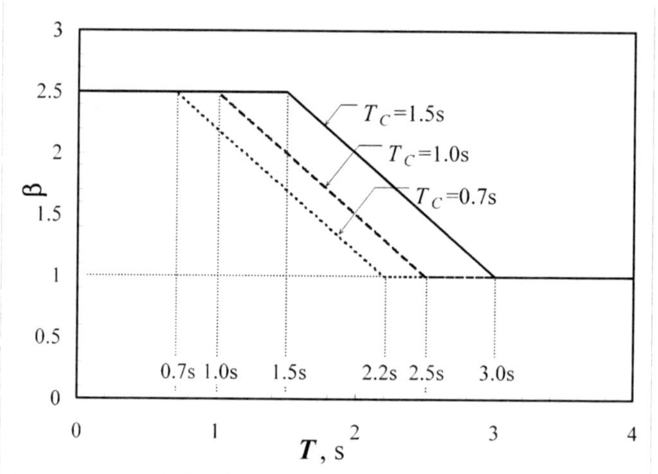

Figure 2: P100-92. Elastic design spectra for different values of corner period T_C.

Figure 3 shows, for illustration, the design spectra normalized by PGA prescribed by P100-92 for reinforced concrete structures.

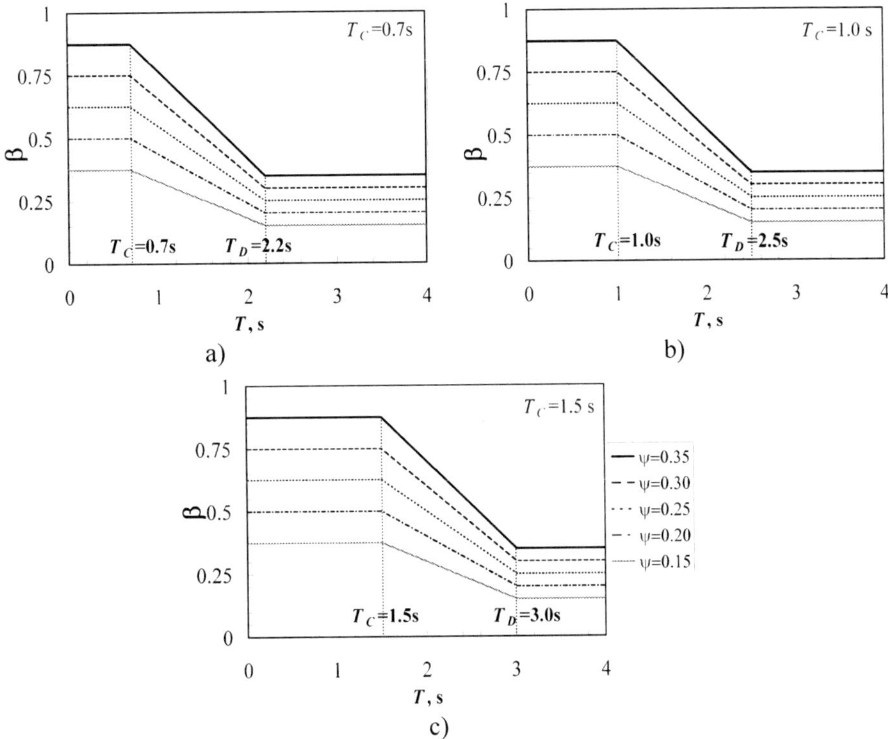

Figure 3: Design spectra normalized by PGA prescribed by the P100-92 code for reinforced concrete structures

2.3. NEW ROMANIAN SEISMIC DESIGN CODE (P100-1/2006)

The structure of the provisions concerning behavior factors in the new Romanian seismic design code is, in general, similar to that of Eurocode 8. However, the values of q differ from those in the European norm and in other provisions and limitations.

The shape of the elastic response spectrum for the horizontal components of seismic action was modified compared with the one in the 1992 version of the code in order to be compatible with that in Eurocode 8.

The new response spectra are specified for zones characterized by three values of the control (corner) period T_C, 0.7, 1.0 and 1.6 s, corresponding to the medium recurrence interval, $MRI = 100$ years considered for Vrancea events. For each of these three zones, a normalized acceleration response spectrum is specified.

The normalized acceleration response spectrum $\beta(T)$ for the horizontal components of seismic action and for 5% damping follows the Eurocode 8 format and is given by the following equations:

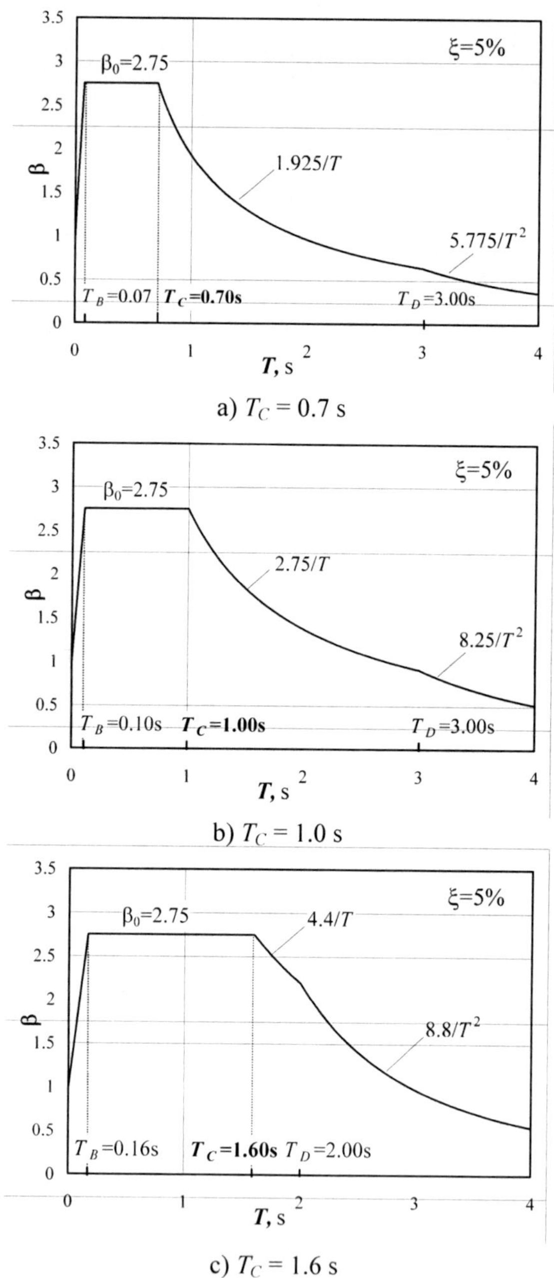

a) $T_C = 0.7$ s

b) $T_C = 1.0$ s

c) $T_C = 1.6$ s

Figure 4: P100-1/2006. Elastic response spectrum normalized by PGA

$$T \leq T_B : \qquad \beta(T) = 1 + \frac{(\beta_0 - 1)}{T_B} T \qquad (6)$$

$$T_B < T \leq T_C : \qquad \beta(T) = \beta_0 \qquad (7)$$

$$T_C < T \leq T_D : \qquad \beta(T) = \beta_0 \frac{T_C}{T} \qquad (8)$$

$$T > T_D : \qquad \beta(T) = \beta_0 \frac{T_C T_D}{T^2} \qquad (9)$$

where:

$\beta(T)$ is the normalized acceleration response spectrum (elastic).

β_0 is the maximum dynamic amplification factor.

T is the fundamental period of vibration of a single-degree-of-freedom structure.

Figure 4 shows the spectra in the new Romanian seismic design code for all specified values of T_C.

The control periods T_B, T_C and T_D are given in Table 1.

TABLE 1: Reference periods of response spectra T_B, T_C, T_D, for horizontal ground motion

Control period	T_B (s)	T_C (s)	T_D (s)
$T_C = 0.7$ s	0.07	0.70	3.00
$T_C = 1.0$ s	0.10	1.00	3.00
$T_C = 1.6$ s	0.16	1.60	2.00

The elastic response spectrum for horizontal ground motion is defined as:

$$S_e(T) = a_g \beta(T). \qquad (10)$$

One should note that the T_C value of 1.5 s in the 1992 code was modified to 1.6 in the 2006 release as a result of extensive studies performed by professor Lungu et al. (1996, 2004). As these studies also suggested that a higher value for the horizontal plateau of the spectra would be more adequate, the plateau was consequently raised to the value of 2.75.

Figure 5 shows an overall comparison between the normalized elastic response spectra in the two releases of P100.

A separate spectrum with $\beta_0 = 3$ and $T_C = 0.7$ s is given in the 2006 release for the crustal sources in the Banat area (in the southwestern part of Romania) where a series of significant seismic events occurred in 1991.

Figure 6 presents, for illustration, design spectra normalized by PGA as prescribed by P100-1/2006 for reinforced concrete structures of uniform elevation of high ductility class (DCH).

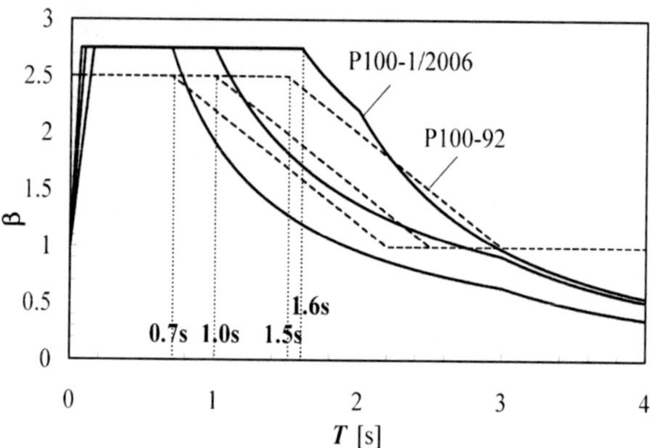

Figure 5: Comparison between the normalized elastic response spectra in the two releases of P100

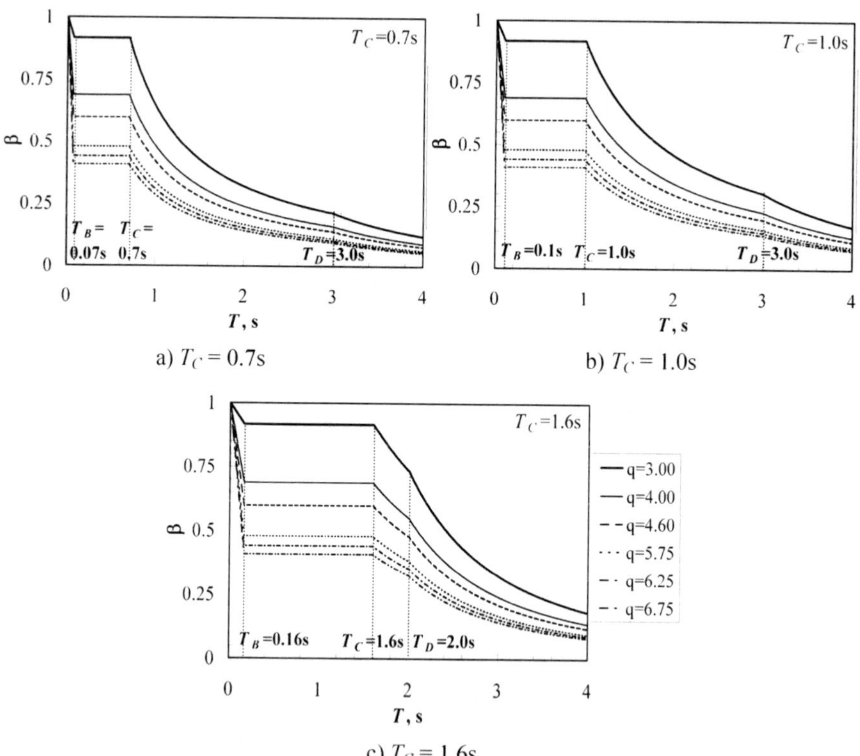

a) $T_C = 0.7$s

b) $T_C = 1.0$s

c) $T_C = 1.6$s

Figure 6: P100-1/2006. Design spectra normalized by PGA for reinforced concrete structures of regular elevation and of high ductility class (DCH)

The display order for the curves in all diagrams is the same as that given in the legend.

A comparison between the requirements of the two versions of the Romanian seismic design code is shown in Figure 7 for two pairs of corresponding values of the ψ coefficient and of the q factor.

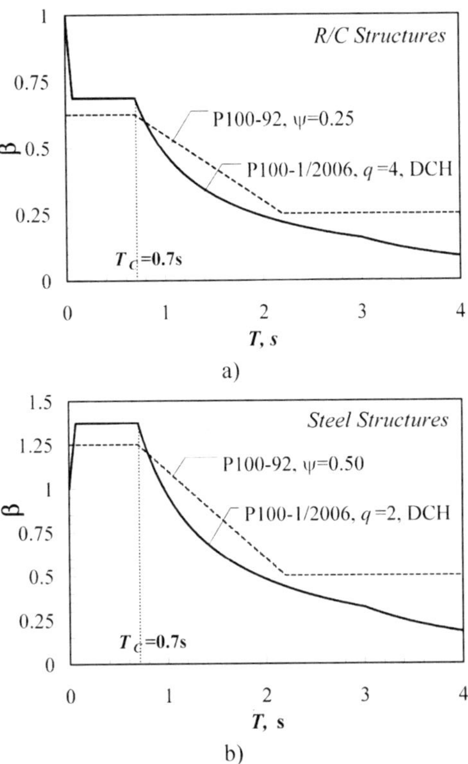

Figure 7: Comparison between the requirements of the two versions of the Romanian seismic design code

3. Structural Overstrength

Due to several factors, the lateral resistance of a structure is usually greater than the lateral force used in seismic design. Among those factors that should be mentioned are the following: higher material strengths than those used in design, structural redundancy, member oversize resulting from story drift limitations or from detailing requirements, strength hardening, multiple load combinations, the effect of non-structural elements, larger reinforcement areas resulting from minimum reinforcement requirements and strain-rate effect. As a result, a structure has a reserve strength that can be mobilized in variable amounts when the structure is subjected to lateral seismic loads.

Moreover, the actual lateral strength is in any case lower than the lateral force that would be induced in the structure if it behaved elastically. This is due to the current design concept that allows, under certain conditions, the occurrence of yielding in structural members.

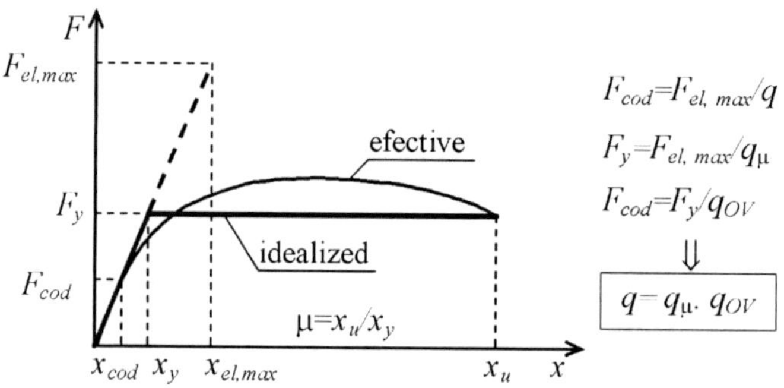

Figure 8: Simplified lateral force – deformation diagram of a structure

A simplified lateral force-deformation diagram of a structure is given in Figure 8 (Uang, 1991). The idealized (bilinear elasto-plastic) diagram corresponds to the significance of the overstrength used in Eurocode 8 and in P100-1/2006 for defining the ratio α_u/α_1. The following notations were used:

F_{code} = design seismic force
F_y = yield strength level
$F_{el,max}$ = maximum base shear that develops in the structure if it remains in the elastic range
x_{code} = deformation at design seismic force
x_y = deformation at yield strength level
$x_{el,max}$ = deformation at $F_{el,max}$
x_u = ultimate displacement (prior to collapse)
μ = x_u/x_y = lateral displacement ductility

With the above notations, it results that the behavior factor q in Eurocode 8 and in P100-1/2006 represents the product of two factors, one containing the effect of inelastic behavior (q_μ), and the other the effect of overstrength (q_{OV}):

$$q = q_\mu \cdot q_{OV}. \tag{11}$$

The overstrength factor q_{OV} is difficult to evaluate analytically. At present, the evaluation of q_{OV} is based mostly on existing experience for different structural types; however, the required overstrength for structures designed according to specified design forces to withstand given seismic actions can be determined quite simply.

The required overstrength is a straightforward criterion for the assessment of the severity of seismic design provisions. In the following, this criterion is used for a comparative evaluation of the requirements of the two versions, old and new, of the P100 code.

In order to perform the assessment, design spectra such as those shown in Figures 3 and 5 were used to represent code requirements. Then, normalized acceleration spectra with 10% probability of exceedance were calculated for q_μ values equal to the q values specified by P100-1/2006 and for q_μ values equal to the $1/\psi$ values specified by P100-92 respectively (Craifaleanu, 2005).

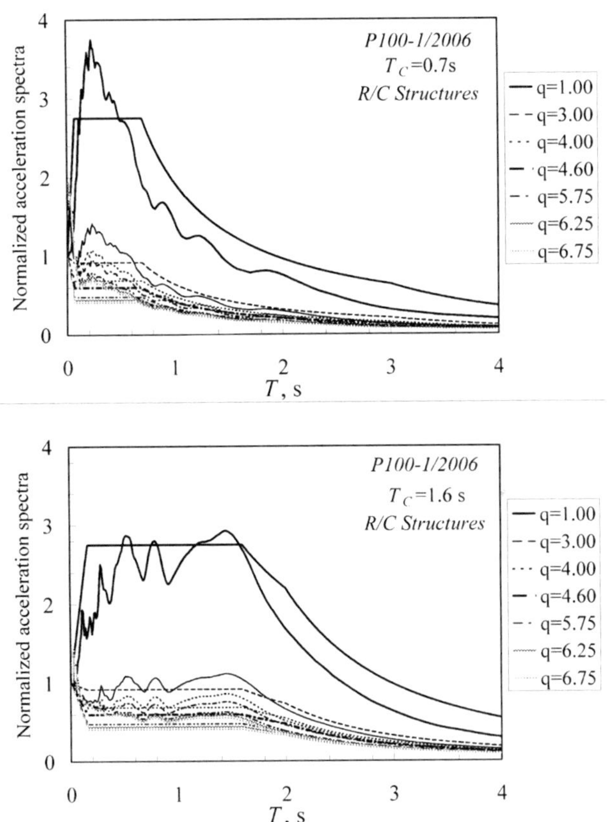

Figure 9: P100-1/2006. Spectral curves used as a basis for the determination of R_{OV} values for reinforced concrete structures of high ductility class (DCH)

Two sets of seismic motions were used to calculate these spectra: the first set consisted of broad frequency band motions recorded in the northeastern part of the country (Moldavia) and the second set consisted of narrow frequency band motions recorded in the soft soil conditions of Bucharest. All motions were recorded during the three strong Vrancea earthquakes that affected Romania in 1977, 1986 and 1990 respectively. The two sets were selected in order to be representative for the seismic zones characterized through the corner periods T_C specified by the above mentioned codes, i.e. $T_C = 0.7$ s and $T_C = 1.6$ s (1.5 s in the old code).

The spectra were determined by considering a 5% damping ratio, an elasto-plastic hysteretic model and a lognormal distribution of the spectral ordinates.

The required overstrength was expressed through the factor R_{OV} calculated as the ratio between the spectral ordinates with 10% probability of exceedance and the design spectra, both determined for the same specified q value. The R_{OV} factor was preferred to q_{OV} since it has a more comprehensive definition as it was obtained directly on the basis of seismic design forces.

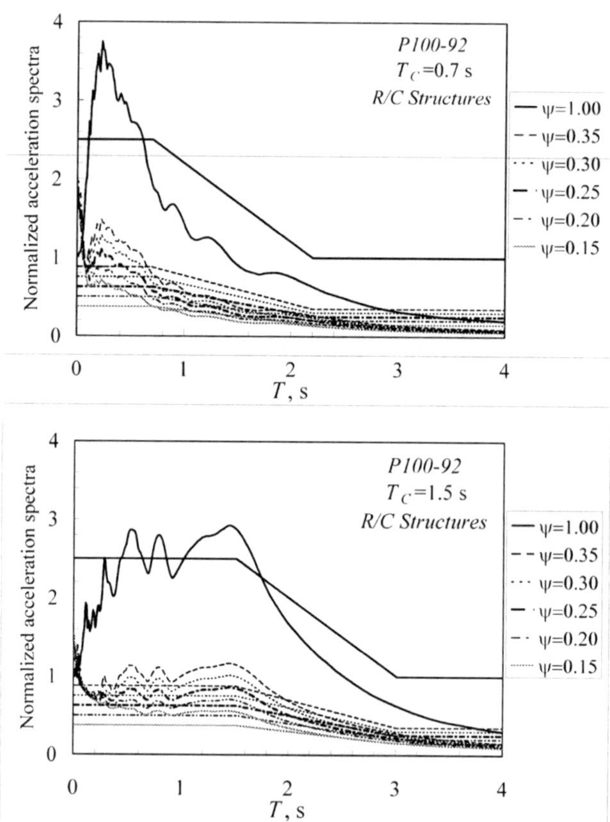

Figure 10: P100-92. Spectral curves used as a basis for the determination of R_{OV} values for reinforced concrete structures

The sets of behavior factors q used in the study correspond to structures of DCH and of regular elevation. The calculations were performed practically for all values given by P100-1/2006 for reinforced concrete structures and for steel structures, respectively (Craifaleanu, 2002).

Figures 9 and 10 show, for illustration, the spectral curves that were used as a basis for the determination of R_{OV} values for reinforced concrete structures. Similar curves were used for steel structures by considering the appropriate values of q and $1/\psi$.

Figures 11 and 12 show the diagrams of the R_{OV} factor for reinforced concrete structures corresponding to the new and to the old code.

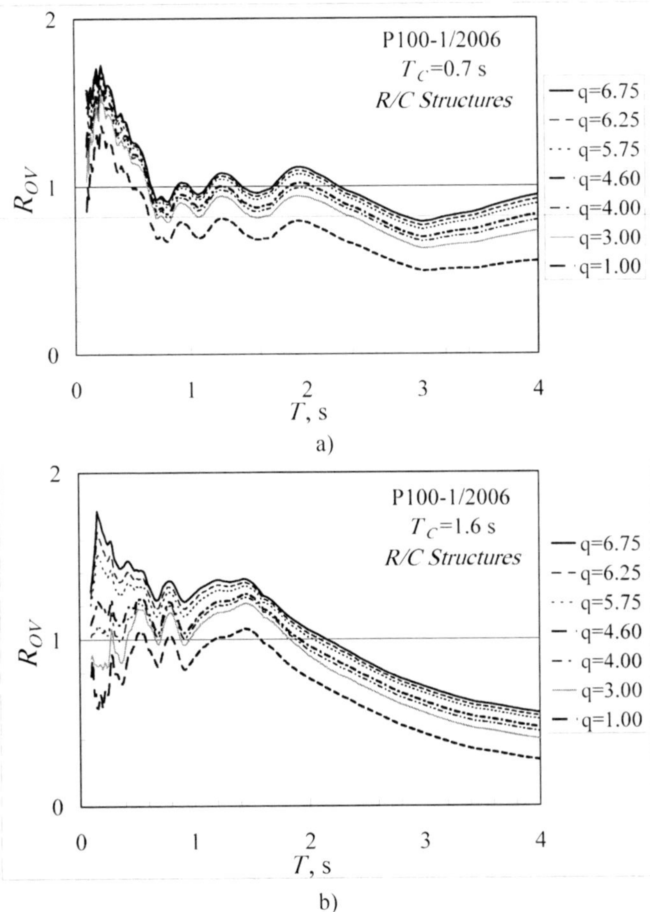

Figure 11: P100-1/2006. R_{OV} values for reinforced concrete structures of high ductility class (DCH)

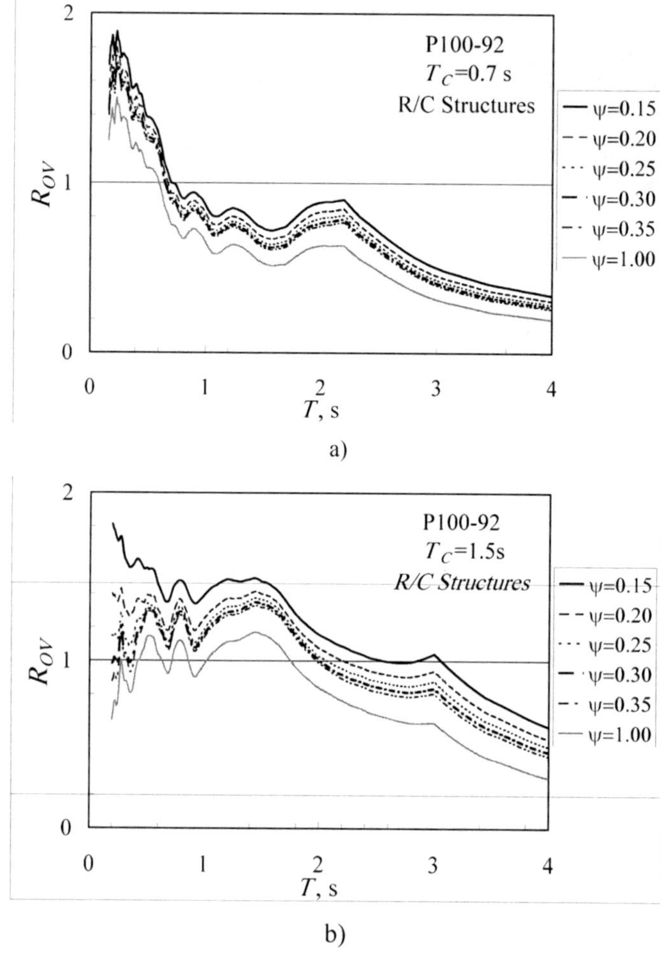

Figure 12: P100-92. R_{OV} values for reinforced concrete structures

Similar diagrams were determined for steel structures by considering the appropriate sets of q and ψ values (Figures 13 and 14).

An important observation to be made is that R_{OV} values also include the difference between the elastic, conventionally established seismic design spectrum and the effectively calculated elastic spectra. For this reason, the curve corresponding to the elastic case ($q = 1$ or ψ = 1) was plotted on every diagram in order to illustrate this difference. A solution to eliminate this effect would have been the use of artificially generated accelerograms compatible with the design spectrum. However, in this case, the specific relevance of natural accelerograms would have been lost.

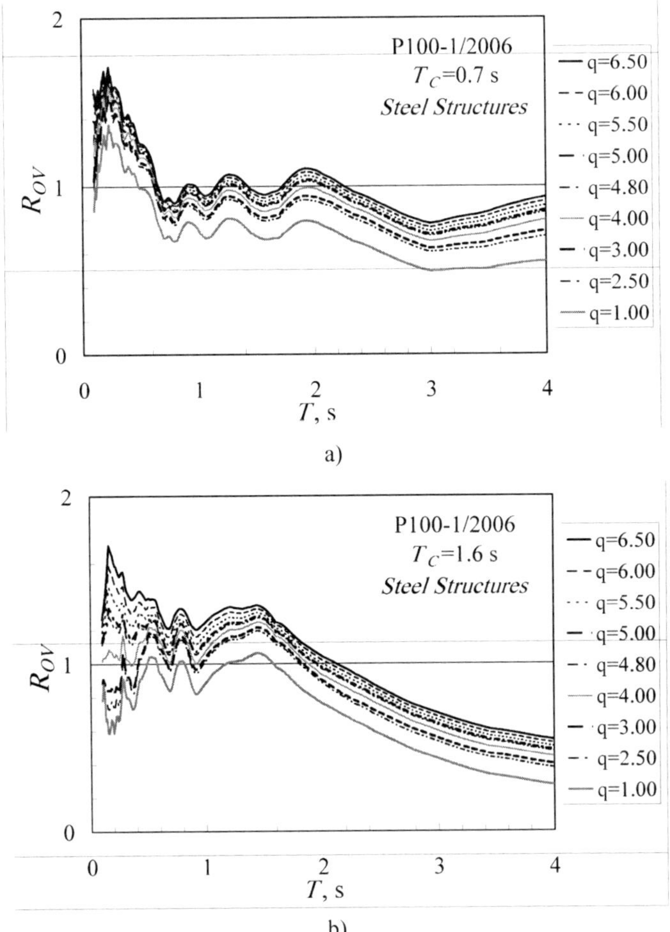

Figure 13: P100-1/2006. R_{OV} values for steel structures of high ductility class (DCH)

Some comments on the obtained R_{OV} values follow.

(a) For the R_{OV} values corresponding to the new code (P100-1/2006)

Substantial differences can be noticed between the values of the diagrams corresponding to the two sets of seismic motions considered. The shapes of the diagrams are also very different, primarily as a consequence of their different spectral contents.

For the diagram corresponding to $T_C = 0.7$ s, the R_{OV} factors reach a maximum at short periods ($T \approx 0.23$ s) after which an abrupt decrease follows so that for periods above T_C, the factors are above unit for only limited period ranges. Even on these ranges they do not exceed 1.1. The maximum value for the case of reinforced concrete structures with $q = 6.75$ is about 1.7.

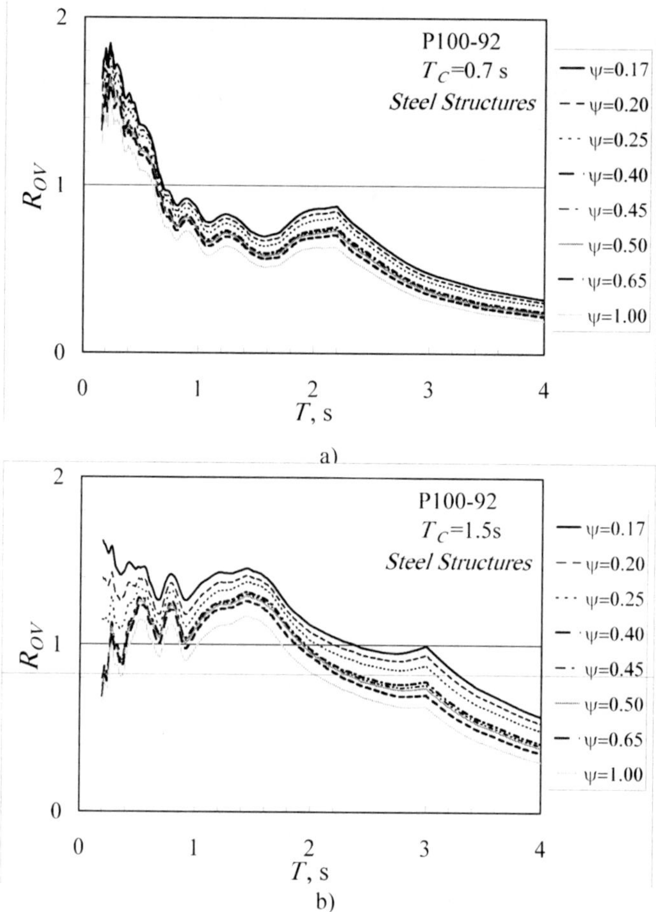

Figure 14: P100-92. R_{OV} values for steel structures of high ductility class (DCH)

For the diagram corresponding to T_C = 1.6 s, after a maximum is reached at a period of about 0.16 s, a decrease can be noticed. It follows then that for periods between 0.5 and 1.2 s, there is a range in which R_{OV} is situated on average between 1.1 and 1.4 for the q values considered. The maximum value for frame or frame equivalent dual concrete structures with q = 6.75 (the largest behavior factor taken into account) is about 1.8. For periods longer than 1.2 s, a marked decrease of R_{OV} can be noticed so that for $T > 2$ s, R_{OV} becomes smaller than 1 which means that theoretically, no overstrength is needed in this case.

Analyzing the R_{OV} values for q = 1.00 (the case of elastic behavior), it can be noticed that while for T_C = 1.6 s R_{OV} exceeds only locally the value of 1.0, for T_C = 0.7 s R_{OV} is greater than 1 on the whole range of periods shorter than T_C, varying between 1.0 and 1.4. This last feature is due to the large values of the normalized acceleration spectra that occur in this period range (see Figure 9).

One should note that when specifying the q values for certain types of structures, the P100-1/2006 code takes into account the possibility of overstrength contributions through the α_u/α_1 ratios that range between 1.0 and 1.3. As a result, by anticipating this phenomenon, structures are designed using larger q factors and consequently smaller seismic forces.

(b) For the R_{OV} values corresponding to the old code (P100-92)

In this case as well, considerable differences appear both in the values and in the shapes of the diagrams determined for the two sets of seismic motions considered.

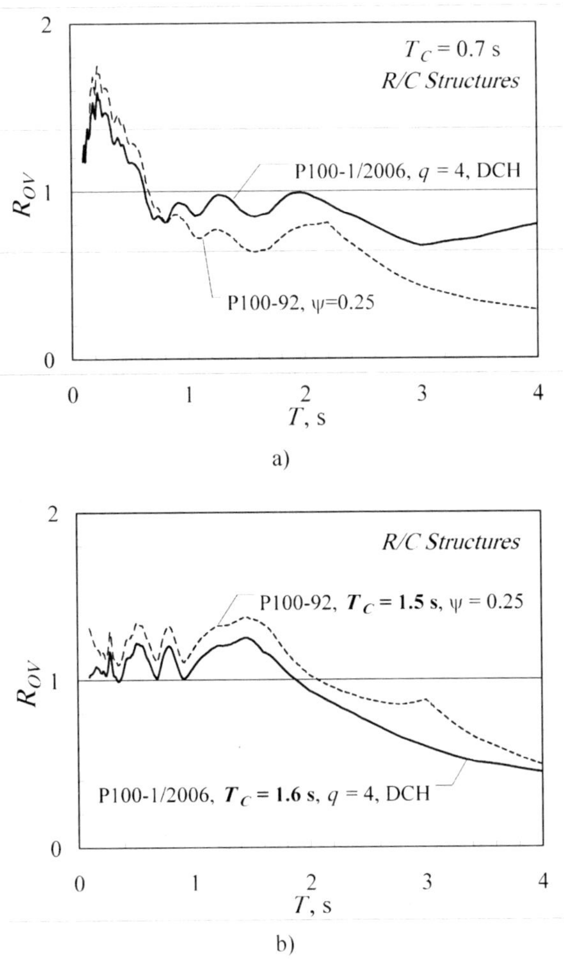

a)

b)

Figure 15: Comparative diagrams of R_{OV}

The differences between the diagrams corresponding to the two codes obviously arise from the different shapes of the design spectra and from the different values of behavior factors.

An overall comparison between the values of the R_{OV} factor in the old and new versions of the P100 code shows the following:

- For $T_C = 1.5$ s (1.6 s), the required overstrength is greater in the old code on the whole period range.
- For $T_C = 0.7$ s, the required overstrength is greater in the old code at periods shorter than T_C and smaller for the rest of the period range.

It must be taken into account that in the second case, the R_{OV} values are in any case below unit for the old code and predominantly below unit for the new code which makes the comparison less significant. Therefore, it can be concluded that taken as a whole, the exigencies of the new code based on an assessment of required structural overstrength are more conservative by comparison with those of the old code.

The above considerations are also illustrated by the comparative diagrams in Figure 15 plotted for two particular cases in which the values of q and $1/\psi$ are equal for a specified structure.

4. Conclusions

1. The required overstrength, expressed by the ratio between the demands imposed by Vrancea earthquakes and those imposed by the Romanian seismic design codes, R_{OV}, varies significantly with the period of vibration and with behavior factors.
2. *In* both codes, required overstrength can reach values up to 1.8 times the seismic design force for behavior factors situated at the limit of the usual range. However, with the exception of periods shorter than 0.5 s, the values of R_{OV} are currently below 1.35–1.40 for the new seismic design code (P100-1/2006) and below 1.5 for the old code (P100-92).
3. By modeling seismic action through two sets of selected accelerograms recorded during strong Vrancea earthquakes in Romania considered as relevant for seismic zones with corner periods T_C of 0.7 and 1.6 (1.5) s, it resulted that the required overstrength for structures designed to resist the seismic forces specified by the new code was lower than that in the old code. This is an indication of the more conservative character of the new provisions.
4. Regarding the values of the behavior and overstrength factors, further studies should be performed for each structural type using detailed structural

models and taking into account all the provisions in the new code in order to validate the currently calculated values of required overstrength.

References

Craifaleanu, I. G. (2002) Aspects concerning the adoption of the Eurocode 8 behavior factors in Romanian seismicity conditions, Bulletin of the Romanian Association of Structural Design Engineers (AICPS), No. 3/2002, pp. 43–48 (in Romanian).

Craifaleanu, I. G. (2005) Nonlinear single-degree-of-freedom models in earthquake engineering, Matrix Rom Publishing House, Bucharest (in Romanian).

Eurocode 8 (2004) Design of structures for earthquake resistance. Part 1: General rules, seismic actions and rules for buildings, EN 1998-1, Dec..

Lungu, D., Cornea, T., Craifaleanu, I., Demetriu, S. (1996) Probabilistic seismic hazard analysis for inelastic structures on soft soils, Proceedings of the Eleventh World Conference on Earthquake Engineering, Acapulco, Mexico, June 23–28, CD-ROM.

Lungu, D., Arion, C., Aldea, A., Vacareanu R. (2004). Representation of seismic action in the new Romanian code for design of earthquake resistant buildings P100-2003, 13th World Conference on Earthquake Engineering, Vancouver, Canada, 15p., CD-ROM.

P100 (1992) Code for the aseismic design of residential, agricultural and industrial structures – P100-92 (in Romanian), Buletinul Constructiilor, INCERC.

P100 (2006) Seismic design code – P100. Part I – P100-1/2006: Design rules for buildings. Ministry of Transport, Constructions and Tourism (in Romanian).

Uang, C.-M. (1991) Establishing R (or R_w) and C_d factors for building seismic provisions, Journal of Structural Engineering, Vol. 117, No. 1, Jan., pp. 19–28.

Printed in the United States
134952LV00002B/40/P